DESIGN AND ANALYSIS OF COMPOSITE STRUCTURES

WITH APPLICATIONS TO AEROSPACE STRUCTURES

Second Edition

Christos Kassapoglou
Delft University of Technology, The Netherlands

WILEY

A John Wiley & Sons, Ltd., Publication

Library of Congress Cataloging-in-Publication Data

A catalogue record for this book is available from the British Library.

ISBN: 9781118401606

Typeset in 10/12pt Times by Aptara Inc., New Delhi, India

Contents

About the Author

Christos Kassapoglou received his BS degree in Aeronautics and Astronautics, and two MS degrees (Aeronautics and Astronautics and Mechanical Engineering) all from the Massachusetts Institute of Technology. He received his Ph.D in Aerospace Engineering from the Delft University of Technology. Since 1984, he has worked in the industry, first at Beech Aircraft on the all-composite Starship I and then at Sikorsky Aircraft in the Structures Research Group specializing on analysis of composite structures of the all-composite Comanche and other helicopters, and leading internally funded research and programs funded by NASA and the US Army. Since 2001, he has been consulting with various companies in the United States on applications of composite structures on airplanes and helicopters. He joined the faculty of the Aerospace Engineering Department of the Delft University of Technology (Aerospace Structures) in 2007 as an Associate Professor. His interests include fatigue and damage tolerance of composites, analysis of sandwich structures, design and optimization for cost and weight and technology optimization. He has over 40 journal papers and three issued or pending patents on related subjects. He is a member of AIAA, AHS and SAMPE.

Series Preface

The field of aerospace is wide ranging and covers a variety of products, disciplines, and domains, not merely in engineering but in many related supporting activities. These combine to enable the aerospace industry to produce exciting and technologically challenging products. A wealth of knowledge is contained by practitioners and professionals in the aerospace fields, which is of benefit to other practitioners in the industry, and to those entering the industry from University.

The *Aerospace Series* aims to be a practical and topical series of books aimed at engineering professionals, operators, users and allied professions such as commercial and legal executives in the aerospace industry. The range of topics is intended to be wide ranging, covering design and development, manufacture, operation and support of aircraft, as well as topics such as infrastructure operations and developments in research and technology. The intention is to provide a source of relevant information that will be of interest and benefit to all those people working in aerospace.

The use of composite materials for aerospace structures has increased dramatically in the last three decades. The attractive strength-to-weight ratios, improved fatigue and corrosion resistance and ability to tailor the geometry and fibre orientations, combined with recent advances in fabrication, have made composites a very attractive option for aerospace applications from both a technical and financial viewpoint. This has been tempered by problems associated with damage tolerance and detection, damage repair, environmental degradation and assembly joints. The anisotropic nature of composites also dramatically increases the number of variables that need to be considered in the design of any aerospace structure.

This book, *Design and Analysis of Composite Structures: With Application to Aerospace Structures*, Second Edition, provides a methodology of various analysis approaches that can be used for the preliminary design of aerospace structures without having to resort to finite elements. Representative types of composite structure are described, along with techniques to define the geometry and lay-up stacking sequence required to withstand the applied loads. The value of such a set of tools is to enable rapid initial trade-off preliminary design studies to be made, before using a detailed finite element analysis on the finalized design configurations.

Allan Seabridge
Jonathan Cooper
Peter Belobaba

Preface to First Edition

This book is a compilation of analysis and design methods for structural components made of advanced composites. The term "advanced composites" is used here somewhat loosely and refers to materials consisting of a high-performance fibre (graphite, glass, Kevlar®, etc.) embedded in a polymeric matrix (epoxy, bismaleimide, PEEK, etc.). The material in this book is the product of lecture notes used in graduate-level classes in Advanced Composites Design and Optimization courses taught at the Delft University of Technology.

The book is aimed at fourth year undergraduate or graduate-level students and starting engineering professionals in the composites industry. The reader is expected to be familiar with classical laminated-plate theory (CLPT) and first ply failure criteria. Also, some awareness of energy methods and Rayleigh–Ritz approaches will make some of the solution methods easier to follow. In addition, basic applied mathematics knowledge such as Fourier series, simple solutions of partial differential equations and calculus of variations are subjects that the reader should have some familiarity with.

A series of attractive properties of composites such as high stiffness and strength-to-weight ratios, reduced sensitivity to cyclic loads, improved corrosion resistance and, above all, the ability to tailor the configuration (geometry and stacking sequence) to specific loading conditions for optimum performance has made them a prime candidate material for use in aerospace applications. In addition, the advent of automated fabrication methods such as advanced fibre/tow placement, automated tape laying, filament winding, has made it possible to produce complex components at costs competitive with if not lower than metallic counterparts. This increase in the use of composites has brought to the forefront the need for reliable analysis and design methods that can assist engineers in implementing composites in aerospace structures. This book is a small contribution towards fulfilling that need.

The objective is to provide methodology and analysis approaches that can be used in preliminary design. The emphasis is on methods that do not use finite elements or other computationally expensive approaches in order to allow the rapid generation of alternative designs that can be traded against each other. This will provide insight in how different design variables and parameters of a problem affect the result.

The approach to preliminary design and analysis may differ according to the application and the persons involved. It combines a series of attributes such as experience, intuition, inspiration and thorough knowledge of the basics. Of these, intuition and inspiration cannot be captured in the pages of a book or itemized in a series of steps. For the first attribute, experience, an attempt can be made to collect previous best practices which can serve as guidelines for future work. Only the last attribute, knowledge of the basics, can be formulated in such a way that the reader can learn and understand them and then apply them to his/her own applications. And doing that is neither easy nor guaranteed to be exhaustive. The wide variety of applications and the peculiarities that each may require in the approach, preclude any complete and in-depth presentation of the material. It is only hoped that the material presented here will serve as a starting point for most types of design and analysis problems.

Given these difficulties, the material covered in this book is an attempt to show representative types of composite structure and some of the approaches that may be used in determining the geometry and stacking sequences that meet applied loads without failure. It should be emphasized that not all methods presented here are equally accurate nor do they have the same range of applicability. Every effort has been made to present, along with each approach, its limitations. There are many more methods than the ones presented here and they vary in accuracy and range of applicability. Additional references are given where some of these methods can be found.

These methods cannot replace thorough finite element analyses which, when properly set up, will be more accurate than most of the methods presented here. Unfortunately, the complexity of some of the problems and the current (and foreseeable) computational efficiency in implementing finite element solutions precludes their extensive use during preliminary design or, even, early phases of the detailed design. There is not enough time to trade hundreds or thousands of designs in an optimization effort to determine the "best" design if the analysis method is based on detailed finite elements. On the other hand, once the design configuration has been finalized or a couple of configurations have been downselected using simpler, more efficient approaches, detailed finite elements can and should be used to provide accurate predictions for the performance, point to areas where revisions of the design are necessary, and, eventually, provide supporting analysis for the certification effort of a product.

Some highlights of composite applications from the 1950s to today are given in Chapter 1 with emphasis on nonmilitary applications. Recurring and nonrecurring cost issues that may affect design decisions are presented in Chapter 2 for specific fabrication processes. Chapter 3 provides a review of CLPT and Chapter 4 summarizes strength failure criteria for composite plates; these two chapters are meant as a quick refresher of some of the basic concepts and equations that will be used in subsequent chapters.

Chapter 5 presents the governing equations for anisotropic plates. It includes the von Karman large deflection equations that are used later to generate simple solutions for post-buckled composite plates under compression. These are followed by a presentation of the types of composite parts found in aerospace structures and the design philosophy typically used to come up with a geometric shape. Design requirements and desired attributes are also discussed. This sets the stage for quantitative requirements that address uncertainties during the design and during service of a fielded structure. Uncertainties in applied loads and variations in usage from one user to another are briefly discussed. A more detailed discussion about uncertainties in material performance (material scatter) leads to the introduction of statistically meaningful (A- and B-basis) design values or allowables. Finally, sensitivity to damage and environmental conditions is discussed and the use of knockdown factors for preliminary design is introduced.

Chapter 6 contains a discussion of buckling of composite plates. Plates are introduced first and beams follow (Chapter 8) because failure modes of beams such as crippling can be introduced more easily as special cases of plate buckling and post-buckling. Buckling under compression is discussed first, followed by buckling under shear. Combined load cases are treated next and a table including different boundary conditions and load cases is provided.

Post-buckling under compression and shear is treated in Chapter 7. For applied compression, an approximate solution to the governing (von Karman) equations for large deflections of plates is presented. For applied shear, an approach that is a modification of the standard approach for metals undergoing diagonal tension is presented. A brief section follows suggesting how post-buckling under combined compression and shear could be treated.

Design and analysis of composite beams (stiffeners, stringers, panel breakers, etc.) are treated in Chapter 8. Calculation of equivalent membrane and bending stiffnesses for cross sections consisting of members with different layups are presented first. These can be used with standard beam design equations and some examples are given. Buckling of beams and beams on elastic foundations is discussed next. This does not differentiate between metals and composites. The standard equations for metals can be used with appropriate (re)definition of terms such as membrane and bending stiffness. The effect of different end conditions is also discussed. Crippling, or collapse after very-short-wavelength buckling, is discussed in detail deriving design equations from plate buckling presented earlier and from semi-empirical approaches. Finally, conditions for inter-rivet buckling are presented.

The two constituents, plates and beams are brought together in Chapter 9 where stiffened panels are discussed. The concept of smeared stiffness is introduced and its applicability is discussed briefly. Then, special design conditions such as the panel breaker condition and failure modes such as skin–stiffener separation are analysed in detail, concluding with design guidelines for stiffened panels derived from the previous analyses.

Sandwich structure is treated in Chapter 10. Aspects of sandwich modelling, in particular, the effect of transverse shear on buckling, are treated first. Various failure modes such as wrinkling, crimping and intracellular buckling are then discussed with particular emphasis on wrinkling with and without waviness. Interaction equations are introduced for analysing sandwich structure under combined loading. A brief discussion on attachments including ramp downs and associated design guidelines close this chapter.

The final chapter, Chapter 11, summarizes design guidelines and rules presented throughout the previous chapters. It also includes some additional rules, presented for the first time in this book, that have been found to be useful in designing composite structures.

To facilitate material coverage and in order to avoid having to read some chapters that may be considered of lesser interest or not directly related to the reader's needs, certain concepts and equations are presented in more than one place. This is minimized to avoid repetition and is done in such a way that reader does not have to interrupt reading a certain chapter and go back to find the original concept or equation on which the current derivation is based.

Specific problems are worked out in detail as examples of applications throughout the book. Representative exercises are given at the end of each chapter. These require the determination of geometry and/or stacking sequence for a specific structure not to fail under certain applied loads. Many of them are created in such a way that more than one answer is acceptable reflecting real-life situations. Depending on the assumptions made and design rules enforced, different but still acceptable designs can be created. Even though low weight is the primary objective of most of the exercises, situations where other issues are important and end up driving the design are also given. For academic applications, experience has shown that students benefit the most if they work out some of these exercises in teams, so design ideas and concepts can be discussed and an approach to a solution formulated. It is recognized that analysis of composite structures is very much in a state of flux, and new and better methods are being developed (e.g. failure theories with and without damage). The present edition includes what are felt to be the most useful approaches at this point in time. As better approaches mature in the future, it will be modified accordingly.

Preface to Second Edition

The first edition of this book met with sufficient interest to justify an improved and enhanced second edition. Feedback from the readers of the first edition led to the addition of two new chapters, one on fittings and one with an example design problem of larger structural parts.

The objective of this book is to introduce basic design and analysis methods of composite structures with sufficient theoretical background for the reader to be able to (a) develop his/her own methods for other situations and (b) extend the present methods to improve their accuracy and applicability. Balancing this objective with the need to provide simple design equations is difficult. Invariably, in some areas the theoretical background will appear too extensive while, in other cases, equations are presented without sufficient derivation. In addition, people who have worked on any of the topics in any detail will, probably, complain that some of the approaches are oversimplified.

Trying to satisfy all the anticipated readers in all aspects covered in this book is simply impossible. If I can get about the same number of complaints at the two extremes of too detailed analysis and oversimplified analysis, I will have achieved my goal. After all, I only hope to give some guidelines for approaches I found useful that the reader can build on and apply to his/her own situations.

The changes from the first edition are (a) addition of more exercises in most chapters, (b) addition of the Puck failure criterion in Chapter 4, (c) addition of a chapter (Chapter 11) on analysis of simple fittings and (d) addition of a chapter (Chapter 13) with a detailed case study for designing a fuselage panel with three different design concepts. Furthermore, a diligent attempt was also made to correct typos of the first edition.

Perhaps, the most important and useful feature of this second edition is that it is accompanied by an App. The App is called CoDeAn (Composites Design and Analysis). Most equations and design procedures in the book have been incorporated in CoDeAn in a user-friendly and efficient manner. This allows iPhone and iPad users to perform analyses and design studies at the touch of a button. A material library with easy addition and modification options is available. Classical laminated-plate theory, first ply failure and buckling analysis are already available with this release. Future releases will include the rest of the book.

1

Applications of Advanced Composites in Aircraft Structures

Some of the milestones in the implementation of advanced composites on aircraft and rotorcraft are discussed in this chapter. Specific applications have been selected that highlight various phases that the composites industry went through while trying to extend the application of composites.

The application of composites in civilian or military aircraft followed the typical stages that every new technology goes through during its implementation. At the beginning, limited application on secondary structure minimized risk and improved understanding by collecting data from tests and fleet experience. This limited usage was followed by wider applications, first in smaller aircraft, capitalizing on the experience gained earlier. More recently, with the increased demand on efficiency and low operation costs, composites are being applied widely on larger aircraft.

Perhaps the first significant application of advanced composites was on the Akaflieg Phönix FS-24 (Figure 1.1) in the late 1950s. What started as a balsa wood and paper sailplane designed by professors at the University of Stuttgart and built by the students was later transformed into a fibreglass/balsa wood sandwich design. Eight planes were eventually built.

The helicopter industry was among the first to recognize the potential of the composite materials and use them on primary structure. The main and tail rotor blades with their beam-like behaviour were one of the major structural parts designed and built with composites towards the end of the 1960s. One such example is the Aerospatiale Gazelle (Figure 1.2). Even though, to first order, helicopter blades can be modelled as beams, the loading complexity and the multiple static and dynamic performance requirements (strength, buckling, stiffness distribution, frequency placement, etc.) make for a very challenging design and manufacturing problem.

In the 1970s, with the composites usage on sailplanes and helicopters increasing, the first all-composite planes appeared. These were small recreational or aerobatic planes. Most notable among them were the Burt Rutan designs such as the Long EZ and Vari-Eze (Figure 1.3). These were largely co-cured and bonded constructions with very limited numbers of fasteners. Efficient aerodynamic designs with mostly laminar flow and light weight led to a combination of speed and agility.

Design and Analysis of Composite Structures: With Applications to Aerospace Structures, Second Edition. Christos Kassapoglou.
© 2013 John Wiley & Sons, Ltd. Published 2013 by John Wiley & Sons, Ltd.

Figure 1.1 Akaflieg Phönix FS-24 (Courtesy of Deutsches Segelflugzeugmuseum; see Plate 1 for the colour figure)

Figure 1.2 Aerospatiale SA 341G Gazelle (Copyright Jenny Coffey, printed with permission; see Plate 2 for the colour figure)

Figure 1.3 Long EZ and Vari-Eze. (Vari-Eze photo: courtesy of Stephen Kearney; Long EZ photo: courtesy of Ray McCrea; see Plate 3 for the colour figure)

Up to that point, usage of composites was limited and/or was applied to small aircraft with relatively easy structural requirements. In addition, the performance of composites was not completely understood. For example, their sensitivity to impact damage and its implications for design only came to the forefront in the late 1970s and early 1980s. At that time, efforts to build the first all-composite airplane of larger size began with the LearFan 2100

Figure 1.4   LearAvia LearFan 2100 (Copyright Thierry Deutsch; see Plate 4 for the colour figure)

(Figure 1.4). This was the first civil aviation all-composite airplane to seek FAA certification (see Section 2.2). It used a pusher propeller and combined high speed and low weight with excellent range and fuel consumption. Unfortunately, while it met all the structural certification requirements, delays in certifying the drive system and the death of Bill Lear the visionary designer and inventor behind the project, kept the LearFan from making it into production and the company, LearAvia, went bankrupt.

The Beech Starship I (Figure 1.5) which followed on the heels of the LearFan in the early 1980s was the first all-composite airplane to obtain FAA certification. It was designed to the new composite structure requirements specially created for it by the FAA. These requirements were the precursor of the structural requirements for composite aircraft as they are today. Unlike the LearFan which was a more conventional skin-stiffened structure with frames and stringers, the Starship fuselage was made of sandwich (graphite/epoxy facesheets with Nomex® core) and had a very limited number of frames, increasing cabin head room for a given cabin diameter and minimizing fabrication cost. It was co-cured in large pieces that were bonded together and, in critical connections such as the wing-box or the main fuselage joints, were also fastened. Designed also by Burt Rutan, the Starship was meant to have mostly laminar

Figure 1.5   Beech (Raytheon Aircraft) Starship I (Courtesy of Brian Bartlett; see Plate 5 for the colour figure)

Figure 1.6   Airbus A-320 (Courtesy of Brian Bartlett; see Plate 6 for the colour figure)

flow and increased range through the use of efficient canard design and blended main wing. Two engines with pusher propellers located at the aft fuselage were to provide enough power for high cruising speed. In the end, the aerodynamic performance was not met and the fuel consumption and cruising speeds missed their targets by a small amount. However, structurally the Starship I proved that the all-composite aircraft could be designed and fabricated to meet the stringent FAA requirements. In addition, invaluable experience was gained in analysis and testing of large composite structures and new low-cost structurally robust concepts were developed for joints and sandwich structure in general.

With fuel prices rising, composites with their reduced weight became a very attractive alternative to the metal structure. Applications in the large civilian transport category started in the early 1980s with the Boeing 737 horizontal stabilizer which was a sandwich construction and continued with larger-scale application on the Airbus A-320 (Figure 1.6). The horizontal and vertical stabilizers as well as the control surfaces of the A-320 are made of composite materials.

The next significant application of composites on primary aircraft structure came in the 1990s with the Boeing 777 (Figure 1.7) where, in addition to the empennage and control surfaces, the main floor beams are also made out of composites.

Figure 1.7   Boeing 777 (Courtesy of Brian Bartlett; see Plate 7 for the colour figure)

Figure 1.8   Airbus A-380 (Courtesy of Bjoern Schmitt, World of Aviation.de; see Plate 8 for the colour figure)

Despite the use of innovative manufacturing technologies which started with early robotics applications on the A320 and continued with significant automation (tape layup) on the 777, the cost of composite structures was not attractive enough to lead to an even larger-scale (e.g. entire fuselage and/or wing structure) application of composites at that time. The Airbus A-380 (Figure 1.8) in the new millennium, was the next major application with glass/aluminium (glare) composites on the upper portion of the fuselage and glass and graphite composites in the centre wing-box, floor beams and aft pressure bulkhead.

Already in the 1990s, the demand for more efficient aircraft with lower operation and maintenance costs made it clear that more usage of composites was necessary for significant reductions in weight in order to gain in fuel efficiency. In addition, improved fatigue lives and improved corrosion resistance compared with aluminium suggested that more composites on aircraft were necessary. This, despite the fact that the cost of composites was still not competitive with aluminium and the stringent certification requirements would lead to increased certification cost.

Boeing was the first to commit to a composite fuselage and wing with the 787 (Figure 1.9) launched in the first decade of the new millennium. Such extended use of composites, about 50% of the structure (combined with other advanced technologies) would give the efficiency improvement (increased range, reduced operation and maintenance costs) needed by the airline operators.

Figure 1.9   Boeing 787 Dreamliner (Courtesy of Agnes Blom; see Plate 9 for the colour figure)

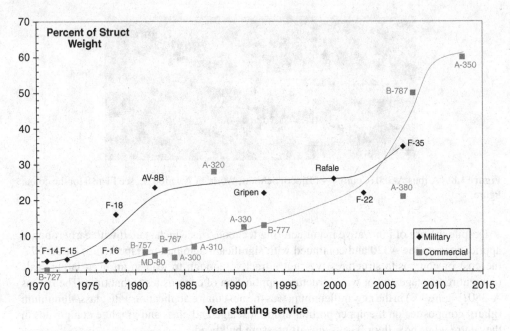

Figure 1.10   Applications of composites in military and civilian aircraft structures

The large number of orders (most successful launch in history) for the Boeing 787 led Airbus to start development of a competing design in the market segment covered by the 787 and the 777. This is the Airbus A-350, with all-composite fuselage and wings.

Another way to see the implementation of composites in aircraft structure over time is by examining the amount of composites (by weight) used in various aircraft models as a function of time. This is shown in Figure 1.10 for some civilian and military aircraft. It should be borne in mind that the numbers shown in Figure 1.10 are approximate as they had to be inferred from open literature data and interpretation of different company announcements [1–8].

Both military and civilian aircraft applications show the same basic trends. A slow start (corresponding to the period where the behaviour of composite structures is still not well understood and limited low risk applications are selected) is followed by rapid growth as experience is gained, reliable analysis and design tools are developed and verified by testing and the need for reduced weight becomes more pressing. After the rapid growth period, the applicability levels off as: (a) it becomes harder to find parts of the structure that are amenable to the use of composites; (b) the cost of further composite implementation becomes prohibitive; and (c) managerial decisions and other external factors (lack of funding, changes in research emphasis, investments already made in other technologies) favour alternatives. As might be expected, composite implementation in military aircraft leads the way. The fact that in recent years civilian applications seem to have overtaken military applications does not reflect true trends as much as lack of data on the military side (e.g. several military programs such as the B-2 have very large composite applications, but the actual numbers are hard to find).

It is still unclear how well the composite primary structures in the most recent programs such as the Boeing 787 and the Airbus A-350 will perform and whether they will meet the design targets. In addition, several areas such as the performance of composites after impact, fatigue and damage tolerance are still the subjects of ongoing research. As our understanding in these areas improves, the development cost, which currently requires a large amount of testing to answer questions where analysis is prohibitively expensive and/or not as accurate as needed to reduce the amount of testing, will drop significantly. In addition, further improvements in robotics technology and integration of parts into larger co-cured structures are expected to make the fabrication cost of composites more competitive compared with metal airplanes.

References

[1] *Jane's All the World's Aircraft 2000–2001*, P. Jackson (ed.), Jane's Information Group, 2000.
[2] *NetComposites News*, 14 December 2000.
[3] *Aerospace America*, May 2001, 45–47.
[4] *Aerospace America,* September 2005.
[5] www.compositesworld.com/articles/boeing-sets-pace-for-composite-usage-in-large-civil-aircraft.
[6] www.boeing.com/commercial/products.html.
[7] Deo, R.B., Starnes, J.H., Holzwarth, R.C. Low-Cost Composite Materials and Structures for Aircraft Applications. Presented at the RTO AVT Specialists' Meeting on Low Cost Composite Structures; May 7-11, 2001 and published in RTO-MP-069(II); Loen, Norway.
[8] Watson, J.C. and Ostrodka, D.L., AV-8B Forward Fuselage Development. Proceedings of 5th Conference on Fibrous Composites in Structural Design; 1981 January; New Orleans, LA.

References

2

Cost of Composites: a Qualitative Discussion

Considering that cost is the most important aspect of an airframe structure (along with the weight), one would expect it to be among the best defined, most studied and most optimized quantities in a design. Unfortunately, it remains one of the least understood and ill-defined aspects of a structure. There are many reasons for this inconsistency, some of which are: (a) cost data for different fabrication processes and types of parts are proprietary and only indirect or comparative values are usually released; (b) there seems to be no well-defined reliable method to relate design properties such as geometry and complexity to the cost of the resulting structure; (c) different companies have different methods of bookkeeping the cost and it is hard to make comparisons without knowing these differences (e.g. the cost of the autoclave can be apportioned to the number of parts being cured at any given time or it may be accounted for as an overhead cost, included in the total overhead cost structure of the entire factory); (d) learning curve effects, which may or may not be included in the cost figures reported, tend to confuse the situation especially since different companies use different production run sizes in their calculations.

These issues are common to all types of manufacturing technologies and not just the aerospace sector. In the case of composites, the situation is further complicated by the relative novelty of the materials and processes being used, the constant emergence of new processes or variations thereof that alter the cost structure and the high nonrecurring cost associated with switching to the new processes that, usually, acts as a deterrent towards making the switch.

The discussion in this chapter attempts to bring up some of the cost considerations that may affect a design. This discussion is by no means exhaustive; in fact, it is limited by the lack of extensive data and generic but accurate cost models. It serves mainly to alert or sensitize a designer to several issues that affect the cost. These issues, when appropriately accounted for, may lead to a robust design that minimizes the weight and is cost competitive with the alternatives.

Design and Analysis of Composite Structures: With Applications to Aerospace Structures, Second Edition. Christos Kassapoglou.
© 2013 John Wiley & Sons, Ltd. Published 2013 by John Wiley & Sons, Ltd.

The emphasis is placed on recurring and nonrecurring costs. The recurring cost is the cost that is incurred every time a part is fabricated. The nonrecurring cost is the cost that is incurred once during the fabrication run.

2.1   Recurring Cost

The recurring cost includes the raw material cost (including scrap) for fabricating a specific part, the labour hours spent in fabricating the part and cost of attaching it to the rest of the structure. The recurring cost is hard to quantify, especially for complex parts. There is no single analytical model that relates specific final part attributes such as geometry, weight, volume, area or complexity to the cost of each process step and through the summation over all process steps to the total recurring cost. One of the reasons for these difficulties and, as a result, the multitude of cost models that have been proposed with varying degrees of accuracy and none of them all-encompassing, is the definition of complexity. One of the most rigorous and promising attempts to define complexity and its effect on the recurring cost of composite parts was by Gutowski et al. [1, 2].

For the case of *hand layup*, averaging over a large quantity of parts of varying complexity ranging from simple flat laminates to compound curvature parts with co-cured stiffeners, the fraction of total cost taken up by the different process steps is shown in Figure 2.1 (taken from Reference 3).

It can be seen from Figure 2.1 that, by far, the costliest steps are locating the plies into the mould (42%) and assembling to the adjacent structure (29%). Over the years, cost-cutting and optimization efforts have concentrated mostly on these two process steps. This is the reason

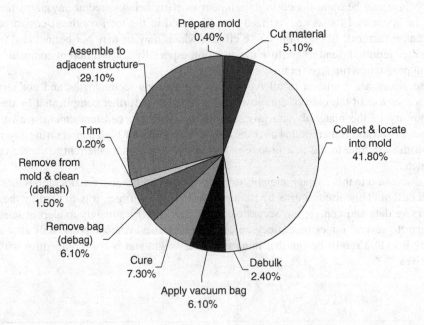

Figure 2.1   Process steps for hand layup and their cost as fractions of total recurring cost [3] (See Plate 10 for the colour figure)

Figure 2.2 Integration of various parts into a single co-cured part to minimize assembly cost (Courtesy of Aurora Flight Sciences)

for introducing automation. Robots, used for example in automated tape layup, take the cut plies and locate them automatically in the mould, greatly reducing the cost associated with that process step, improving the accuracy and reducing or eliminating human error, thereby increasing consistency and quality. Since assembly accounts for about one-third of the total cost, increasing the amount of co-curing where various components are cured at the same time reduces drastically the assembly cost. An example of this integration is shown in Figure 2.2.

These improvements as well as others associated with other process steps such as automated cutting (using lasers or water jets), trimming and drilling (using numerically controlled equipment) have further reduced the cost and improved quality by reducing the human involvement in the process. Hand layup and its automated or semi-automated variations can be used to fabricate just about any piece of airframe structure. An example of a complex part with compound curvature and parts intersecting in different directions is shown in Figure 2.3.

Further improvements have been brought to bear by taking advantage of the experience acquired in the textile industry. By working with fibres alone, several automated techniques such as knitting, weaving, braiding and stitching can be used to create a preform, which is then injected with resin. This is the *resin transfer moulding (RTM) process*. The raw material cost can be less than half the raw material cost of preimpregnated material (prepreg) used in hand layup or automated tape layup because the impregnation step needed to create the prepreg used in those processes is eliminated. On the other hand, ensuring that the resin fully wets all fibres everywhere in the part and that the resin content is uniform and equal to the desired resin content can be hard for complex parts and may require special tooling, complex design of injection and overflow ports and use of high pressure. It is not uncommon, for complex RTM parts to have 10–15% less strength (especially in compression and shear) than their equivalent prepreg parts due to reduced resin content. Another problem with matched metal

Figure 2.3   Portion of a three-dimensional composite part with compound curvature fabricated using hand layup

moulding RTM is the high nonrecurring cost associated with the fabrication of the moulds. For this reason, variations of the RTM process such as *vacuum-assisted RTM (VARTM)* where one of the tools is replaced by a flexible caul plate whose cost is much lower than an equivalent matched metal mould or *resin film infusion (RFI)* where the resin is drawn into dry fibre preforms from a pool or film located under it and/or from staged plies that already have resin in them, have been used successfully in several applications (Figure 2.4). Finally, due to the fact that the process operates with resin and fibres separately, the high amounts of scrap associated with hand layup can be significantly reduced.

Introduction of more automation led to the development of *automated fibre or tow placement*. This was a result of trying to improve filament winding (see below). Robotic heads can each dispense material as narrow as 3 mm and as wide as 100 mm by manipulating individual strips (or tows) each 3 mm wide. Tows are individually controlled so the amount of material laid down in the mould can vary in real time. Starting and stopping individual tows also allows the creation of cutouts 'on the fly'. The robotic head can move in a straight line at very high rates (as high as 30 m/min). This makes automated fibre placement an ideal process for laying material down to create parts with large surface area and small variations in thickness or with a limited number of cutouts. For maximum efficiency, structural details (e.g. cutouts) that require starting and stopping the machine or cutting material while laying it down should be avoided. Material scrap is very low. Convex as well as concave tools can be used since the machine does not rely on constant fibre tension, as in filament winding, to lay the material down. There are limitations with the process associated with the accuracy of starting and stopping when material is laid down at high rates and the size and shape of the tool when concave tools are used (in order to avoid interference of the robotic head with the tool). The ability to steer fibres on prescribed paths (Figure 2.5) can also be used as an advantage by

Figure 2.4 Curved stiffened panels made with the RTM process

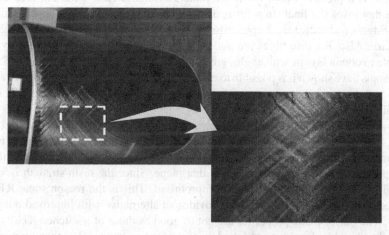

Figure 2.5 Composite cylinder with steered fibres fabricated by automated fibre placement (made in a collaborative effort by TUDelft and NLR; see Plate 11 for the colour figure)

transferring the loads efficiently across the part. This results in laminates where stiffness and strength are a function of location and provides an added means for optimization [4, 5].

Automated fibre placement is most efficient when making large parts. Parts such as stringers, fittings, small frames that do not have at least one sizeable side where the advantage of the high lay-down rate of material by the robotic head can be brought to bear, are hard to make and/or not cost competitive. In addition, skins with large amounts of taper and number of cutouts may also not be amenable to this process.

In addition to the above processes that apply to almost any type of part (with some exceptions already mentioned for automated fibre placement) specialized processes that are very efficient for the fabrication of specific part types or classes of parts have been developed. The most common of these are filament winding, pultrusion and press moulding using long discontinuous fibres and sheet moulding compounds.

Filament winding, as already mentioned, is the precursor to advanced fibre or tow placement. It is used to make pressure vessels and parts that can be wound on a convex mandrel. The use of a convex mandrel is necessary in order to maintain tension on the filaments being wound. The filaments are drawn from a spool without resin and are driven through a resin bath before they are wound around the mandrel. Due to the fact that tension must be maintained on the filaments, their paths can only be geodetic paths on the surface of the part being woven. This means that, for a cylindrical part, if the direction parallel to the cylinder axis is denoted as the zero direction, winding angles between 15° and 30° are hard to maintain (filaments tend to slide) and angles less than 15° cannot be wound at all. Thus, for a cylindrical part with conical closeouts at its ends, it is impossible to include 0° fibres using filament winding. 0° plies can be added by hand if necessary at a significant increase in cost. Since the material can be dispensed at high rates, filament winding is an efficient and low-cost process. In addition, fibres and matrix are used separately and the raw material cost is low. Material scrap is very low.

Pultrusion is a process where fibres are pulled through a resin bath and then through a heated die that gives the final shape. It is used for making long constant-cross-section parts such as stringers and stiffeners. Large cross-sections, measuring more than 25 × 25 cm are hard to make. Also, because fibres are pulled, if the pulling direction is denoted by 0°, it is not possible to obtain layups with angles greater than 45° (or more negative than –45°). Some recent attempts have shown it is possible to obtain longitudinal structures with some taper. The process is very low cost. Long parts can be made and then cut at the desired length. Material scrap is minimal.

With *press moulding* it is possible to create small three-dimensional parts such as fittings. Typically, composite fittings made with hand layup or RTM without stitching suffer from low out-of-plane strength. There is at least one plane without any fibres crossing it and thus only the resin provides strength perpendicular to that plane. Since the resin strength is very low, the overall performance of the fitting is compromised. This is the reason some RTM parts are stitched. Press moulding (Figure 2.6) provides an alternative with improved out-of-plane properties. The out-of-plane properties are not as good as those of a stitched RTM structure, but better than hand laid-up parts and the low cost of the process makes them very attractive for certain applications. The raw material is essentially a slurry of randomly oriented long discontinuous fibres in the form of chips. High pressure applied during cure forces the chips to completely cover the tool cavity. Their random orientation is, for the most part, maintained. As a result, there are chips in every direction with fibres providing extra strength. Besides

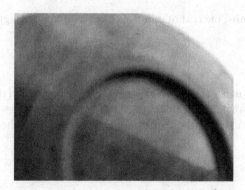

Figure 2.6 Portion of a composite fitting made by press moulding

three-dimensional fittings, the process is also very efficient and reliable for making clips and shear ties. Material scrap is minimal. The size of the parts to be made is limited by the press size and the tool cost. If there are enough parts to be made, the high tooling cost is offset by the low recurring cost.

There are other fabrication methods or variations within a fabrication process that specialize in certain types of parts and/or part sizes. The ones mentioned above are the most representative. There is one more aspect that should be mentioned briefly; the effect of learning curves. Each fabrication method has its own learning curve which is specific to the process, the factory and equipment used and the skill level of the personnel involved. The learning curve describes how the recurring cost for making the same part multiple times decreases as a function of the number of parts. It reflects the fact that the process is streamlined and people find more efficient ways to do the same task. Learning curves are important when comparing alternative fabrication processes. A process with a steep learning curve can start with a high unit cost but, after a sufficiently large number of parts, can yield unit costs much lower than another process, which starts with lower unit cost, but has a shallower learning curve. As a result, the first process may result in lower average cost (total cost over all units divided by the number of units) than the first.

As a rule, fabrication processes with little or no automation have steeper learning curves and start with higher unit cost. This is because an automated process has fixed throughput rates while human labour can be streamlined and become more efficient over time as the skills of the people involved improve and ways of speeding up some of the process steps used in making the same part are found. The hand layup process would fall in this category with, typically, an 85% learning curve. An 85% learning curve means that the cost of unit $2n$ is 85% of the cost of unit n. Fabrication processes involving a lot of automation have shallower learning curves and start at lower unit cost. One such example is the automated fibre/tow placement process with, typically, a 92% learning curve. A discussion of some of these effects and the associated tradeoffs can be found in Reference 3.

An example comparing a labour intensive process with 85% learning curve and cost of unit one 40% higher than an automated fabrication process with 92% learning curve is given here to highlight some of the issues that are part of the design phase, in particular at early stages when the fabrication process or processes have not been finalized yet.

Assuming identical units, the cost of unit n, $C(n)$, is assumed to be given by a power law:

$$C(n) = \frac{C(1)}{n^r} \tag{2.1}$$

where $C(1)$ is the cost of unit 1 and r is an exponent that is a function of the fabrication process, factory capabilities, personnel skill, etc.

If $p\%$ is the learning curve corresponding to the specific process, then

$$p = \frac{C(2n)}{C(n)} \tag{2.2}$$

Using Equation (2.1) to substitute in Equation (2.2) and solving for r, it can be shown that,

$$r = -\frac{\ln p}{\ln 2} \tag{2.3}$$

For our example, with process A having $p_A = 0.85$ and process B having $p_B = 0.92$, substituting in Equation (2.3) gives $r_A = 0.2345$ and $r_B = 0.1203$. If the cost of unit 1 of process B is normalized to 1, $C_B(1) = 1$, then the cost of unit 1 of process A will be 1.4, based on our assumption stated earlier, so $C_A(1) = 1.4$. Putting it all together,

$$C_A(n) = \frac{1.4}{n^{0.2345}} \tag{2.4}$$

$$C_B(n) = \frac{1}{n^{0.1203}} \tag{2.5}$$

The cost as a function of n for each of the two processes can now be plotted in Figure 2.7. A logarithmic scale is used on the x axis to better show the differences between the two curves.

It can be seen from Figure 2.7 that a little after the 20th part, the unit cost of process A becomes less than that of process B suggesting that for sufficiently large runs, process A may

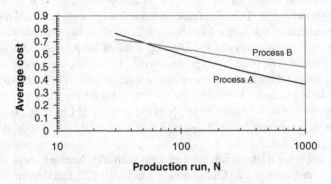

Figure 2.7　Unit recurring cost for a process with no automation (process A) and an automated process (process B)

be competitive with process B. To investigate this further, the average cost over a production run of N units is needed. If N is large enough, the average cost can be accurately approximated by:

$$C_{av} = \frac{1}{N} \sum_{n=1}^{N} C(n) \approx \frac{1}{N} \int_{1}^{N} C(n)\,dn \qquad (2.6)$$

and using Equation (2.1),

$$C_{av} = \frac{1}{N} \int_{1}^{N} \frac{C(1)}{n^r}\,dn = \frac{C(1)}{1-r}\left(\frac{1}{N^r} - \frac{1}{N}\right) \qquad (2.7)$$

Note that to derive Equation (2.7) the summation was approximated by an integral. This gives accurate results for $N > 30$. For smaller production runs ($N < 30$) the summation in Equation (2.6) should be used. Equation (2.7) is used to determine the average cost for Process A and Process B as a function of the size of the production run N. The results are shown in Figure 2.8.

As can be seen from Figure 2.8, Process B, with automation, has lower average cost as long as less than approximately 55 parts are made ($N < 55$). For $N > 55$, the steeper learning curve of Process A leads to lower average cost for that process. Based on these results, the less-automated process should be preferred for production runs with more than 50–60 parts. However, these results should be viewed only as preliminary, as additional factors that play a role were neglected in the above discussion. Some of these factors are briefly discussed below.

Process A, which has no automation, is prone to human errors. This means that: (a) the part consistency will vary more than in Process B; and (b) the quality and accuracy may not always be satisfactory requiring repairs or scrapping of parts.

In addition, process improvements, which the equations presented assume to be continuous and permanent, are not always possible. It is likely that after a certain number of parts, all possible improvements have been implemented. This would suggest that the learning curves typically reach a plateau after a while and the cost cannot be reduced below that plateau without

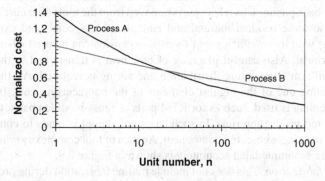

Figure 2.8 Average recurring cost for a process with no automation (process A) and a fully automated process (process B)

major changes in the process (new equipment, new process steps, etc.). These drastic changes are more likely in automated processes where new equipment is developed regularly than in a nonautomated process. Therefore, while the conclusion that a less-automated process will give lower average cost over a sufficiently large production run is valid, in reality, it may only occur under very special circumstances favouring continuous process improvement, consistent high part quality and part accuracy, etc. In general, automated processes are preferred because of their quality, consistency and potential for continuous improvement.

The above is a very brief reference to some of the major composite fabrication processes. It serves to bring some aspects to the forefront as they relate to design decisions. More in-depth discussion of some of these processes and how they relate to the design of composite parts can be found in References 6 and 7.

2.2   Nonrecurring Cost

The main components of nonrecurring cost follow the phases of the development of a program and are the following.

Design. Typically divided in stages (e.g. conceptual, preliminary and detail) it is the phase of creating geometry of the various parts and coming up with the material(s) and fabrication processes (see Sections 5.1.1 and 5.1.2 for a more detailed discussion). For composites it is more involved than for metals because it includes detailed definition of each ply in a layup (material, orientation, location of boundaries, etc.). The design of press-moulded parts would take less time than other fabrication processes as definition of the boundaries of each ply is not needed. The material under pressure fills the mould cavity and the concept of a ply is more loosely used.

Analysis. In parallel with the design effort, it determines applied loads for each part and comes up with the stacking sequence and geometry to meet the static and cyclic loads without failure and with minimum weight and cost. The multitude of failure modes specific to composites (delamination, matrix failure, fibre failure, etc.) makes this an involved process that may require special analytical tools and modelling approaches.

Tooling. This includes the design and fabrication of the entire tool string needed to produce the parts: moulds, assembly jigs and fixtures, etc. For composite parts cured in the autoclave, extra care must be exercised to account for thermal coefficient mismatch (when metal tools are used) and spring-back phenomena where parts removed from the tools after cure tend to deform slightly to release some residual thermal and cure stresses. Special (and expensive) metal alloys (e.g. invar) with low coefficients of thermal expansion can be used where dimensional tolerances are critical. Also careful planning of how heat is transmitted to the parts during cure for more uniform temperature distribution and curing is required. All these add to the cost, making tooling one of the biggest elements of the nonrecurring cost. In particular, if matched metal tooling is used, such as for RTM parts or press-moulded parts, the cost can be prohibitive for short production runs. In such cases an attempt is made to combine as many parts as possible in a single co-cured component. An idea of tool complexity when local details of a wing-skin are accommodated accurately is shown in Figure 2.9.

Nonrecurring fabrication. This does not include routine fabrication during production that is part of the recurring cost. It includes: (a) one-off parts made to toolproof the tooling concepts; (b) test specimens to verify analysis and design and provide the database needed to support

Figure 2.9 Co-cure of large complex parts (Courtesy of Aurora Flight Sciences; see Plate 12 for the colour figure)

design and analysis; and (c) producibility specimens to verify the fabrication approach and avoid surprises during production. This can be costly when large co-cured structures are involved with any of the processes already mentioned. It may take the form of a building-block approach where fabrication of subcomponents of the full co-cured structure is done first to check different tooling concepts and verify part quality. Once any problems (resin-rich, resin-poor areas, locations with insufficient degree of cure or pressure during cure, voids, local anomalies such as 'pinched' material, fibre misalignment), are resolved, more complex portions leading up to the full co-cured structure are fabricated to minimize the risk and verify the design.

Testing. During this phase, the specimens fabricated during the previous phase are tested. This includes the tests needed to verify analysis methods and provide missing information for various failure modes. This does not include testing needed for certification (see next item). If the design has opted for large co-cured structures to minimize recurring cost, the cost of testing can be very high since it, typically, involves testing of various subcomponents first and then testing the full co-cured component. Creating the right boundary conditions and applying the desired load combinations in complex components results in expensive tests.

Certification. This is one of the most expensive nonrecurring cost items. Proving that the structure will perform as required and providing relevant evidence to certifying agencies requires a combination of testing and analysis [8–10]. The associated test program can be extremely broad (and expensive). For this reason, a building-block approach is usually followed where tests of increasing complexity, but reduced in numbers follow simpler more numerous tests, each time building on the previous level in terms of information gained, increased confidence in the design performance and reduction of risk associated with the full-scale article. In a broad level description going from the simplest to the most complex: (a) material qualification where thousands of coupons with different layups are fabricated and tested under different applied loads and environmental conditions with and without damage to provide statistically meaningful values (see Sections 5.1.3–5.1.5) for strength and stiffness of the material and stacking sequences to be used; (b) element tests of specific structural details isolating failure modes or interactions; (c) subcomponent and component tests verifying how the elements come together and providing missing (or hard to otherwise accurately quantify) information on failure loads and modes; (d) full-scale test. Associated with each test level, analysis is used to reduce test data, bridge structural performance from one level to the next and justify the reduction of specimens at the next level of higher complexity. The tests include static and fatigue tests leading to the flight test program that is also part of the certification effort. When new fabrication methods are used, it is necessary to prove that they will generate parts of consistently high quality. This, sometimes, along with the investment in equipment purchasing and training, acts as a deterrent in switching from a proven method (e.g. hand layup) with high comfort level to a new method some aspects of which may not be well known (e.g. automated fibre placement).

The relative cost of each of the different phases described above is a strong function of the application, the fabrication process(es) selected and the size of the production run. It is, therefore, hard to create a generic pie chart that would show how the cost associated with each compares. In general, it can be said that certification tends to be most expensive followed by tooling, nonrecurring fabrication and testing.

2.3   Technology Selection

The discussion in the two previous sections shows that there is a wide variety of fabrication processes, each with its own advantages and disadvantages. Trading these and calculating the recurring and nonrecurring costs associated with each selection is paramount in coming up with the best choice. The problem becomes very complex when one considers large components such as the fuselage or the wing or entire aircraft. At this stage, it is useful to define the term 'technology' as referring to any combination of material, fabrication process and design concept. For example, graphite/epoxy skins using fibre placement would be one technology. Similarly, sandwich skins with a mixture of glass/epoxy and graphite/epoxy plies made using hand layup would be another technology.

In a large-scale application such as an entire aircraft, it is extremely important to determine the optimum technology mix, i.e. the combination of technologies that will minimize weight and cost. This can be quite complicated since different technologies are more efficient for different types of part. For example, fibre-placed skins might give the lowest weight and recurring cost, but assembling the stringers as a separate step (bonding or fastening) might make the resulting skin/stiffened structure less cost competitive. On the other hand, using resin transfer moulding to co-cure skin and stringers in one step might have lower overall recurring cost at a slight increase in weight (due to reduced strength and stiffness) and a significant increase in nonrecurring cost due to increased tooling cost. At the same time, fibre placement may require significant capital outlays to purchase automated fibre/tow placement machines. These expenditures require justification accounting for the size of the production run, availability of capital and the extent to which capital investments already made on the factory floor for other fabrication methods have been amortized or not.

These tradeoffs and final selection of optimum technology mix for the entire structure of an aircraft are done early in the design process and 'lock in' most of the cost of an entire program. For this reason it is imperative that the designer be able to perform these trades in order to come up with the 'best alternatives'. As will be shown in this section these 'best alternatives'

Table 2.1 Part families of an airframe

Part family	Description
Skins and covers	Two-dimensional parts with single curvature
Frames, bulkheads, beams, ribs, intercostals	Two-dimensional flat parts
Stringers, stiffeners, breakers	One-dimensional (long) parts
Fittings	Three-dimensional small parts connecting other parts
Decks and floors	Mostly flat parts
Doors and fairings	Parts with compound curvature
Miscellaneous	Seals, etc.

are a function of the amount of risk one is willing to take, the amount of investment available and the relative importance of recurring, nonrecurring cost and weight [11–14].

In order to make the discussion more tractable, the airframe (load-bearing structure of an aircraft) is divided into part families. These are families of parts that perform the same function, have approximately the same shapes, are made with the same material(s) and can be fabricated by the same manufacturing process. The simplest division into part families is shown in Table 2.1. In what follows the discussion will include metals for comparison purposes.

The technologies that can be used for each part family are then determined. This includes the material (metal or composite and, if composite, the type of composite), fabrication process (built-up sheet metal, automated fibre placement, resin transfer moulding, etc.) and design concept (e.g. stiffened skin versus sandwich). In addition, the applicability of each technology to each part family is determined. This means determining what portion in the part family can be made by the technology in question. Usually, as the complexity of the parts in a part family increases, a certain technology becomes less applicable. For example, small skins with large changes in thickness across their length and width cannot be made by fibre placement and have low cost; or pultrusion cannot be used (efficiently) to make tapering beams. A typical breakdown by part family and applicability by technology is shown in Table 2.2. For convenience, the following shorthand designations are used: SMT = (built-up) sheet metal, HSM = high-speed-machined aluminium, HLP = hand layup, AFP = automated fibre placement, RTM = resin transfer moulding, ALP = automated (tape) layup, PLT = pultrusion. The numbers in Table 2.2 denote the percentage of the parts in the part family that can be made by the selected process and have acceptable (i.e. competitive) cost.

Table 2.2 Applicability of fabrication processes by part family

Part family	SMT (%)	HLP (%)	HSM (%)	AFP (%)	RTM (%)	PLT (%)	ALP (%)
Skins and covers	100	100	15	80	100	0	50
Frames, etc.	100	100	65	55	100	10	30
Stringers, etc.	100	100	5	0	100	90	0
Fittings	100	85	5	0	100	0	0
Decks and floors	90	100	35	40	90	10	20
Doors and fairings	80	100	5	35	90	5	10

It is immediately obvious from Table 2.2 that no single technology can be used to make an entire airframe in the most cost-effective fashion. There are some portions of certain part families that are more efficiently made by another technology. While the numbers in Table 2.2 are subjective, they reflect what is perceived to be the reality of today and they can be modified according to specific preferences or expected improvements in specific technologies.

Given the applicabilities of Table 2.2, recurring and nonrecurring cost data are obtained or estimated by part family. This is done by calculating or estimating the average cost for a part of medium complexity in the specific part family made by a selected process and determining the standard deviation associated with the distribution of cost around that average as the part complexity ranges from simple to complex parts. This can be done using existing data as is shown in Figure 2.10, for technologies already implemented such as HLP or by extrapolating and approximating limited data from producibility evaluations, vendor information and anticipated improvements for new technologies or technologies with which a particular factory has not had enough experience.

In the case of the data shown in Figure 2.10, data over 34 different skin parts made with hand layup show an average (or mean) cost of 14 h/kg of finished product and a standard deviation around that mean of about 11 h/kg (the horizontal arrows in Figure 2.10 cover approximately two standard deviations). This scatter around the mean cost is mostly due to variations in complexity. A simple skin (flat, constant thickness, no cutouts) can cost little as 1 h/kg while a complex skin (curved, with ply dropoffs, with cutouts) can cost as high as 30 h/kg. In addition to part complexity, there is a contribution to the standard deviation due to uncertainty. This uncertainty results mainly from two sources [12]: (a) not having enough experience with the process and applying it to types of part to which it has not been applied before; this is referred to as production-readiness; and (b) operator or equipment variability.

Determining the portion of the standard deviation caused by uncertainty is necessary in order to proceed with the selection of the best technology for an application. One way to separate uncertainty from complexity is to use a reliable cost model to predict the cost of parts of different complexity for which actual data are available. The difference between the predictions and the actual data is attributed to uncertainty. By normalizing the prediction by the actual cost for all parts available, a distribution is obtained the standard deviation of which is a measure of the uncertainty associated with the process in question. This standard deviation (or its square, the variance) is an important parameter because it can be associated with the

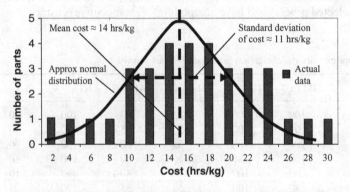

Figure 2.10 Distribution of recurring cost of HLP skins

risk. If the predicted cost divided by actual cost data were all in a narrow band around the mean, the risk in using this technology (e.g. HLP) for this part family (e.g. skins) would be very low since the expected cost range would be narrow. Since narrow distributions have low variances, the lower the variance, the lower the risk.

It is more convenient, instead of using absolute cost numbers to use cost savings numbers obtained by comparing each technology of interest with a baseline technology. In what follows, SMT is used as the baseline technology. Positive cost savings numbers denote cost reduction below SMT cost and negative cost savings numbers denote cost increase above SMT costs. Also, generalizing the results from Figure 2.10, it will be assumed that the cost savings for a certain technology applied to a certain part family is normally distributed. Other statistical distributions can be used and, in some cases, will be more accurate. For the purposes of this discussion, the simplicity afforded by assuming a normal distribution is sufficient to show the basic trends and draw the most important conclusions.

By examining data published in the open literature, inferring numbers from trend lines and using experience, the mean cost savings and variances associated with the technologies given in Table 2.2 can be compiled. The results are shown in Table 2.3. Note that these results reflect a specific instant in time and they comprise the best estimate of current costs for a given technology. This means that some learning curve effects are already included in the numbers. For example, HLP and RTM parts have been used fairly widely in industry and factories have come down their respective learning curves. Other technologies such as AFP have not been used as extensively and the numbers quoted are fairly high up in the respective learning curves.

For each technology/part family combination in Table 2.3, two numbers are given. The first is the cost savings as a fraction (i.e. 0.17 implies 17% cost reduction compared to SMT) and the second is the variance (square of standard deviation) of the cost savings population. Negative cost savings numbers imply increase in cost over SMT. They are included here because the weight savings may justify the use of the technology even if, on average, the cost is higher. For SMT and some HLP cases, the variance is set to a very low number, 0.0001 to reflect the fact that the cost for these technologies and part families is well understood and there is little uncertainty associated with it. This means the technology has already been in use for

Table 2.3 Typical cost data by technology by part family

Part family	SMT (%)	HLP (%)	HSM (%)	AFP (%)	RTM (%)	PLT (%)	ALP (%)
Skins and covers	0.0	**0.17**	0.2	0.25	0.08	0.08	**0.32**
	0.0001	0.0061	0.02	0.009	0.003	0.06	0.01
Frames, etc.	0.0	0.1	0.28	0.1	0.18		0.40
	0.0001	0.0001	0.006	0.06	0.008		0.08
Stringers, etc.	0.0	−0.05	(in skins)		0.05	0.40	**0.35**
	0.0001	0.0001			0.002	0.001	0.09
Fittings	0.0		0.2		−0.10		
	0.0001		0.005		0.015		
Decks and floors	0.0	−0.01		0.15	−0.15		**0.20**
	0.0001	0.0001		0.01	0.008		**0.02**
Doors and fairings	0.0		0.1	0.25	−0.10		0.35
	0.0001	0.0021		**0.026**	0.01		**0.05**

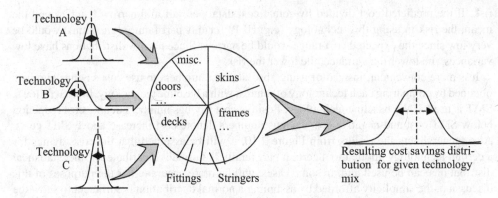

Figure 2.11   Combining different technologies to an airframe and expected cost distribution

that part family for some time. Some of the data in Table 2.3 are highlighted to show some of the implications: (a) HLP skins have 17% lower cost than SMT skin mostly due to co-curing large pieces and eliminating or minimizing assembly; (b) ALP has the lowest cost numbers, but limited applicability (see Table 2.2); (c) the variance in some cases such as ALP decks and floors or AFP doors and fairings is high because for many parts in these families additional nonautomated steps are necessary to complete fabrication. This is typical of sandwich parts containing core where core processing involves manual labour and increases the cost. Manual labour increases the uncertainty due to the operator variability already mentioned.

Given the data in Tables 2.2 and 2.3, one can combine different technologies to make a part family. Doing that over all part families results in a technology mix. This technology mix has an overall mean cost savings and variance associated with it that can be calculated using the data from Tables 2.2 and 2.3 and using the percentages of how much of each part family is made by each technology [12, 13]. This process is shown in Figure 2.11. Obviously, some technology mixes are better than others because they have lower recurring cost and/or lower risk. An optimization scheme can then be set up [13] that aims at determining the technology mix that minimizes the overall recurring cost savings (below the SMT baseline) keeping the associated variance (and thus the risk) below a preselected value. By changing that preselected value from very small (low risk) to very high (high risk) different optimum mixes can be obtained. A typical result of this process is shown in Figure 2.12 for the case of a fuselage and wing of a 20-passenger commuter plane.

The risk is shown in Figure 2.12 on the x axis as the square root of the variance or standard deviation of the cost savings of the resulting technology mix. For each value of risk, the optimization process results in a technology mix that maximizes cost savings. Assuming that the cost savings of each technology mix is normally distributed, the corresponding probabilities that the cost savings will be lower than a specified value can be determined [13]. These different probabilities trace the different curves shown in Figure 2.12. For example, if the risk is set at 0.05 on the x axis, the resulting optimum mix has 1% probability of not achieving 11.5% savings, 2.5% probability of not achieving 13.5% savings, 5% probability of not achieving 15% savings and so on. Note that all curves, except the 50% probability curve go through a maximum. This maximum can be used for selecting the optimum technology mix to be used. For example, if a specific factory/management team is risk averse it would probably go with

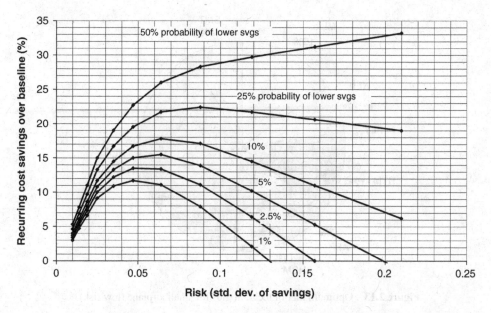

Figure 2.12   Recurring cost savings as a function of risk

the 1% curve which goes through a maximum at a risk value slightly less than 0.05. The team would expect savings of at least 11.5%. A more aggressive team might be comfortable with 25% probability that the cost savings is lower and would use the 25% curve. This has a maximum at a risk value of 0.09 with corresponding savings of 22.5%. However, there is a 25% probability that this level of savings will not be met. That is, if this technology mix were to be implemented a large number of times, it would meet or exceed the 22.5% savings target only 75% of the time. It is up to the management team and factory to decide which risk level and curve they should use. It should be noted that for very high risk values, beyond 0.1, the cost saving curves eventually become negative. For example the 1% curve becomes negative at a risk value of 0.13. This means that the technology mix corresponding to a risk value of 0.13 has so much uncertainty that there is 99% probability that the cost savings will be negative, i.e. the cost will be higher than the SMT baseline.

Once a risk level is selected from Figure 2.12, the corresponding technology mix is known from the optimization process. Examples for low and high risk values are shown in Figures 2.13 and 2.14.

For the low-risk optimum mix of Figure 2.13, there is a 10% probability of not achieving 12.5% cost savings. For the high-risk optimum mix of Figure 2.14 there is a 10% probability of not achieving 7% cost savings. The only reason to go with the high-risk optimum mix is that, at higher probability values (greater than 25%) it exceeds the cost savings of the low-risk optimum mix.

A comparison of Figures 2.14 and 2.13 shows that as the risk increases, the percentage usage of baseline SMT and low-risk low-return HLP and RTM decreases while the usage of higher-risk high-return AFP and ALP increases. ALP usage doubles from 6% to 12% and AFP usage increases by a factor of almost 7, from 3% to 20%. The amount of PLT also increases (in fact doubles), but since PLT is only limited to stringers in this example, the overall impact

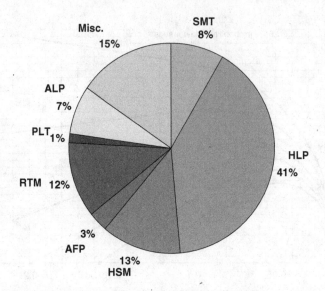

Figure 2.13   Optimum mix of technologies for small airplane (low risk)

of using PLT is quite small. It should be noted that there is a portion of the airframe denoted by 'Misc'. These are miscellaneous parts such as seals or parts for which applicability is unclear and mixing technologies (e.g. pultruded stringers co-bonded on fibre-placed skins) might be a better option, but no data were available for generating predictions.

Finally, the breakdown by part family for one of the cases, the low-risk optimum mix of Figure 2.13 is shown in Table 2.4. For example, 21.1% of the frames are made by HLP, 32.4%

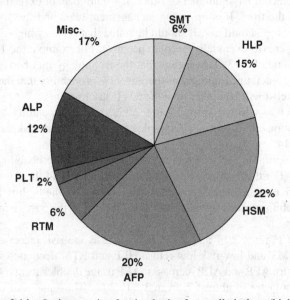

Figure 2.14   Optimum mix of technologies for small airplane (high risk)

Table 2.4 Low-risk technology mix by part family and technology

	SMT (%)	%HLP (%)	%HSM (%)	%AFP (%)	%RTM (%)	%PLT (%)	%ALP (%)
Skins + …	0	81	0	4.9	3.6	0	10.5
Frames + …	0	21.1	32.4	0	40.5	0	6
Stringers	79.3	0	0	0	0	20.7	0
Fittings	44.1	0	55.5	0	0	0	0.4
Decks + …	0	76.5	0	8.4	3.1	0	12
Doors + …	0	81	0	10	0	0	9

by HSM, 40.5% by RTM and 6% by ALP. Note that SMT is only used for three quarters of the stringers and almost half the fittings. Note that these percentages are the results of the optimization mentioned earlier and do not exactly determine which parts will be made with what process, only that a certain percentage of parts for each part family is made by a certain process. It is up to the designers and manufacturing personnel to decide how these percentages can be achieved or, if not possible, determine what the best compromise will be. For example, 6% of the frames and bulkheads made by ALP would probably correspond to the pressure bulkheads and any frames with deep webs where automated layup can be used effectively.

The above discussion focused on recurring cost as a driver. The optimum technology mixes determined have a certain weight and nonrecurring cost associated with them. If weight or nonrecurring cost were the drivers, different optimum technology mixes would be obtained. Also, the optimized results are frozen in time in the sense that the applicabilities of Table 2.2 and the cost figures of Table 2.3 are assumed constant. Over time, as technologies improve, these data will change and the associated optimum technology mixes will change. Results for the time-dependent problem with different drivers such as nonrecurring cost or optimum return on investment can be found in References 11 and 14.

It should be kept in mind that some of the data used in this section are subjective or based on expectations of what certain technologies will deliver in the future. As such, the results should be viewed as trends that will change with different input data. What is important here is that an approach has been developed that can be used to trade weight, recurring cost and nonrecurring cost and determine the optimum mix of technologies given certain cost data. The interested user of the approach can use his/her own data and degree of comfort in coming up with the optimum mix of technologies for his/her application.

2.4 Summary and Conclusions

An attempt to summarize the above discussion by fabrication process and to collect some of the qualitative considerations that should be taken into account during the design and analysis phases of a program using composite materials is shown in Table 2.5. For reference, sheet metal built-up structure and high-speed-machining (aluminium or titanium) are also included. This table is meant to be a rough set of guidelines and it is expected that different applications and manufacturing experiences can deviate significantly from its conclusions.

As shown in the previous section, there is no single process that can be applied to all types of parts and result in the lowest recurring and/or nonrecurring costs. A combination of processes is necessary. In many cases, combining two or more processes in fabricating a single part, thus

Table 2.5 Qualitative cost considerations affecting design/analysis decisions

Process	Application	Comments
Sheet metal	All airframe structure	Assembly intensive, relatively heavy. Moderate tooling costs including fit-out and assembly jigs
High-speed machining	Frames, bulkheads, ribs, beams, decks and floors. In general, parts with one flat surface that can be created via machining	Very low tooling cost. Very low recurring cost. Can generate any desired thickness greater than 0.6 mm. Moderate raw material cost due to the use of special alloys. Extremely high scrap rate (more than 99% of the raw material ends up recycled as machined chips). Limited due to vibrations to part thicknesses greater than 0.6–0.7 mm. Issues with damage tolerance (no built-in crack stoppers) repair methods and low damping; size of billet limits size of part that can be fabricated
Hand layup	All airframe structure	Weight reductions over equivalent metal of at least 15%. Recurring cost competitive with sheet metal when large amount of co-curing is used. Moderate scrap. High raw material cost. High tooling cost. Hard to fabricate 3-D fittings. Reduced out-of-plane strength (important in fittings and parts with out-of-plane loading)
Automated fibre/tow placement	Skins, decks, floors, doors, fairings, bulkheads, large ribs and beams. In general, parts with large surface area	Weight reductions similar to hand layup. Recurring cost can be less than metal baseline if the number of starts and stops for the machine are minimized (few cutouts, plydrops, etc.). Less scrap than hand layup. High tooling cost. For parts made on concave tools, limited by size of robotic head (interference with tool). Fibre steering is promising for additional weight savings but is limited by a maximum radius of curvature the machine can turn without buckling the tows
RTM	All airframe structure	Weight reductions somewhat less than hand layup due to decreased fibre volume for complex parts. Combined with automated preparation of fibre performs it can result in low recurring fabrication cost. Relatively low scrap rate. Very high tooling cost if matched metal tooling is used. Less so for vacuum-assisted RTM (half of the tool is a semi-rigid caul plate) or resin film infusion. To use unidirectional plies, some carrier or tackifier is needed for the fibres, increasing the recurring cost somewhat

Pultrusion	Constant cross-section parts: stiffeners, stringers, small beams	Weight reductions somewhat less than hand layup due to the fact that not all layups are possible (plies with 45° orientation or higher when 0 is aligned with the long axis of the part). Very low recurring cost and relatively low tooling cost compared with other fabrication processes. Reduced strength and stiffness in shear and transverse directions due to inability to generate any desired layup
Filament winding	Concave parts wound on a rotating mandrel: pressure vessels, cylinders, channels (wound and then cut)	Weight reductions somewhat less than hand layup due to difficulty in achieving the required fibre volume and due to inability to achieve certain stacking sequences. Low scrap rate, low raw material cost. Low recurring fabrication cost. Moderate tooling cost. Only convex parts wound on a mandrel where the tension in the fibres can be maintained during fabrication. Cannot wind angles shallower than geodetic lines (angles less than 15° not possible for long slender parts with 0 aligned with the long axis of the part). Reduced strength and stiffness
Press moulding	Fittings, clips, shear ties, small beams, ribs, intercostals	Weight reductions in the range 10–20% over aluminium baseline (weight savings potential limited due to the use of discontinuous fibres). Very low recurring cost with very short production cycle (minutes to a couple of hours). Low material scrap. Limited by the size of the press. Very high tooling cost for the press mould. Reduced strength due to the use of long discontinuous fibres, but good out-of-plane strength due to 'interlocking' of fibres

creating a hybrid process (e.g. automated fibre-placed skins with staged pultruded stiffeners, all co-cured in one cure cycle) appears to be the most efficient approach. In general, co-curing as large parts as possible and combining with as much automation as possible seems to have the most promise for parts of low cost, high quality and consistency. Of course, the degree to which this can be done depends on how much risk is considered acceptable in a specific application and to what extent the investment required to implement more than one fabrication processes is justified by the size of the production run. These combinations of processes and process improvements have already started to pay off and, for certain applications [15], the cost of composite airframe is comparable if not lower than that of equivalent metal structure.

Exercises

2.1 Hand layup, resin transfer moulding and press moulding are considered as the candidate processes for the following part:

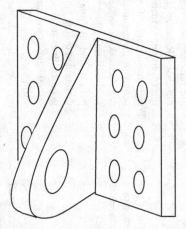

Discuss qualitatively how each choice may affect the structural performance and the weight of the final product. Include size effects, out-of-plane load considerations, load path continuity around corners, etc.

2.2 Hand layup, automated fibre placement and filament winding are proposed as candidate processes for the following part:

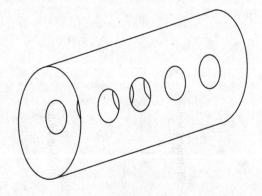

Discuss qualitatively how each choice may affect the structural performance and the weight of the final product. Include size effects, load path continuity, etc. in your discussion. Assume there are no local reinforcements (e.g. around window cutouts) or attachments to adjacent structure.

2.3 A certain composites technology is considered for implementation in a factory for the first time to make composite parts for airplanes. This technology requires new equipment that costs $1.5 million. In addition, experts estimate that the nonrecurring cost for each new part the factory makes with the new technology (design, analysis, ... all the way to certification) is, on the average, $20,000/kg of structure replaced. The technology has 25% applicability (by weight) over a structure that weighs 42,000 kg. The experts also expect that the new technology will save 15% of the weight of the structure it replaces (which is the reason for considering the switch to the new technology; it makes it more attractive to the customer) and it will start two times more expensive than the structure it replaces which, currently, costs 3.5 h/kg. It will also go down a 90.5% learning curve. How many aircraft must they sell to get their investment back if the hourly rate (including overhead) for the factory is 150$/h in one plant and 95$/h in another? (Give two answers; one for implementing the technology in one plant and one in the other). Make a plot showing the total $ "savings" as a function of number of aircraft sold for the two different factory hourly rates. Assume that the baseline technology is already far enough down its own learning curve so its cost will not change over time. Also assume that for every kilogram saved, the customer is willing to pay an extra $250. (Note that in this simplification of the problem, the time element, inflation, cost escalation, return on investment calculations, etc. are neglected.)

References

[1] Gutowski, T.G., Hoult, D., Dillon, G., Mutter, S., Kim, E., Tse, M., and Neoh, E.T., Development of a Theoretical Cost Model for Advanced Composite Fabrication, *Proc. 4th NASA/DoD Advanced Composites Technology Conf.*, Salt Lake City, UT, 1993.

[2] Gutowski, T.G., Neoh, E.T., and Dillon, G., Scaling Laws for Advanced Composites Fabrication Cost, *Proc. 5th NASA/DoD Advanced Composites Technology Conf.*, Seattle, WA, 1994, pp. 205–233.

[3] Apostolopoulos, P., and Kassapoglou, C., Cost Minimization of Composite Laminated Structures – Optimum Part Size as a Function of Learning Curve Effects and Assembly, *Journal of Composite Materials*, **36**(4), 501–518 (2002).

[4] Tatting, B., and Gürdal, Z., Design and Manufacture of Elastically Tailored Tow-Placed Plates, NASA CR 2002 211919, August 2002.

[5] Jegley, D., Tatting, B., and Gürdal, Z., Optimization of Elastically Tailored Tow-Placed Plates with Holes, *44th AIAA/ASME/ASCE/AHS Structures, Structural Dynamics and Materials Conf.*, Norfolk VA, 2003, also paper AIAA-2003-1420.

[6] *1st NASA Advanced Composites Technology Conf.*, NASA CP 3104, Seattle WA, 1990.

[7] *2nd NASA Advanced Composites Technology Conf.*, NASA CP 3154, Lake Tahoe, NV, 1991.

[8] Abbott, R., Design and Certification of the All-Composite Airframe. *Society of Automotive Engineers Technical Paper Series – Paper No 892210*; 1989.

[9] Whitehead, R.S., Kan, H.P., Cordero, R. and Saether, E.S., Certification Testing Methodology for Composite Structures: Volume I – Data Analysis. *Naval Air Development Center Report 87042-60(DOT/FAA/CT-86-39)*; 1986.

[10] Whitehead, R.S., Kan, H.P., Cordero, R. and Saether, E.S., Certification Testing Methodology for Composite Structures: Volume II – Methodology Development. *Naval Air Development Center Report 87042-60(DOT/FAA/CT-86-39)*; 1986.

[11] Kassapoglou, C., Determination of the Optimum Implementation Plan for Manufacturing Technologies – The case of a Helicopter Fuselage, *Journal of Manufacturing Systems*, **19**, 121–133 (2000).

[12] Kassapoglou, C., Selection of Manufacturing Technologies for Fuselage Structures for Minimum Cost and Low Risk: Part A – Problem Formulation, *Journal of Composites Technology and Research*, **21**, 183–188 (1999).

[13] Kassapoglou, C., Selection of Manufacturing Technologies for Fuselage Structures for Minimum Cost and Low risk: Part B – Solution and Results, *Journal of Composites Technology and Research*, **21**, 189–196 (1999).

[14] Sarkar, P. and Kassapoglou, C., An ROI-Based Strategy for Implementation of Existing and Emerging Technologies, *Spring Research Conf. on Statistics in Industry and Technology*, Minneapolis-St Paul, MN, June 1999. Also *IEEE Transactions on Engineering Management*, **48**, 414–427 (1999).

[15] McGettrick, M., and Abbott, R., To MRB or not to be: Intrinsic Manufacturing Variabilities and Effects on Load Carrying Capacity, *Proc. 10th DoD/NASA/FAA Conf. on Fibrous Composites in Structural Design*, Hilton Head, SC, 1993, pp. 5–39.

3

Review of Classical Laminated Plate Theory

This chapter gives some basic laminate definitions and a brief summary of the classical laminated-plate theory (CLPT). Aspects of CLPT, in particular, the laminate stiffness matrices are used throughout the remainder of this book.

3.1 Composite Materials: Definitions, Symbols and Terminology

A composite material is any material that consists of at least two constituents. In this book, the term 'composite material' refers to a mixture of fibres and matrix resulting in a configuration that combines some of the best characteristics of the two constituents. There is a large variety of possible combinations. For fibres, some of the options include, E- or S-glass, quartz, graphite, Kevlar®, boron, silicon, etc., appearing in long continuous or short discontinuous form. The matrix materials cover a wide range of thermoset (epoxy, polyester, phenolics, polyimides, bismaleimids) or thermoplastic resins or metals such as aluminium or steel. The building block of a composite material is the ply or lamina. Plies or laminae are stacked together (different orientations and materials can be combined) to make a laminate.

The most common plies used are unidirectional plies (where all fibres are aligned in one direction) or fabric plies (plain weave, satins, etc.) where fibres are oriented in two mutually perpendicular directions. If each ply in the stacking sequence or layup making up a laminate is denoted by its orientation θ (in degrees) relative to a reference axis ($-90° < \theta \leq +90°$), as shown, for example, in Figure 3.1, then a laminate can be denoted by its stacking sequence (or layup):

$$[\theta_1/\theta_2^{\frac{1}{2}}/\theta_3 \ldots]$$

where θ_1, θ_2, etc. are the angles of successive plies starting from the top of the laminate.

Design and Analysis of Composite Structures: With Applications to Aerospace Structures, Second Edition. Christos Kassapoglou.
© 2013 John Wiley & Sons, Ltd. Published 2013 by John Wiley & Sons, Ltd.

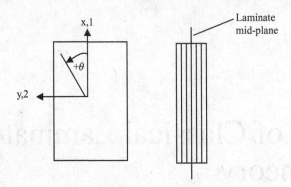

Figure 3.1 Laminate axes and definition of positive θ orientation

If more than one material type is used in the same laminate the angular orientation can be followed by a symbol that denotes the material type. For example, in the following stacking sequence,

$$[\theta_1(T)/\theta_2(F)/\theta_3(T)\ldots]$$

the first and third plies are made with unidirectional tape material and the second with fabric material.

When fabric material is used, it is also common to indicate the two orientations in each ply in parentheses such as

$$(0/90), (\pm45), (20/-70)$$

where the first denotes a fabric ply with fibres oriented in the $0°$ and $90°$ directions, the second denotes a ply with fibres in the $+45°$ and $-45°$ directions and the third a ply with fibres in the $+20°$ and $-70°$ directions.

There are several special laminate types often encountered in practice some of which are: (a) symmetric, (b) balanced, (c) cross-ply, (d) angle-ply and (e) quasi-isotropic laminates.

Symmetric laminates are laminates that have a symmetric stacking sequence with respect to the laminate mid-plane (see Figure 3.1). This means that the material, thickness and orientation of each pair of plies located symmetrically with respect to the laminate mid-plane are the same. A symmetric stacking sequence is usually denoted by writing half of it and using the subscript s:

$$[35/20/40]_S \text{ is the same as } [35/20/40/40/20/35]$$

This contracted notation has the advantage of simplicity, but requires caution when the total number of plies is odd. In such a case, the centre ply, half of which lies on one side of the mid-plane and half on the other, is denoted with an overbar:

$$[35/20/\overline{40}]_S \text{ is the same as } [35/20/40/20/35]$$

Balanced laminates are laminates in which for each $+\theta$ ply there is a $-\theta$ ply (of the same material and thickness) somewhere in the stacking sequence. Special properties of balanced and/or symmetric laminates related to their structural response will be presented in subsequent sections.

Cross-ply laminates consist only of $0°$ and $90°$ plies. Angle-ply laminates do not contain any $0°$ or $90°$ plies.

Finally, quasi-isotropic laminates have the same stiffness in any direction in their plane (xy plane in Figure 3.1). One way to create a quasi-isotropic stacking sequence of n plies is to require that there is no direction that has more fibres than any other direction. A simple procedure to accomplish this is to divide the range of angles from $0°$ to $180°$ in n equal segments and, assign to each ply one angle increment corresponding to these segments.

For example, if there are eight plies, the angle increment is $180/8 = 22.5°$. Then, mixing the following angles in any order creates a quasi-isotropic laminate:

$$0, 22.5, 45, 67.5, 90, 112.5(\text{or} -67.5), 135(\text{or} -45), 157.5(\text{or} -22.5)$$

Taking this one step further, for a symmetric laminate, the rule is only applied to half the laminate since the other half is automatically created by symmetry. For the same case of $n = 8$, the angle increment is now $180/(8/2) = 45°$. The following angles, in any order and repeated symmetrically give a quasi-isotropic, symmetric, 8-ply laminate.

$$0°, 45°, 90°, 135°(\text{or} -45°)$$

Some possible quasi-isotropic stacking sequences in this case are:

$$[0/45/90/-45]_S, [45/-45/0/90]_S, [45/90/0/-45]_S, \text{etc.}$$

To complete the discussion of stacking sequence notation, other shorthand methods include the use of parentheses with subscripts to denote a repeating pattern within the stacking sequence and the use of numerical subscripts or superscripts outside the brackets. Examples of these are:

$[(15/-15)_3/0/30]_S$ is the same as $[15/-15/15/-15/15/-15/0/30/30/0/-15/15/-15/15/-15/15]$

$[15/-15/0/30]_{2S}$ is the same as $[15/-15/0/30/15/-15/0/30/30/0/-15/15/30/0/-15/15]$

$[15/-15/0/30]_{S2}$ is the same as $[15/-15/0/30/30/0/-15/15/15/-15/0/30/30/0/-15/15]$

3.2 Constitutive Equations in Three Dimensions

Composite materials are, by their nature, anisotropic. In three dimensions, the engineering stresses and strains describing completely the state of deformation in a composite are denoted in matrix form, respectively:

$$\left[\sigma_x \ \sigma_y \ \sigma_z \ \tau_{yz} \ \tau_{xz} \ \tau_{xy} \right]$$
$$\left[\varepsilon_x \ \varepsilon_y \ \varepsilon_z \ \gamma_{yz} \ \gamma_{xz} \ \gamma_{xy} \right]$$

The first three are the normal stresses (strains) and the last three are the shear stresses (strains). It is customary for two-dimensional problems to use x and y as the in-plane coordinates (see Figure 3.1) and z as the out-of-plane coordinate (perpendicular to the plane of Figure 3.1).

Stresses and strains are related through the generalized stress–strain relations (Hooke's law) [1–5]:

$$\begin{Bmatrix} \sigma_x \\ \sigma_y \\ \sigma_z \\ \tau_{yz} \\ \tau_{xz} \\ \tau_{xy} \end{Bmatrix} = \begin{bmatrix} E_{11} & E_{12} & E_{13} & E_{14} & E_{15} & E_{16} \\ E_{21} & E_{22} & E_{23} & E_{24} & E_{25} & E_{26} \\ E_{31} & E_{32} & E_{33} & E_{34} & E_{35} & E_{36} \\ E_{41} & E_{42} & E_{43} & E_{44} & E_{45} & E_{46} \\ E_{51} & E_{52} & E_{53} & E_{54} & E_{55} & E_{56} \\ E_{61} & E_{62} & E_{63} & E_{64} & E_{65} & E_{66} \end{bmatrix} \begin{Bmatrix} \varepsilon_x \\ \varepsilon_y \\ \varepsilon_z \\ \gamma_{yz} \\ \gamma_{xz} \\ \gamma_{xy} \end{Bmatrix} \qquad (3.1)$$

Note that there is an apparent mix-up of subscripts in Equation (3.1) where the stiffness components E_{ij} have numerical indices while the stress and strain components have letter indices. This is done on purpose to keep the engineering notations for stresses and strains and the usual (contracted tensor) notation for the stiffness terms, which uses numbers instead of letters.

Equation (3.1) relates the strains to stresses through the fourth order elasticity tensor $\underline{E}$. It can be shown, based on energy considerations [6] that the elasticity tensor is symmetric, i.e. $E_{ij} = E_{ji}$. Thus, for a general anisotropic body, there are 21 independent elastic constants, as highlighted by the dashed line in Equation (3.1a).

$$\begin{Bmatrix} \sigma_x \\ \sigma_y \\ \sigma_z \\ \tau_{yz} \\ \tau_{xz} \\ \tau_{xy} \end{Bmatrix} = \begin{bmatrix} E_{11} & E_{12} & E_{13} & E_{14} & E_{15} & E_{16} \\ E_{12} & E_{22} & E_{23} & E_{24} & E_{25} & E_{26} \\ E_{13} & E_{23} & E_{33} & E_{34} & E_{35} & E_{36} \\ E_{14} & E_{24} & E_{34} & E_{44} & E_{45} & E_{46} \\ E_{15} & E_{25} & E_{35} & E_{45} & E_{55} & E_{56} \\ E_{16} & E_{26} & E_{36} & E_{46} & E_{56} & E_{66} \end{bmatrix} \begin{Bmatrix} \varepsilon_x \\ \varepsilon_y \\ \varepsilon_z \\ \gamma_{yz} \\ \gamma_{xz} \\ \gamma_{xy} \end{Bmatrix} \quad \begin{array}{l} \text{Independent elastic} \\ \text{constants} \end{array} \qquad (3.1a)$$

The discussion in this book is further confined to orthotropic materials. These are materials that possess two planes of symmetry. In such a case, some of the coupling terms in Equation (3.1a) are zero:

$$E_{14} = E_{15} = E_{16} = E_{24} = E_{25} = E_{26} = E_{34} = E_{35} = E_{36} = 0 \qquad (3.2)$$

In addition, for an orthotropic body, shear stresses in one plane do not cause shear strains in another. Thus,

$$E_{45} = E_{46} = E_{56} = 0 \qquad (3.3)$$

With these simplifications, the stress–strain relations for an orthotropic material have the form:

$$
\begin{Bmatrix} \sigma_x \\ \sigma_y \\ \sigma_z \\ \tau_{yz} \\ \tau_{xz} \\ \tau_{xy} \end{Bmatrix} =
\begin{bmatrix}
E_{11} & E_{12} & E_{13} & 0 & 0 & 0 \\
E_{12} & E_{22} & E_{23} & 0 & 0 & 0 \\
E_{13} & E_{23} & E_{33} & 0 & 0 & 0 \\
0 & 0 & 0 & E_{44} & 0 & 0 \\
0 & 0 & 0 & 0 & E_{55} & 0 \\
0 & 0 & 0 & 0 & 0 & E_{66}
\end{bmatrix}
\begin{Bmatrix} \varepsilon_x \\ \varepsilon_y \\ \varepsilon_z \\ \gamma_{yz} \\ \gamma_{xz} \\ \gamma_{xy} \end{Bmatrix}
\tag{3.4}
$$

A ply of unidirectional composite material, with the x axis of Figure 3.1 aligned with the fibre direction and the y axis transverse to it, possesses two planes of symmetry and is thus described by Equation (3.4). The same holds true for a fabric ply with the x axis aligned with one fibre direction and the y axis aligned with the other. Such plies form the building blocks for composite parts discussed in this book. Note that, in the laminate coordinate system, different plies stacked together, which are not 0, 90 or (0/90) will no longer possess two planes of symmetry and some of the coupling terms in Equation (3.1a) are nonzero. However, it is always possible to find an axis system (principal axes), in general not coinciding with the laminate axes, in which the laminate is orthotropic. In general, the entire laminate can be described by a stress–strain relation of the form:

$$
\begin{Bmatrix} \sigma_x \\ \sigma_y \\ \sigma_z \\ \tau_{yz} \\ \tau_{xz} \\ \tau_{xy} \end{Bmatrix} =
\begin{bmatrix}
E_{11} & E_{12} & E_{13} & 0 & 0 & E_{16} \\
E_{12} & E_{22} & E_{23} & 0 & 0 & E_{26} \\
E_{13} & E_{23} & E_{33} & 0 & 0 & E_{36} \\
0 & 0 & 0 & E_{44} & E_{45} & 0 \\
0 & 0 & 0 & E_{45} & E_{55} & 0 \\
E_{16} & E_{26} & E_{36} & 0 & 0 & E_{66}
\end{bmatrix}
\begin{Bmatrix} \varepsilon_x \\ \varepsilon_y \\ \varepsilon_z \\ \gamma_{yz} \\ \gamma_{xz} \\ \gamma_{xy} \end{Bmatrix}
\tag{3.5}
$$

where E_{ij} are now laminate and not ply quantities.

The inverse of Equation (3.5), expressing the strains in terms of the stresses via the compliance tensor S_{ij} is also often used:

$$
\begin{Bmatrix} \varepsilon_x \\ \varepsilon_y \\ \varepsilon_z \\ \gamma_{yz} \\ \gamma_{xz} \\ \gamma_{xy} \end{Bmatrix} =
\begin{bmatrix}
S_{11} & S_{12} & S_{13} & 0 & 0 & S_{16} \\
S_{12} & S_{22} & S_{23} & 0 & 0 & S_{26} \\
S_{13} & S_{23} & S_{33} & 0 & 0 & S_{36} \\
0 & 0 & 0 & S_{44} & S_{45} & 0 \\
0 & 0 & 0 & S_{45} & S_{55} & 0 \\
S_{16} & S_{26} & S_{36} & 0 & 0 & S_{66}
\end{bmatrix}
\begin{Bmatrix} \sigma_x \\ \sigma_y \\ \sigma_z \\ \tau_{yz} \\ \tau_{xz} \\ \tau_{xy} \end{Bmatrix}
\tag{3.6}
$$

where the compliance matrix is the inverse of the stiffness matrix:

$$
[S] = [E]^{-1}
\tag{3.7}
$$

Note that Equations (3.5), (3.6) and (3.7) refer to laminate quantities while Equation (3.4) refers to an orthotropic material such as a ply. The underlying assumptions are that: (a) at the laminate and, often, the ply scales, the fibre/matrix combination can be treated as a homogeneous material with smeared properties; (b) plane sections remain plane during deformation; (c) there is a perfect bond between fibres and matrix; and (d) there is a perfect bond between plies.

3.2.1   Tensor Transformations

If the stiffness (or compliance) properties are known in one coordinate system, they can be obtained in any other coordinate system through standard tensor transformations. These can be expressed concisely if the tensor notation is used (each index ranges from 1 to 3 and repeating indices sum). Defining ℓ_{ij} to be the (direction) cosine of the angle between axes i and j, the compliance tensor S_{mnpq} in one coordinate system is obtained in terms of the compliance tensor S_{ijkr} in another via the relation:

$$S_{mnpq} = \ell_{mi}\ell_{nj}\ell_{pk}\ell_{qr}S_{ijkr} \tag{3.8}$$

with an analogous relation for the stiffness E_{mnpq}.

If the two coordinate systems have the z axis (out-of-plane in the case of a laminate) in common, Equation (3.8) simplifies and can be expanded relatively easily. If the original coordinate system coincides with the ply axis system (x along fibres, y perpendicular to the fibres, as shown in Figure 3.2), then the compliance matrix in a coordinate system whose axis

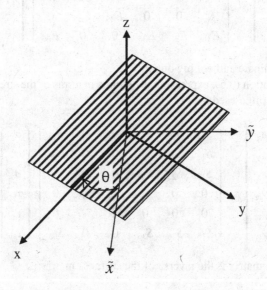

Figure 3.2   Coordinate system transformation

forms an angle θ with the x axis of the original coordinate system can be shown to be given by Equations (3.9).

$$S_{11} = S_{11}^0 \cos^4 \theta + \left(2S_{12}^0 + S_{66}^0\right) \sin^2 \theta \cos^2 \theta + S_{22}^0 \sin^4 \theta$$

$$S_{12} = \left(S_{11}^0 + S_{22}^0 - S_{66}^0\right) \sin^2 \theta \cos^2 \theta + S_{22}^0 \left(\sin^4 \theta + \cos^4 \theta\right)$$

$$S_{13} = S_{13}^0 \cos^2 \theta + S_{23}^0 \sin^2 \theta$$

$$S_{22} = S_{11}^0 \sin^4 \theta + \left(2S_{12}^0 + S_{66}^0\right) \sin^2 \theta \cos^2 \theta + S_{22}^0 \cos^4 \theta$$

$$S_{23} = S_{13}^0 \sin^2 \theta + S_{23}^0 \cos^2 \theta$$

$$S_{33} = S_{33}^0$$

$$S_{16} = 2S_{11}^0 \cos^3 \theta \sin \theta - 2S_{22}^0 \cos \theta \sin^3 \theta + \left(2S_{22}^0 + S_{66}^0\right) \left(\cos \theta \sin^3 \theta - \cos^3 \theta \sin \theta\right) \quad (3.9)$$

$$S_{26} = 2S_{11}^0 \cos \theta \sin^3 \theta - 2S_{22}^0 \cos^3 \theta \sin \theta + \left(2S_{22}^0 + S_{66}^0\right) \left(\cos^3 \theta \sin \theta - \cos \theta \sin^3 \theta\right)$$

$$S_{36} = 2 \left(S_{13}^0 - S_{23}^0\right) \cos \theta \sin \theta$$

$$S_{44} = S_{55}^0 \sin^2 \theta + S_{44}^0 \cos^2 \theta$$

$$S_{45} = \left(S_{55}^0 - S_{44}^0\right) \sin \theta \cos \theta$$

$$S_{55} = S_{55}^0 \cos^2 \theta + S_{44}^0 \sin^2 \theta$$

$$S_{66} = 4 \left(S_{11}^0 + S_{22}^0 - 2S_{12}^0\right) \sin^2 \theta \cos^2 \theta + S_{66}^0 \left(\sin^4 \theta + \cos^4 \theta - 2 \sin^2 \theta \cos^2 \theta\right)$$

where the quantities in the xyz coordinate system (basic ply) have a superscript 0 and are given in terms of the corresponding stiffnesses of the basic ply by:

$$S_{11}^0 = \frac{1}{E_{11}}$$

$$S_{12}^0 = -\frac{v_{12}}{E_{11}}$$

$$S_{66}^0 = \frac{1}{G_{12}}$$

$$S_{22}^0 = \frac{1}{E_{22}}$$

$$S_{13}^0 = -\frac{v_{13}}{E_{11}} \qquad (3.10)$$

$$S_{23}^0 = -\frac{v_{23}}{E_{22}}$$

$$S_{33}^0 = \frac{1}{E_{33}}$$

$$S_{44}^0 = \frac{1}{G_{23}}$$

$$S_{55}^0 = \frac{1}{G_{13}}$$

where E_{ij} are stiffnesses of the basic $(0°)$ ply with subscripts 1, 2 and 3 corresponding to the coordinates x, y and z.

3.3   Constitutive Equations in Two Dimensions: Plane Stress

When dealing with thin composites, where the thickness of the laminate is much smaller than the other dimensions of the structure, the laminate is often assumed to be in a state of plane stress. This is usually the case of a composite plate that is thin compared with its in-plane dimensions. Then, the out-of-plane stresses σ_z, τ_{yz} and τ_{xz} are negligible compared to the in-plane stresses:

$$\sigma_z \approx \tau_{yz} \approx \tau_{xz} \approx 0 \tag{3.11}$$

For an orthotropic material such as a single ply in the ply axes or a symmetric and balanced laminate in the laminate axes, placing Equation (3.11) in Equation (3.5) gives:

$$
\begin{aligned}
\sigma_x &= E_{11}\varepsilon_x + E_{12}\varepsilon_y + E_{13}\varepsilon_z \\
\sigma_y &= E_{12}\varepsilon_x + E_{22}\varepsilon_y + E_{23}\varepsilon_z \\
0 &= E_{13}\varepsilon_x + E_{23}\varepsilon_y + E_{33}\varepsilon_z \\
0 &= E_{44}\gamma_{yz} \\
0 &= E_{55}\gamma_{xz} \\
\tau_{xy} &= E_{66}\gamma_{yz}
\end{aligned}
\tag{3.12a–f}
$$

From Equations (3.12d) and (3.12e),

$$\gamma_{yz} = \gamma_{xz} = 0 \tag{3.13}$$

Equation (3.12c) can be solved for ε_z and the result substituted in Equations (3.12a) and (3.12b). This gives the equations

$$
\begin{aligned}
\sigma_x &= E_{11}\varepsilon_x + E_{12}\varepsilon_y + E_{13}\left(-\frac{E_{13}}{E_{33}}\varepsilon_x - \frac{E_{23}}{E_{33}}\varepsilon_y \right) \\
\sigma_y &= E_{12}\varepsilon_x + E_{22}\varepsilon_y + E_{23}\left(-\frac{E_{13}}{E_{33}}\varepsilon_x - \frac{E_{23}}{E_{33}}\varepsilon_y \right)
\end{aligned}
\tag{3.14}
$$

which, upon collecting terms can be rewritten as:

$$
\begin{aligned}
\sigma_x &= \left(E_{11} - \frac{E_{13}^2}{E_{33}} \right)\varepsilon_x + \left(E_{12} - \frac{E_{13}E_{23}}{E_{33}} \right)\varepsilon_y \\
\sigma_y &= \left(E_{12} - \frac{E_{13}E_{23}}{E_{33}} \right)\varepsilon_x + \left(E_{22} - \frac{E_{23}^2}{E_{33}} \right)\varepsilon_y
\end{aligned}
\tag{3.15}
$$

Equations (3.14) and (3.15) along with Equation (3.12f) form the constitutive relations (stress–strain equation) for composite materials undergoing plane stress. Redefining

$$Q_{xx} = E_{11} - \frac{E_{13}^2}{E_{33}}$$

$$Q_{xy} = E_{12} - \frac{E_{13}E_{23}}{E_{33}}$$

$$Q_{yy} = E_{22} - \frac{E_{23}^2}{E_{33}}$$

$$Q_{SS} = E_{66}$$

(3.16)

the equations for plane stress can be rewritten in matrix form:

$$\left\{\begin{array}{c} \sigma_x \\ \sigma_y \\ \tau_{xy} \end{array}\right\} = \left[\begin{array}{ccc} Q_{xx} & Q_{xy} & 0 \\ Q_{xy} & Q_{yy} & 0 \\ 0 & 0 & Q_{SS} \end{array}\right] \left\{\begin{array}{c} \varepsilon_x \\ \varepsilon_y \\ \gamma_{xy} \end{array}\right\}$$

(3.17)

It should be emphasized that the form of Equations (3.17) is the same irrespective of whether one deals with a single ply or a laminate, provided that the coordinate system is such that both the ply and the laminate are orthotropic. However, the values of the stiffnesses Q_{xx}, Q_{xy}, etc., differ between ply and laminate.

The easiest way to use Equations (3.17) is to start from basic ply properties as measured from simple coupon tests, calculate the values for Q_{xx}, Q_{xy}, etc., then determine the corresponding values for any (rotated) ply and, finally, an entire laminate.

Let E_L, E_T, G_{LT} and ν_{LT} be the Young's modulus along the fibres (longitudinal direction), Young's modulus transverse to the fibres, shear modulus and (major) Poisson's ratio, respectively. These values can all be obtained from standard coupon tests.

Now in a uniaxial tension test (see Figure 3.3), where the applied load is parallel to the fibres of a unidirectional ply (which define the x direction), the slope of the applied stress σ_x versus longitudinal strain ε_x is the Young's modulus E_L and the slope of the transverse strain ε_y versus longitudinal strain ε_x is the Poisson's ratio ν_{LT}:

$$\nu_{LT} = -\frac{\varepsilon_y}{\varepsilon_x}$$

(3.18)

Using this to substitute for ε_y in the first of Equations (3.17) gives:

$$\sigma_x = Q_{xx}\varepsilon_x - Q_{xy}\nu_{LT}\varepsilon_x$$

(3.19)

For the same uniaxial tension test, $\sigma_y = 0$ and the second of Equations (3.17) gives

$$0 = Q_{xy}\varepsilon_x + Q_{yy}\varepsilon_y \Rightarrow \frac{Q_{xy}}{Q_{yy}} = -\frac{\varepsilon_y}{\varepsilon_x}$$

(3.20)

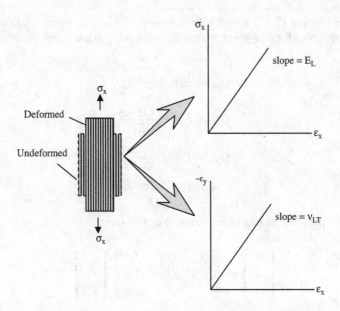

Figure 3.3   Quantities measured in a uniaxial tension test of a single unidirectional ply

Comparing, Equations (3.18) and (3.20)

$$\nu_{LT} = \frac{Q_{xy}}{Q_{yy}} \tag{3.21}$$

and substituting in Equation (3.19),

$$\sigma_x = \left(Q_{xx} - \frac{Q_{xy}^2}{Q_{yy}} \right) \varepsilon_x \tag{3.22}$$

Equation (3.22) implies that the slope E_L of the σ_x versus ε_x curve (see Figure 3.3) is given by:

$$E_L = Q_{xx} - \frac{Q_{xy}^2}{Q_{yy}} \tag{3.23}$$

In a completely analogous fashion, but now considering a uniaxial tension test transverse to the fibres and noticing that ν_{TL} is the Poisson's ratio that describes contraction in the x direction when a tension load is applied in the y direction, Equations (3.24) and (3.25) are obtained as analogues to Equations (3.21) and (3.23):

$$\nu_{TL} = \frac{Q_{xy}}{Q_{xx}} \tag{3.24}$$

$$E_T = Q_{yy} - \frac{Q_{xy}^2}{Q_{xx}} \tag{3.25}$$

Note that Equations (3.21) and (3.24) imply that

$$v_{LT} Q_{yy} = v_{TL} Q_{xx} \tag{3.26}$$

Equations (3.23), (3.25) and (3.26) form a system of three equations in the three unknowns Q_{xx}, Q_{xy} and Q_{yy}. Solving gives,

$$Q_{xx} = \frac{E_L}{1 - v_{LT} v_{TL}} \tag{3.27}$$

$$Q_{yy} = \frac{E_T}{1 - v_{LT} v_{TL}} \tag{3.28}$$

$$Q_{xy} = \frac{v_{LT} E_T}{1 - v_{LT} v_{TL}} = \frac{v_{TL} E_L}{1 - v_{LT} v_{TL}} \tag{3.29}$$

Considering now a pure shear test of a unidirectional ply where G_{LT} is the slope of the shear stress (τ_{xy}) versus the shear strain (γ_{xy}) curve, the last of Equations (3.17) implies that

$$Q_{ss} = G_{LT} \tag{3.30}$$

Equations (3.27), (3.28), (3.29) and (3.30) can be used to substitute in Equations (3.17) to obtain the final form of the stress–strain equations for an orthotropic ply under plane stress:

$$\begin{Bmatrix} \sigma_x \\ \sigma_y \\ \tau_{xy} \end{Bmatrix} = \begin{bmatrix} \dfrac{E_L}{1 - v_{LT} v_{TL}} & \dfrac{v_{LT} E_T}{1 - v_{LT} v_{TL}} & 0 \\ \dfrac{v_{LT} E_T}{1 - v_{LT} v_{TL}} & \dfrac{E_T}{1 - v_{LT} v_{TL}} & 0 \\ 0 & 0 & G_{LT} \end{bmatrix} \begin{Bmatrix} \varepsilon_x \\ \varepsilon_y \\ \gamma_{xy} \end{Bmatrix} \tag{3.31}$$

The next step is to obtain the stress–strain relations for any ply rotated by an angle θ. In general, the stress–strain Equations (3.17) now become:

$$\begin{Bmatrix} \sigma_1 \\ \sigma_2 \\ \tau_{12} \end{Bmatrix} = \begin{bmatrix} Q_{11} & Q_{12} & Q_{16} \\ Q_{12} & Q_{22} & Q_{26} \\ Q_{16} & Q_{26} & Q_{66} \end{bmatrix} \begin{Bmatrix} \varepsilon_1 \\ \varepsilon_2 \\ \gamma_{12} \end{Bmatrix} \tag{3.32}$$

To relate these quantities to the corresponding ones for an orthotropic ply requires transforming stresses, strains and stiffnesses by the angle θ.

The stiffness transformation follows the standard tensor transformation (Equations 3.8), which, for a ply rotated in its plane, as is the case of interest here, are simplified to the equations analogous to Equations (3.9) that were obtained for the compliances. For the plane

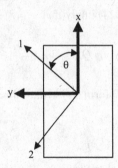

Figure 3.4   Coordinate system for ply rotated by an angle θ

stress case, Equations (3.8) or (3.9) applied to the stiffness tensor give the stiffnesses in the 1–2 coordinate system of Figure 3.4 as:

$$
\begin{aligned}
Q_{11}^{(\theta)} &= m^4 Q_{xx} + n^4 Q_{yy} + 2m^2 n^2 Q_{xy} + 4m^2 n^2 Q_{ss} \\
Q_{22}^{(\theta)} &= n^4 Q_{xx} + m^4 Q_{yy} + 2m^2 n^2 Q_{xy} + 4m^2 n^2 Q_{ss} \\
Q_{12}^{(\theta)} &= m^2 n^2 Q_{xx} + m^2 n^2 Q_{yy} + (m^4 + n^4) Q_{xy} - 4m^2 n^2 Q_{ss} \\
Q_{66}^{(\theta)} &= m^2 n^2 Q_{xx} + m^2 n^2 Q_{yy} - 2m^2 n^2 Q_{xy} + (m^2 - n^2)^2 Q_{ss} \\
Q_{16}^{(\theta)} &= m^3 n Q_{xx} - mn^3 Q_{yy} + (mn^3 - m^3 n) Q_{xy} + 2(mn^3 - m^3 n) Q_{ss} \\
Q_{26}^{(\theta)} &= mn^3 Q_{xx} - m^3 n Q_{yy} + (m^3 n - mn^3) Q_{xy} + 2(m^3 n - mn^3) Q_{ss}
\end{aligned}
\tag{3.33}
$$

where $m = \cos\theta$ and $n = \sin\theta$.

The stresses and strains transform using second-order tensor transformation equations instead of the fourth-order tensor transformation for stiffnesses and compliances given by Equation (3.8). Using ℓ_{ij} to denote the direction cosines between axes i and j, the stress transformation equations can be written as:

$$
\sigma_{mn} = \ell_{mp} \ell_{nq} \sigma_{pq}
\tag{3.34}
$$

which, expanded out for the case shown in Figure 3.4 reads:

$$
\begin{Bmatrix} \sigma_1 \\ \sigma_2 \\ \tau_{12} \end{Bmatrix} =
\begin{bmatrix}
\cos^2\theta & \sin^2\theta & 2\sin\theta\cos\theta \\
\sin^2\theta & \cos^2\theta & -2\sin\theta\cos\theta \\
-\sin\theta\cos\theta & \sin\theta\cos\theta & (\cos^2\theta - \sin^2\theta)
\end{bmatrix}
\begin{Bmatrix} \sigma_x \\ \sigma_y \\ \tau_{xy} \end{Bmatrix}
\tag{3.35}
$$

An analogous expression is obtained for the strain transformation. However, since here engineering notation is used (instead of tensor notation) the form of Equation (3.35) for the (engineering) strains is:

$$
\begin{Bmatrix} \varepsilon_1 \\ \varepsilon_2 \\ \gamma_{12} \end{Bmatrix} =
\begin{bmatrix}
\cos^2\theta & \sin^2\theta & \sin\theta\cos\theta \\
\sin^2\theta & \cos^2\theta & -\sin\theta\cos\theta \\
-2\sin\theta\cos\theta & 2\sin\theta\cos\theta & (\cos^2\theta - \sin^2\theta)
\end{bmatrix}
\begin{Bmatrix} \varepsilon_x \\ \varepsilon_y \\ \gamma_{xy} \end{Bmatrix}
\tag{3.36}
$$

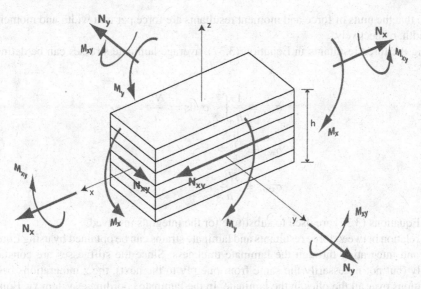

Figure 3.5 Force and moment resultants applied to a laminate (arrows indicate sign convention for positive values) (Figure courtesy of Imco van Gent)

Note the changes in the factors of 2 in the last row and column of the transformation matrix. These come from the fact that the engineering shear strain is twice the tensor strain: $\gamma_{xy} = 2\varepsilon_{12}$.

While the equations in terms of stresses and strains can be (and often are) used, in practice it is convenient to define force and moment resultants by integrating through the thickness of a laminate. For a laminate of thickness h as shown in Figure 3.5, the following quantities are defined:

$$N_x = \int_{-\frac{h}{2}}^{\frac{h}{2}} \sigma_x \, dz$$

$$N_y = \int_{-\frac{h}{2}}^{\frac{h}{2}} \sigma_y \, dz \tag{3.37}$$

$$N_{xy} = \int_{-\frac{h}{2}}^{\frac{h}{2}} \tau_{xy} \, dz$$

which are the force resultants and

$$M_x = \int_{-\frac{h}{2}}^{\frac{h}{2}} \sigma_x z \, dz$$

$$M_y = \int_{-\frac{h}{2}}^{\frac{h}{2}} \sigma_y z \, dz \tag{3.38}$$

$$M_{xy} = \int_{-\frac{h}{2}}^{\frac{h}{2}} \tau_{xy} z \, dz$$

which are the moment resultants.

Note that the units of force and moment resultants are force per unit width and moment per unit width, respectively.

Using the force resultants in Equation (3.37), average laminate stresses can be defined as follows:

$$\sigma_{xav} = \frac{1}{h} \int_{-\frac{h}{2}}^{\frac{h}{2}} \sigma_x \mathrm{d}z = \frac{N_x}{h}$$

$$\sigma_{yav} = \frac{N_y}{h} \tag{3.39}$$

$$\sigma_{xyav} = \frac{N_{xy}}{h}$$

where Equations (3.37) are used to substitute for the integrals involved.

The relation between force resultants and laminate strains can be obtained by using Equation (3.32) and integrating through the laminate thickness. Since the stiffnesses are constant in each ply (but not necessarily the same from one ply to the next), the z integrations become summations over all the plies in the laminate. In the laminate coordinate system xy, Equation (3.32) integrated with respect to z gives

$$\begin{Bmatrix} N_x \\ N_y \\ N_{xy} \end{Bmatrix} = \begin{bmatrix} A_{11} & A_{12} & A_{16} \\ A_{12} & A_{22} & A_{26} \\ A_{16} & A_{26} & A_{66} \end{bmatrix} \begin{Bmatrix} \varepsilon_x \\ \varepsilon_y \\ \gamma_{xy} \end{Bmatrix} \tag{3.40}$$

where

$$A_{ij} = \sum_{k=1}^{n} Q_{ij}(z_k - z_{k-1}) \tag{3.41}$$

where $i,j = 1,2,6$, the summation is carried over all n plies of the laminate and z_k, z_{k-1} are the upper and lower z coordinates of the kth ply, as shown in Figure 3.6.

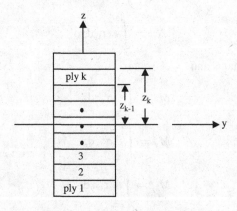

Figure 3.6 Ply numbering system

Equations (3.40) and (3.41) describe the membrane deformations of a laminate under in-plane loads. For a laminate under bending loads, the standard Kirchhoff plate theory assumptions are used: plane sections remain plane and perpendicular to the neutral axis. Denoting the out-of-plane displacement by w, the curvatures κ_x, κ_y and κ_{xy} are defined as:

$$\kappa_x = -\frac{\partial^2 w}{\partial x^2}$$

$$\kappa_y = -\frac{\partial^2 w}{\partial y^2} \qquad (3.42)$$

$$\kappa_{xy} = -2\frac{\partial^2 w}{\partial x \partial y}$$

It can be shown that κ_x and κ_y are inversely proportional to the local radii of curvature in the x and y directions, respectively. Note that w is only a function of the in-plane coordinates x and y and is not a function of the out-of-plane coordinate z.

In a pure bending situation with small deflections, the strains are proportional to the curvatures and are assumed to vary linearly through the laminate thickness (see also Section 5.2.2). They are then given by

$$\varepsilon_x = z\kappa_x$$
$$\varepsilon_y = z\kappa_y \qquad (3.43)$$
$$\gamma_{xy} = z\kappa_{xy}$$

Now writing the first of Equations (3.32) in the laminate coordinate system xy, multiplying both sides by z and integrating through the thickness of the laminate, gives:

$$\int_{-\frac{h}{2}}^{\frac{h}{2}} \sigma_x z dz = \int_{-\frac{h}{2}}^{\frac{h}{2}} Q_{11} z^2 \kappa_x dz + \int_{-\frac{h}{2}}^{\frac{h}{2}} Q_{12} z^2 \kappa_y dz + \int_{-\frac{h}{2}}^{\frac{h}{2}} Q_{16} z^2 \kappa_{xy} dz$$

According to the first of Equations (3.38), the left-hand side of Equation (3.44) is M_x. Denoting

$$D_{11} = \int_{-\frac{h}{2}}^{\frac{h}{2}} Q_{11} z^2 dz, \quad D_{12} = \int_{-\frac{h}{2}}^{\frac{h}{2}} Q_{12} z^2 dz \text{ and } D_{16} = \int_{-\frac{h}{2}}^{\frac{h}{2}} Q_{16} z^2 dz \qquad (3.44)$$

Equation (3.44) can be rewritten in the form:

$$M_x = D_{11}\kappa_x + D_{12}\kappa_y + D_{16}\kappa_{xy} = -D_{11}\frac{\partial^2 w}{\partial x^2} - D_{12}\frac{\partial^2 w}{\partial y^2} - 2D_{16}\frac{\partial^2 w}{\partial x \partial y} \qquad (3.45)$$

where D_{11}, D_{12} and D_{16} are laminate bending stiffnesses.

Operating on the second and third of Equations (3.32) in an analogous fashion, the following constitutive equations for pure bending of a laminate can be obtained:

$$\begin{Bmatrix} M_x \\ M_y \\ M_{xy} \end{Bmatrix} = \begin{bmatrix} D_{11} & D_{12} & D_{16} \\ D_{12} & D_{22} & D_{26} \\ D_{16} & D_{26} & D_{66} \end{bmatrix} \begin{Bmatrix} \kappa_x \\ \kappa_y \\ \kappa_{xy} \end{Bmatrix} \tag{3.46}$$

where

$$D_{ij} = \sum_{k=1}^{n} \frac{Q_{ij}}{3} \left(z_k^3 - z_{k-1}^3 \right) \tag{3.47}$$

with $i,j = 1,2,6$, the summation carried over all plies n of the laminate and z_k, z_{k-1} the upper and lower z coordinates of the kth ply, as shown in Figure 3.6.

Equations (3.40) describe the pure membrane deformations of a laminate and Equations (3.46) the pure bending deformations. In this decoupled form, in-plane strains ε_x, ε_y and γ_{xy} can only be caused by in-plane loads N_x, N_y and N_{xy}, while curvatures κ_x, κ_y and κ_{xy} can only be caused by bending moments M_x, M_y and M_{xy}. However, for a general laminate, it is possible to have coupling between the membrane and bending behaviours, with strains caused by bending moments and/or curvatures caused by in-plane loads. In such a case, the strains are given by a superposition of the membrane strains and the curvatures. The membrane strains are constant through the thickness of the laminate and equal to the mid-plane strains ε_{xo}, ε_{yo} and γ_{xyo}. Therefore,

$$\begin{aligned} \varepsilon_x &= \varepsilon_{xo} + z\kappa_x \\ \varepsilon_y &= \varepsilon_{yo} + z\kappa_y \\ \gamma_{xy} &= \gamma_{xyo} + z\kappa_{xy} \end{aligned} \tag{3.48}$$

Reducing Equations (3.32) to a format in terms of force and moment resultants and combining Equations (3.40), (3.46) and (3.48) the generalized constitutive relations for any laminate (including membrane-bending coupling) have the form:

$$\begin{Bmatrix} N_x \\ N_y \\ N_{xy} \\ M_x \\ M_y \\ M_{xy} \end{Bmatrix} = \begin{bmatrix} A_{11} & A_{12} & A_{16} & B_{11} & B_{12} & B_{16} \\ A_{12} & A_{22} & A_{26} & B_{12} & B_{22} & B_{26} \\ A_{16} & A_{26} & A_{66} & B_{16} & B_{26} & B_{66} \\ B_{11} & B_{12} & B_{16} & D_{11} & D_{12} & D_{16} \\ B_{12} & B_{22} & B_{26} & D_{12} & D_{22} & D_{26} \\ B_{16} & B_{26} & B_{66} & D_{16} & D_{26} & D_{66} \end{bmatrix} \begin{Bmatrix} \varepsilon_{xo} \\ \varepsilon_{yo} \\ \gamma_{xyo} \\ \kappa_x \\ \kappa_y \\ \kappa_{xy} \end{Bmatrix} \tag{3.49}$$

where A_{ij} and D_{ij} are defined by Equations (3.41) and (3.47) and

$$B_{ij} = \sum_{k=1}^{n} \frac{Q_{ij}}{2} \left(z_k^2 - z_{k-1}^2 \right) \tag{3.50}$$

with $i,j = 1,2,6$, the summation carried over all plies n of the laminate and z_k, z_{k-1} the upper and lower z coordinates of the kth ply, as shown in Figure 3.6.

It is important to note that if the order of plies in a stacking sequence is changed the A matrix remains unaffected but the B and D matrices change. This can be of particular importance for buckling-critical designs and provides an option of optimizing a layup by reordering the plies without increasing its weight.

If the mid-plane strains and curvatures of a laminate are known, direct substitution in Equations (3.49) will give the applied forces and moments. Usually, however, the forces and moments are known and the strains and curvatures are sought for. They can be obtained by inverting relations (3.49). The result is [4,5]:

$$
\begin{Bmatrix} \varepsilon_{xo} \\ \varepsilon_{yo} \\ \gamma_{xyo} \\ \kappa_x \\ \kappa_y \\ \kappa_{xy} \end{Bmatrix} =
\begin{bmatrix}
\alpha_{11} & \alpha_{12} & \alpha_{16} & \beta_{11} & \beta_{12} & \beta_{16} \\
\alpha_{12} & \alpha_{22} & \alpha_{26} & \beta_{21} & \beta_{22} & \beta_{26} \\
\alpha_{16} & \alpha_{26} & \alpha_{66} & \beta_{61} & \beta_{62} & \beta_{66} \\
\beta_{11} & \beta_{21} & \beta_{61} & \delta_{11} & \delta_{12} & \delta_{16} \\
\beta_{12} & \beta_{22} & \beta_{62} & \delta_{12} & \delta_{22} & \delta_{26} \\
\beta_{16} & \beta_{26} & \beta_{66} & \delta_{16} & \delta_{26} & \delta_{66}
\end{bmatrix}
\begin{Bmatrix} N_x \\ N_y \\ N_{xy} \\ M_x \\ M_y \\ M_{xy} \end{Bmatrix}
\tag{3.51}
$$

$$
[\alpha] = [A]^{-1} + [A]^{-1}[B]\big[[D] - [B][A]^{-1}[B]\big]^{-1}[B][A]^{-1}
\tag{3.52}
$$

$$
[\beta] = -[A][B]\big[[D] - [B][A]^{-1}[B]\big]^{-1}
\tag{3.53}
$$

$$
[\delta] = \big[[D] - [B][A]^{-1}[B]\big]^{-1}
\tag{3.54}
$$

with square brackets denoting a matrix and the exponent -1 denoting the inverse of a matrix. Note that the β matrix at the top right of Equation (3.51) need not be symmetric. Its transpose appears at the lower left of the matrix in the right-hand side of Equation (3.51).

The most important laminate layup is that of a symmetric laminate (see also Section 3.1). For such a laminate, the coupling matrix B is zero. This can be seen from Equation (3.50) where the contributions to each entry of the matrix coming from two plies located symmetrically with respect to the mid-plane subtract each other (Q_{ij} are the same because the laminate is symmetric and the coefficients $z_k^2 - z_{k-1}^2$ are equal and opposite). With the B matrix zero, there is no membrane-stretching coupling in the laminate behaviour. Also, Equations (3.52), (3.53) and (3.54) simplify. Denoting the inverse of the A matrix by a and the inverse of the D matrix by d, Equations (3.52), (3.53) and (3.54) become:

$$
[\alpha] = [A]^{-1} = [a]
\tag{3.52a}
$$

$$
[\beta] = 0
\tag{3.53a}
$$

$$
[\delta] = [D]^{-1} = [d]
\tag{3.54a}
$$

and substituting in Equation (3.51)

$$
\begin{Bmatrix} \varepsilon_{xo} \\ \varepsilon_{yo} \\ \gamma_{xyo} \\ \kappa_x \\ \kappa_y \\ \kappa_{xy} \end{Bmatrix} =
\begin{bmatrix}
a_{11} & a_{12} & a_{16} & 0 & 0 & 0 \\
a_{12} & a_{22} & a_{26} & 0 & 0 & 0 \\
a_{16} & a_{26} & a_{66} & 0 & 0 & 0 \\
0 & 0 & 0 & d_{11} & d_{12} & d_{16} \\
0 & 0 & 0 & d_{12} & d_{22} & d_{26} \\
0 & 0 & 0 & d_{16} & d_{26} & d_{66}
\end{bmatrix}
\begin{Bmatrix} N_x \\ N_y \\ N_{xy} \\ M_x \\ M_y \\ M_{xy} \end{Bmatrix}
\tag{3.51a}
$$

valid for a symmetric laminate.

Laminate symmetry will be invoked often in subsequent chapters. It should be emphasized here that, in designing composite structures, symmetric and balanced laminates are preferred. They decouple membrane from bending behaviour and stretching from shearing deformations, thus avoiding unwanted failure modes that may occur under some loading conditions.

Very often used in design are the so-called engineering constants. These are stiffness properties that can be measured in the laboratory using simple tests. For example, a uniaxial test of a symmetric and balanced laminate would provide a value for the membrane stiffness E_{xm}^L (or E_{1m} if 1–2 are the laminate axes) of the laminate. For such a laminate under uniaxial loading N_x (with $N_y = 0$), the first two of Equations (3.49) read as

$$N_x = A_{11}\varepsilon_{xo} + A_{12}\varepsilon_{yo}$$
$$0 = A_{12}\varepsilon_{xo} + A_{22}\varepsilon_{yo}$$

The second equation can be used to solve for ε_{yo}:

$$\varepsilon_{yo} = -\frac{A_{12}}{A_{22}}\varepsilon_{xo} \tag{3.55}$$

which, substituted in the first equation, gives

$$N_x = \left(A_{11} - \frac{A_{12}^2}{A_{22}}\right)\varepsilon_{xo} \tag{3.56}$$

Now using the first of Equations (3.39) to substitute for the stress σ_{xav} measured in a uniaxial test, the following relation is obtained.

$$\sigma_{xav} = \frac{1}{h}\left(A_{11} - \frac{A_{12}^2}{A_{22}}\right)\varepsilon_{xo} \tag{3.57}$$

It can also be shown that the 11 entry of the inverse of the A matrix for a symmetric and balanced laminate is

$$a_{11} = \frac{A_{22}}{A_{11}A_{22} - A_{12}^2} \tag{3.58}$$

Equations (3.37) and (3.38) imply that the laminate Young's modulus for membrane deformations of a symmetric and balanced laminate is given by:

$$E_{1m} = \frac{1}{h}\frac{A_{11}A_{22} - A_{12}^2}{A_{22}} = \frac{1}{ha_{11}} \tag{3.59}$$

Also, from Equations (3.18) and (3.55), the Poisson's ratio for a symmetric and balanced laminate undergoing membrane deformations is

$$\nu_{12m} = \frac{A_{12}}{A_{22}} \tag{3.60}$$

An analogous expression can be derived for the bending modulus E_{1b} of a laminate. For the special case of a laminate with B matrix zero and $D_{16} = D_{26} = 0$, the fourth and fifth equations of relations (3.49) and (3.51) can be used to eliminate κ_y and obtain the relation

$$M_x = \left(D_{11} - \frac{D_{12}^2}{D_{22}}\right)\kappa_x \tag{3.61}$$

In a pure bending test, such as a four-point bending test, the moment curvature relation has the form:

$$M_x = \frac{M}{b} = \frac{1}{b}E_{1b}I\kappa_x \tag{3.62}$$

where I is the moment of inertia $bh^3/12$.

Comparing Equations (3.61) and (3.62) it can be seen that

$$E_{1b} = \frac{12}{h^3}\left(D_{11} - \frac{D_{12}^2}{D_{22}}\right) \tag{3.63}$$

which, using d_{11}, can be shown to be

$$E_{1b} = \frac{12}{h^3}\frac{D_{11}D_{22} - D_{12}^2}{D_{22}} = \frac{12}{h^3 d_{11}} \tag{3.64}$$

In general, the stiffness calculated by Equation (3.59), corresponding to stretching of a laminate, is not the same as that calculated by Equation (3.64), which corresponds to bending of a laminate. This will be shown later on to cause some problems on the selection of the stiffness value to be used for certain problems (see, for example, Section 8.2). In general, for bending problems the bending stiffnesses are used and for stretching problems the membrane stiffnesses are used. However, in situations where both behaviours occur simultaneously it is not always clear what values should be used and it is not uncommon to use the values that give the most conservative results.

Relations (3.59) and (3.64) were derived for special laminates to avoid algebraic complexity and to emphasize the underlying physical models. In general, the laminate stiffness properties in all directions for symmetric laminates can be found to be [5]:

$$
\begin{aligned}
E_{1m} &= \frac{1}{ha_{11}} & E_{1b} &= \frac{12}{h^3 d_{11}} \\
E_{2m} &= \frac{1}{ha_{22}} & E_{2b} &= \frac{12}{h^3 d_{22}} \\
G_{12m} &= \frac{1}{ha_{66}} & G_{12b} &= \frac{12}{h^3 d_{66}} \\
\nu_{12m} &= -\frac{a_{12}}{a_{11}} & \nu_{12b} &= -\frac{d_{12}}{d_{11}} \\
\nu_{21m} &= -\frac{a_{12}}{a_{22}} & \nu_{21b} &= -\frac{d_{12}}{d_{22}}
\end{aligned}
\tag{3.65}
$$

Exercises

3.1 Assume a layup consists of n plies of the same material, all at the same orientation (not necessarily $0°$). Let E be the Young's modulus of a single ply at that orientation, G the corresponding shear modulus and v_{12}, v_{21} the two Poisson's ratios. Derive analytical expressions for $A_{11}, A_{12}, A_{22}, A_{66}, D_{11}, D_{12}, D_{22}, D_{66}$ as functions of E, G, v_{12}, v_{21} and the thickness h of the laminate (still having all plies with the same fibre orientation).

3.2 By mistake, the layup of a specific laminate fabricated in the factory was not labelled and the stacking sequence is unknown. The laminate was fabricated using a graphite/epoxy material with the following basic ply properties:

$$E_x = 131 \text{ GPa}$$
$$E_y = 11.7 \text{ GPa}$$
$$G_{xy} = 4.82 \text{ GPa}$$
$$v_{xy} = 0.29$$
$$t_{ply} = 0.3048 \text{ mm}$$

To avoid throwing the expensive laminate away an engineer cuts a small strip of material from the edge. The strip is 152.4 mm long by 25.4 mm wide and has a thickness of 1.83 mm. First he/she tests this in a three-point bending configuration and then in tension as shown in the Figure below:

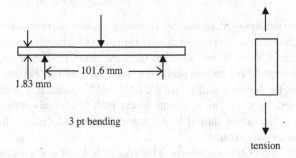

3 pt bending   tension

In the three-point bending test he/she notices that the specimen undergoes pure bending and in the tension test the specimen only elongates (and contracts by a small amount transversely to the load).

Using the results of the three-point bending test, the engineer notices that when plotting the centre deflection as a function of the applied load at the centre, he/she obtains (for low loads) an (almost) straight line with slope 0.03826 mm/N. Unfortunately, this information is not sufficient to determine the stacking sequence conclusively. Part of the problem is that it is hard to measure the centre deflection of the three-point bending test accurately. During the uniaxial tension test the engineer notices that a maximum load of 2225 N results in a specimen elongation of 0.0941 mm. Now the engineer is confident he/she knows the stacking sequence. What is the stacking sequence? (In this factory only laminates with 0, 45, −45 and 90 plies are used.)

3.3 In a pure bending test of a balanced and symmetric laminate, the following data are collected:

(a) When the applied M_y is 64.635 N mm/mm and M_x is 109.1 N mm/mm, the maximum axial strain $\varepsilon_{x\max}$ is 3000 microstrain and the minimum axial strain $\varepsilon_{x\min}$ is -3000 microstrain.

(b) The maximum axial strain $\varepsilon_{y\max}$ is zero microstrain and the minimum axial strain $\varepsilon_{y\min}$ is zero microstrain. These strains are the same everywhere on the laminate.

(c) The laminate thickness is 2.1336 mm.

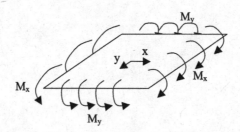

Draw an approximate picture of how the laminate deforms in this situation.

In a second test, only $M_x = 44.45$ N mm/mm is applied and $\kappa_x = 0.002$/mm (κ_y was not measured).

In a third test, only $M_y = 44.45$ N mm/mm is applied and $\kappa_y = 0.0026$/mm (κ_x was not measured.

Determine D_{11}, D_{12} and D_{22} for this laminate. Show the mathematical expressions you used to determine them (closed form solutions). Assume that, for this laminate, $D_{16} = D_{26} = 0$.

For the material under consideration, the ply thickness is known to be 0.3048 mm.

References

[1] Tsai, S.W., Mechanics of Composite Materials Part I — Introduction, *Report AFML-TR-66-149, Part I, Air Force Materials Laboratory Research and Technology Division*, Wright-Patterson Air Force Base, Ohio, June 1966.

[2] Tsai, S.W., Mechanics of Composite Materials, Part II —Theoretical Aspects, *Report AFML-TR-66-149, Part II, Air Force Materials Laboratory Research and Technology Division*, Wright-Patterson Air Force Base, Ohio, November 1966.

[3] Ashton, J.E., Halpin, J.C. and Petit, P.H., *Primer on Composite Materials: Analysis*, Technomic, Westport, CT, 1969.

[4] Jones, R.M, *Mechanics of Composite Materials*, McGraw-Hill, Washington DC, 1975.

[5] Tsai, S.W. and Hahn, H.T., *Introduction to Composite Materials*, Technomic Publishing Co, Westport, CT, 1980.

[6] Love, A.E.H., *A Treatise on the Mathematical Theory of Elasticity*, 4th edn, Dover Publications, NY, 1944, Article 66.

4

Review of Laminate Strength and Failure Criteria

If the loads applied to a laminate are sufficiently high then the strength of the material is exceeded and the laminate fails. It is, therefore, very important to be able to use the stresses and/or strains calculated in the previous chapter to predict failure. This, however, is complicated by the fact that final failure of a laminate does not always coincide with the onset of damage. Depending on the laminate lay-up and loading, damage may start at a load significantly lower than the load at which final failure occurs. Being able to predict when damage starts and how it evolves requires individual modelling of the matrix and fibres. Usually, damage starts in the form of matrix cracks between fibres in plies transverse to the primary load direction. As the load increases the crack density increases and the cracks may coalesce into delaminations (where plies locally separate from one another) or branch out to adjacent plies [1]. In addition, local stress concentrations may lead to failure of the fibre–matrix interphase. Further increase of the load accumulates this type of damage and causes some fibres to fail until the laminate can no longer sustain the applied load and fails catastrophically. The detailed analysis of damage creation and evolution accounting for the individual constituents of a ply is the subject of micromechanics [2, 3].

In an alternative simplified approach, each ply is modelled as homogeneous, having specific failure modes which are characterized by tests. For a unidirectional ply the following failure modes are usually recognized:

Tension failure along the fibres with strength symbol X^t
Compression failure along the fibres with strength symbol X^c
Tension failure transverse to the fibres with strength symbol Y^t
Compression failure transverse to the fibres with strength symbol Y^c
Pure shear failure of a ply with strength symbol S

These strength values, obtained experimentally, are already one step away from the individual failures of fibre and matrix and their interphase. The details of damage onset, such as matrix cracks leading to fibre failure or failure of the fibre–matrix interphase leading to fibre

Design and Analysis of Composite Structures: With Applications to Aerospace Structures, Second Edition. Christos Kassapoglou.
© 2013 John Wiley & Sons, Ltd. Published 2013 by John Wiley & Sons, Ltd.

failure, are lumped into a single experimentally measured value. This value is a macroscopic value that describes when a single ply will fail catastrophically given a specific loading.

In parallel to, or instead of, the five strength values just mentioned, ultimate strain values can be used for the same loading situations, again obtained experimentally. Using ultimate stress values is interchangeable with ultimate strain values (in terms of obtaining the same failure load at which the ply fails) only for loading situations for which the stress–strain curve is linear to failure or very nearly so. This means that for tension and compression along the fibres, going from predictions obtained with a strength-based model to predictions from a strain model requires only the use of a constant of proportionality which is the Young's modulus (in the direction of the load) divided by a Poisson's ratio term. For shear loading and transverse tension or compression, where the stress–strain curves are, usually, nonlinear, simply multiplying strain-based predictions with a constant of proportionality does not give the correct strength failure values. A model that accounts for nonlinearities in the stress–strain curve must be used.

Consider now the case of a laminate in which all the plies are the same with the same arbitrary orientation θ. An arbitrary in-plane loading applied to this laminate results in the same combined state of stress (and strain) in each ply. This state of stress or strain must be transformed to the principal axes for the ply, which are the ply axes (one axis parallel to the fibres and one transverse to them). The resulting principal stresses (or strains) are compared with their respective maximum values (strength or ultimate strain). Obviously, in this special case, all plies fail simultaneously. The approach where the principal stresses in a ply are compared with the ultimate strength values in the respective directions is the maximum stress theory. The approach where principal strains in a ply are compared with the ultimate strain values in the respective directions is the maximum strain theory. Note that, for generalized loading, even if all stress–strain curves are linear, the predictions from the two methods will differ slightly due to a Poisson's ratio effect.

The situation becomes more complicated when the plies in a laminate do not all have the same ply orientation. The procedure is as follows:

1. Given the applied loads, the corresponding laminate mid-plane strains and curvatures are computed using Equations (3.51).
2. These are then used along with Equations (3.48) to determine the individual strains within each ply in the laminate axes (see Figure 4.1).
3. Ply strains in the laminate axes can be translated to ply stresses in the laminate axes using Equations (3.32).
4. Depending on the type of failure criterion used (stress- or strain-based) the ply stresses and/or strains in the laminate axes are transformed to ply stresses and/or strains in the ply axes (see Figure 4.1) using Equations (3.35) and/or (3.36). For each ply, the ply axis system has one axis parallel to the fibres and the other perpendicular to them.
5. Using the results of the previous step, a failure criterion is applied to determine which ply fails. This determines first-ply failure.
6. If desired, post-first-ply failure analysis can follow. The stiffness and strength properties of the failed ply are adjusted accounting for the type of failure that occurred and steps 1–5 are repeated until the next ply fails.
7. Step 6 is repeated until all plies in the laminate have failed.

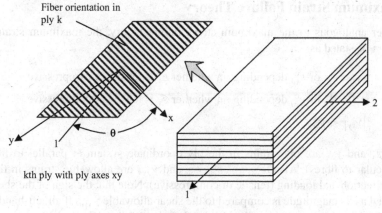

Fiber orientation in ply k

kth ply with ply axes xy

θ

x

y

1

1

2

Laminate with laminate axes 1-2

Figure 4.1 Laminate and ply axes systems

The failure predictions obtained with the procedure just described may vary significantly depending on the failure criterion used. There are a large number of failure criteria, stress-based, strain-based or energy-based. A few representative ones are briefly discussed in subsequent sections.

4.1 Maximum Stress Failure Theory

In this case, the principal stresses in each ply are compared with their corresponding strength values X^t, X^c, Y^t, Y^c and S. In a design situation these strength values are adjusted through statistical analysis (see Sections 5.1.3–5.1.6) to obtain reduced values that account for material scatter and adverse environmental effects. In some cases, the effect of damage is also included in these reduced strength values. These reduced values are also termed allowables. The maximum stress failure criterion can be expressed as:

$$\sigma_x < X^t \text{ or } X^c \text{ depending on whether } \sigma_x \text{ is tensile or compressive}$$

$$\sigma_y < Y^t \text{ or } Y^c \text{ depending on whether } \sigma_y \text{ is tensile or compressive} \qquad (4.1)$$

$$|\tau_{xy}| < S$$

where σ_x, σ_y and τ_{xy} are ply stresses in the ply coordinate system (x parallel to fibres and y perpendicular to fibres). Note that the sign of the shear stress is immaterial as its magnitude is compared with the shear allowable S. If all left-hand sides of Equation (4.1) are less than the right-hand sides there is no failure. Failure occurs as soon as one (or more) of the left-hand sides equals the right-hand side. The failure mode is the one for which Equation (4.1) is met. For example, if σ_x is compressive and the first of Equations (4.1) is met, then the failure mode is compressive failure along the fibres.

4.2   Maximum Strain Failure Theory

In a manner analogous to the maximum stress failure theory, the maximum strain failure criterion can be stated as:

$$\varepsilon_x < \varepsilon_{xu}^t \text{ or } \varepsilon_{xu}^c \text{ depending on whether } \varepsilon_x \text{ is tensile or compressive}$$

$$\varepsilon_y < \varepsilon_{yu}^t \text{ or } \varepsilon_{yu}^c \text{ depending on whether } \varepsilon_y \text{ is tensile or compressive} \qquad (4.2)$$

$$|\gamma_{xy}| < \gamma_{xyu}$$

where ε_x, ε_y and γ_{xy} are ply strains in the ply coordinate system (x parallel to fibres and y perpendicular to fibres). Also, ε_{xu}^t, ε_{xu}^c, ε_{yu}^t, ε_{yu}^c and γ_{xyu} are allowable strains in the corresponding direction and loading (tensile or compressive). Note that the sign of the shear strain is immaterial as its magnitude is compared to the shear allowable γ_{xyu}. If all left-hand sides of Equation (4.2) are less than the right-hand sides there is no failure. Failure occurs, in a specific failure mode, as soon as one (or more) of the left-hand sides equals the right-hand side.

4.3   Tsai–Hill Failure Theory

In the two previous failure criteria, each stress or strain is individually compared with its respective allowable. In general, however, stresses (or strains) may interact with each other and lead to failure, even if each compared individually with its respective allowable suggests that there is no failure. Hill [4] was among the first to propose a combined failure criterion for composite materials. For a single ply under plane stress, with ply axes xy as shown in Figure 4.1, the criterion has the form:

$$F_x \sigma_x^2 + F_y \sigma_y^2 + F_{xy} \sigma_x \sigma_y + F_S \tau_{xy}^2 = 1 \qquad (4.3)$$

The form of Equation (4.3) is exactly analogous to the von Mises yield criterion in isotropic materials:

$$\frac{\sigma_x^2}{\sigma_{yield}^2} + \frac{\sigma_y^2}{\sigma_{yield}^2} - \frac{\sigma_x \sigma_y}{\sigma_{yield}^2} + \frac{3\tau_{xy}^2}{\sigma_{yield}^2} = 1 \qquad (4.4)$$

with σ_{yield} the yield stress of the material. In fact, Equation (4.3) was proposed by Hill (for a three-dimensional state of stress) as a model of yielding in anisotropic materials. For composite materials, where the concept of macroscopic yielding (at the laminate or the ply level) is not really valid, failure replaces yielding.

Equation (4.3) recognizes the fact that the failure strengths of a composite ply are different in different directions. Tsai [5] determined the stress coefficients in Equation (4.3) by considering three simple loading situations: (a) only σ_x acts on a ply with corresponding strength X; (b) only σ_y acts with corresponding strength Y; and (c) only τ_{xy} acts with corresponding strength S.

For example, if only σ_x acts, Equation (4.3) reads:

$$F_x \sigma_x^2 = 1 \qquad (4.5)$$

It is also known that if only σ_x acts, which is parallel to the fibres, failure will occur when σ_x equals X or

$$\sigma_x^2 = X^2 \tag{4.6}$$

Comparing Equations (4.5) and (4.6) it can be seen that:

$$F_x = \frac{1}{X^2} \tag{4.7}$$

Considering the remaining two load cases would give another two conditions to determine two of the three remaining unknowns F_y, F_{xy} and F_s. One more condition is obtained by considering the original three-dimensional form of the Hill yield criterion [4] in which F_x, F_y and F_{xy} are interdependent through distortional deformations of a representative volume of material. This gives one additional equation. The final form of the Tsai–Hill failure criterion is:

$$\frac{\sigma_x^2}{X^2} - \frac{\sigma_x \sigma_y}{X^2} + \frac{\sigma_y^2}{Y^2} + \frac{\tau_{xy}^2}{S^2} = 1 \tag{4.8}$$

4.4 Tsai–Wu Failure Theory

The Tsai-Wu failure criterion [6] was a result of an attempt to mathematically generalize the Tsai–Hill failure criterion creating a curve fit based on tensor theory and accounting for the fact that composites have different strengths in tension and compression. This means that the Tsai–Wu failure theory is not entirely based on physical phenomena, but includes a curve-fitting aspect. In fact, one of the unknown coefficients in the criterion is obtained by requiring that the von Mises yield criterion be recovered if the material were isotropic. As was mentioned in the previous section, yielding and, more so, distortional energy theory on which the von Mises criterion is based, are not applicable to composites so the Tsai–Wu criterion should be viewed as a convenient (and useful) curve fit more than a physics-based model of failure. The form of the criterion is:

$$\frac{\sigma_x^2}{X^t X^c} + \frac{\sigma_y^2}{Y^t Y^c} - \sqrt{\frac{1}{X^t X^c} \frac{1}{Y^t Y^c}} \sigma_x \sigma_y + \left(\frac{1}{X^t} - \frac{1}{X^c} \right) \sigma_x + \left(\frac{1}{Y^t} - \frac{1}{Y^c} \right) \sigma_y + \frac{\tau_{xy}^2}{S^2} = 1 \tag{4.9}$$

Note that tensile and compressive strengths are input as positive values (magnitudes) in the above equation. With the exception of biaxial compression situations where the predictions are, at best, unrealistic, the Tsai–Wu criterion gives predictions that range from acceptable to excellent when compared with test results.

4.5 Puck Failure Theory

Most failure criteria give reasonable to excellent predictions (depending on the criterion) when a single stress acts on a simple geometry such as uniaxial tension. It is under combined states of stress that most failure criteria partially or completely break down. Attempting to account for the interaction of different stress states can be difficult and it is based on assumptions the validity of which can be questionable and/or hard to prove. The Tsai–Hill and Tsai–Wu criteria described in the two previous sections were based on observations made on ductile materials adapted for composite materials. As such, the failure criteria recover well-established conditions for yielding when they are applied to metals. However, one major issue with this approach is that failure of composites is governed primarily by brittle fracture. Therefore, criteria based on the failure behaviour of brittle materials are expected to more accurately describe fracture of composite materials.

The formulation of failure of composites by Puck [7] is one of the best attempts to combine phenomenological observations with thorough understanding of the behaviour of composite materials during fracture. The main characteristics of the Puck failure theory are the replacement of interaction curves by physically based reasoning of how combined stresses promote or delay fracture and the premise that failure at the critical failure plane is caused exclusively by the stresses acting on that plane. For example, a tensile normal stress on the fracture plane reduces the failure load while a compressive normal stress on the fracture plane increases the failure load. Note that the inclination (and location) of the fracture plane are not known in advance. Considering failure of a unidirectional ply, six different loading types, each with its associated corresponding failure strength are distinguished: (1) Tension loading along the fibres, (2) Compression loading along the fibres, (3) Tension loading perpendicular to the fibres, (4) Compression loading perpendicular to the fibres, (5) Shear 12 or 13 loading with the fibres aligned with the 1 direction and (6) Shear 23 loading with the fibres still in the 1 direction. It is important to note that the corresponding failure strength values are not the same as the traditional strength values X^t, X^c, Y^t, Y^c and S (see Section 4.1) but some of them can be related to these ply-level strength properties.

Focusing on the individual constituents, the following failure conditions are established [7, 8]:

Fibre tension or compression failure

$$\frac{E_L}{X}\left|\left[\varepsilon_L + \frac{\nu_{LTf}}{E_f}m_\sigma \sigma_T\right]\right| = 1 \tag{4.10}$$

where X should be replaced with X^t or X^c depending on whether the quantity in brackets is positive or negative, respectively. Also, the subscripts 'L' and 'T' refer to longitudinal and transverse directions, see Section 3.3. The subscript 'f' denotes fibre quantities and m_σ is a stress magnification factor that accounts for the difference in transverse stiffnesses between fibre and matrix. For glass fibres, $m_\sigma \approx 1.3$ and for carbon fibres, $m_\sigma \approx 1.1$.

Matrix tension failure (when $\sigma_T \geq 0$):

$$\sqrt{\left(\frac{\tau_{TL}}{S}\right)^2 + \left(1 - s_{nnL}\frac{Y^T}{S}\right)^2 \left(\frac{\sigma_T}{Y^T}\right)^2} + s_{nnL}\frac{\sigma_T}{S^{TZ}} = 1 \tag{4.11}$$

where s_{nnL} is the slope of the failure envelope evaluated with $\sigma_n = 0$ when σ_n and τ_{nL} are the stresses acting on the failure plane and n is the normal direction to that plane. Also, S^{TZ} is the shear strength of a unidirectional ply loaded in the 23 plane with the fibres aligned in the 1 direction.

Matrix compression failure with fracture plane perpendicular to the loading and $\sigma_T < 0 \leq |\sigma_T/\tau_{TL}| \leq R/S_c$:

$$\sqrt{\left(\frac{\tau_{TL}}{S}\right)^2 + \left(s_{nnL}\frac{Y^T}{S}\right)^2} + s_{nnL}\frac{\sigma_T}{S} = 1 \qquad (4.12)$$

with

$$R = \frac{Y^c}{2\left(1 + s_{nnT}\right)}$$

$$S_c = S\sqrt{1 + 2s_{nnT}}$$

and s_{nnT} the slope of the failure surface at $\sigma_n = 0$ when σ_n and τ_{nT} are the stresses acting on the failure plane with n the normal to that plane.

Finally, matrix compression failure with fracture plane at an angle to the loading (dominated by local shear):

$$-\frac{Y^c}{\sigma_T}\left[\left(\frac{\tau_{TL}}{Y^c}\frac{R}{S}\right)^2 + \left(\frac{\sigma_T}{Y^c}\right)^2\right] = 1 \qquad (4.13)$$

valid when $\sigma_T < 0 \leq |\tau_{TL}/\sigma_T| \leq S_c/R$.

Equations (4.11), (4.12) and (4.13) refer to different matrix failure modes and account for the, usually nonlinear, effect of matrix cracking on final failure. Despite its appeal, the Puck criterion involves parameters that require significant extra effort to determine such as the slopes of the failure surfaces, the transverse stress magnification factors and the location and inclination of the fracture plane.

4.6 Other Failure Theories

In the discussion of some of the failure criteria presented in the previous sections some of the shortfalls of these failure theories were mentioned. Many attempts have been made in the past to propose improved failure criteria [9–11] that do not suffer from the shortfalls mentioned and are in closer agreement with experimental results. This is still an open subject of research and the sometimes heated discussion [12, 13] has yet to reach definitive conclusions.

One of the major problems of interaction failure criteria such as the Tsai–Wu and Tsai–Hill is the physical meaning of interaction terms (in addition to difficulty of experimentally obtaining them). Related to that, but more generic as a problem of some of the failure criteria already presented, is the smearing of properties and treating each ply as homogeneous with single values to represent failure strengths in different directions [12]. Failure theories [14] that account for the individual failure modes of fibre and matrix are more promising in that respect. Specifically, the Hashin–Rotem failure criterion has the form [14]:

$$\left.\begin{array}{ll} \dfrac{\sigma_x}{X^{\mathrm{t}}} = 1 & \text{when } \sigma_x \text{ is tensile} \\[1.5em] \dfrac{\sigma_x}{X^{\mathrm{c}}} = 1 & \text{when } \sigma_x \text{ is compressive} \end{array}\right\} \text{fibre failure}$$

$$\left.\begin{array}{ll} \dfrac{\sigma_y^2}{(Y^{\mathrm{t}})^2} + \dfrac{\tau_{xy}^2}{S^2} = 1 & \text{when } \sigma_y \text{ is tensile} \\[1.5em] \dfrac{\sigma_y^2}{(Y^{\mathrm{c}})^2} + \dfrac{\tau_{xy}^2}{S^2} = 1 & \text{when } \sigma_y \text{ is compressive} \end{array}\right\} \text{matrix failure}$$
(4.14)

More recently (see for example [15]), failure criteria based on micromechanics analysis of the composite constituents under different loading situations have emerged and appear to be the most promising, but at a significant increase in complexity and computational cost.

In view of the difficulties of failure theories to accurately predict first-ply failure, extending to subsequent ply failure and final laminate collapse is even harder. In fact, other than disregarding very early failures of plies with fibres transverse to the main tensile load in a laminate, there is no reliable method for performing post-first-ply failure analysis other than the approach by Dávila *et al.* [15] implemented in a finite element environment. Several attempts have been made [16, 17] with varying degrees of success.

An attempt to evaluate all previous work on the subject and determine which criterion or criteria are the most accurate to use is still ongoing with the worldwide failure exercise [18]. While it has identified some criteria that are more promising than others, one of them being the Puck criterion discussed in the previous section, it has yet to conclusively determine the best one.

In the general case where a laminate is under a three-dimensional state of stress, modified criteria accounting for out-of-plane stresses and their interaction with in-plane stresses have to be used, or, in some cases, individual criteria for in-plane and out-of-plane loads are used [19].

Which failure criterion or criteria will be used in a specific application is very much a matter of preference, available resources and test data. The simpler failure criteria such as maximum strain or Tsai–Hill and Tsai–Wu (despite their shortcomings) can be very useful for preliminary design if supported by test data covering the load situations of interest. In other cases, emphasis is placed on test data and laminate strength is obtained from test rather than failure criteria. In what follows in this book, wherever laminate strength is needed (for example for crippling calculations in Section 8.5) it is assumed that the reader will use whichever method to predict laminate strength that he/she considers more reliable and accurate.

References

[1] Williams, J.G., Fracture Mechanics of Composites Failure, Proceedings of the Institution of Mechanical Engineers C, *Journal of Mechanical Engineering Sciences*, **204**, 209–218 (1990).

[2] Gotsis, P.K., Chamis, C. and Minnetyan, L., Prediction of Composite Laminate Fracture Micromechanics and Progressive Failure, *Composites Science and Technology*, **58**, 1137–1149 (1998).

[3] Aboudi, J., Micromechanical Analysis of the Strength of Unidirectional Fiber Composites, *Composites Science and Technology*, **33**, 79–46 (1988).

[4] Hill, R., *The Mathematical Theory of Plasticity*, Oxford University Press, London, 1950.

[5] Tsai, S.W., *Strength Theories of Filamentary Structures, in Fundamental Aspects of Fiber Reinforced Plastic Composites*, Wiley, New York, 1968, pp 3–11.

[6] Tsai, S.W. and Wu, E.M., A General Theory of Strength for Anisotropic Materials, *Journal of Composite Materials*, **5**, 58–80 (1971).

[7] Puck, A. and Schürmann, H, Failure Analysis of FRP Laminates by Means of Physically Based Phenomenological Models, *Composites Science and Technology*, **58**, 1045–1067 (1998).

[8] Laš, V, Zemčík, R., Kroupa, T. and Kottner, R., Failure Prediction of Composite Materials, *Bulletin of Applied Mechanics*, **4**, 81–87 (2008).

[9] Nahas, M.N., Survey of Failure and Post-Failure Theories of Laminated Fiber-Reinforced Composites, *Journal of Composites Technology and Research*, **8**, 138–153 (1986).

[10] Quinn, B.J. and Sun, C.T., A Critical Evaluation of Failure Analysis Methods for Composite Laminates, Proc. 10th DoD/NASA/FAA Conf. on Fibrous Composites in Structural Design, vol. V, 1994, pp V21–V37.

[11] Hart-Smith, L.J., A Re-examination of the Analysis of In-plane Matrix Failures in Fibrous Composite Laminates, *Composites Science and Technology*, **56**, 107–121 (1996).

[12] Hart-Smith, L.J., The Role of Biaxial Stresses in Discriminating Between Meaningful and Illusory Composite Failure Theories, *Composite Structures*, **25**, 3–20 (1993).

[13] Hart-Smith, L.J., An Inherent Fallacy in Composite Interaction Failure Criteria, *Composites*, **24**, 523–524 (1993).

[14] Hashin, Z. and Rotem, A., A Fatigue Failure Criterion for Fiber Reinforced Materials, *Journal of Composite Materials*, **7**, 448–464 (1973).

[15] Dávila, C., Camanho, P.P., and Rose, C.A., Failure Criteria for FRP Laminates, *Journal of Composite Materials*, **39**, 323–345 (2005).

[16] Nuismer, R.J., Continuum Modeling of Damage Accumulation and Ultimate Failure in Fiber Reinforced Laminated Composite Materials, Research Workshop, Mechanics of Composite Materials, Duke University, Durham NC, 1978, pp 55–77.

[17] Reddy, Y.S. and Reddy, J.N., Three-Dimensional Finite Element Progressive Failure Analysis of Composite Laminates Under Axial Extension, *Journal of Composites Technology and Research*, **15**, 73–87 (1993).

[18] Hinton, M, Kaddour, S., Smith, P., Li, S. and Soden, P, "Failure Criteria in Fiber Reinforced Polymer Composites: Can Any of the Predictive Theories be Trusted?" presentation to NAFEMS World Congress, Boston, May 23–26, 2011.

[19] Brewer, J.C. and Lagace, P.A., Quadratic Stress Criterion for Initiation of Delamination, *Journal of Composite Materials*, **22**, 1141–1155 (1988).

5

Composite Structural Components and Mathematical Formulation

5.1 Overview of Composite Airframe

A section of a fuselage structure, showing some of the typical parts that make it up, is shown in Figure 5.1. Similar part types are used to make up a wing structure.

The types of parts that make up an airframe (fuselage and/or wing) are the same for metal and composite structures. In fact, it is possible to replace, part for part, an aluminium airframe by an equivalent composite airframe. This would typically be a skin-stiffened built-up structure with fasteners connecting the different parts. In general, such a one-for-one replacement does not make full use of composite capabilities and results in minor weight reductions (<15%) with relatively high fabrication cost because the different parts are made separately and assembled together with fasteners. Such a construction, especially when the skin lay-up is quasi-isotropic is referred to as 'black aluminium' to emphasize the fact that the design imitates or closely matches the aluminium design and little or no attempt is made to use composites to their fullest potential.

Each part or component in an airframe structure serves a specific purpose (or, sometimes, multiple purposes) so that the ensemble is as efficient as possible. Efficiency typically refers to the lowest weight, given a set of applied loads, but it can be any combination of desired attributes such as weight, cost, natural frequency, etc. With reference to Figure 5.1, the parts used in a composite (also metal) airframe can be broken into the part families or types shown in Table 5.1. Note that a less-detailed breakdown was given in Chapter 2, Table 2.1 for the purposes of cost discussion.

Each part must be designed so that it does not fail under the applied loads and it meets all other design requirements (see Section 5.1.1). Usually, the main objective is to keep the weight as low as possible but, as already mentioned, additional objectives such as minimum cost (see Chapter 2) are also incorporated in the design process.

Design and Analysis of Composite Structures: With Applications to Aerospace Structures, Second Edition. Christos Kassapoglou.
© 2013 John Wiley & Sons, Ltd. Published 2013 by John Wiley & Sons, Ltd.

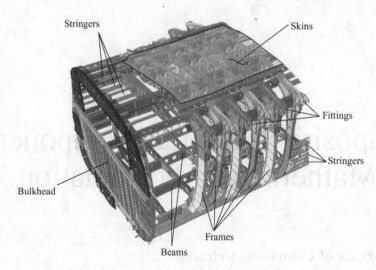

Figure 5.1   Typical fuselage part break-down

5.1.1   The Structural Design Process: The Analyst's Perspective

The objective of the structural design process is to create a structure that meets specific requirements and has certain desirable attributes. The typical design requirements can be summarized into: (1) fit, form and function; (2) applied loads; (3) corrosion resistance and

Table 5.1   Part families in a composite airframe

Part family	Description	Usage	
Skins	Two-dimensional thin structures covering the outside of fuselage or wing; usually single curvature	Fuselage	Wing
Stringers, stiffeners, panel breakers	One-dimensional beam-like structures	Fuselage	Wing
Frames, bulkheads	Two-dimensional ring-like structures at specific intervals along fuselage	Fuselage	
Beams	Two-dimensional plate-like structures	Fuselage	Wing
Spars	Two-dimensional plate-like structures along the length of the wing		Wing
Ribs	Two-dimensional plate-like structures at specific intervals along the wingspan		Wing
Intercostals	Two-dimensional plate-like structures acting as supports	Fuselage	
Fittings	Three-dimensional structures connecting adjacent parts	Fuselage	Wing
Decks, floors	Two-dimensional flat structures	Fuselage	
Doors, fairings	Two-dimensional structures usually with compound curvature	Fuselage	Wing

resistance to fluids; (4) thermal expansion coefficient placement; (5) frequency placement. Briefly, each of these is discussed below.

Fit, form and function. The structure to be designed must fit within the allowable envelope, that is, it must avoid interference with the adjacent structure, must have the appropriate material and generic shape so it performs optimally and must perform the assigned function without flaws. The latter includes providing attachment points for other structures as needed and access (through-paths for example) for intersecting parts such as electrical and hydraulic equipment or ducts.

Applied loads. The structure to be designed must not fail under the applied static loads and must have the desired life under the applied fatigue loads. In addition, the structure must be able to withstand certain static and fatigue loads in the presence of damage without jeopardizing the operation of the remainder of the structure (e.g. if the structure is damaged and the load is transferred to the adjacent structure the adjacent structure should still be able to perform without failure).

Corrosion resistance, resistance to fluids (jet fuel, etc.). Exposure to water vapours or water or other fluids such as fuel and hydraulic fluids is unavoidable during the service life of many parts. The amount of corrosion and/or the associated reduction of strength or stiffness must be minimized.

Thermal expansion coefficient placement. Airframe structures are exposed to wide variations in ambient temperature, either due to their location on the airframe (e.g. parts near the exhaust of an engine) or due to the environment (e.g. satellites). Such structures must be designed to have low thermal expansion coefficients so that any deformations resulting from temperature changes do not compromise the performance of the structure and do not lead to premature failure.

Frequency placement. Airframe structures operate in a vibration environment with specific driving frequencies (from the engines) or random vibrations (gusts, etc.). The natural frequencies of the main structural modes such as the first few bending and torsional modes must be sufficiently far from the driving frequencies to avoid large deflections and premature failure.

Depending on the application, some or all of these design requirements must be met. In some cases additional requirements may be imposed. What makes the problem more challenging is that these requirements must be met while specific desirable attributes are also achieved. The most common desirable attributes are: (1) minimum weight; (2) minimum cost; (3) low maintenance; (4) replaceability across assemblies, etc. These are discussed briefly below.

Minimum weight. Minimizing the structural weight increases the amount of payload or weight of fuel that can be carried (for a given gross weight). Or, if the weight reduction is not translated to payload or fuel increase, it translates to overall size reductions (engines are smaller, wings are smaller, etc.), which, in turn, reduces fuel consumption and acquisition and maintenance cost.

Minimum cost. This can be: (a) the recurring fabrication cost (labour and materials to build each part or aircraft); (b) the nonrecurring cost which is the cost incurred once in each program and includes development/research cost, tooling cost, cost for testing and certification, cost for developing drawings and doing analysis, etc.; (c) acquisition cost (cost incurred by the customer in purchasing the part or aircraft); (d) operating cost, etc. See also Chapter 2 for a brief discussion on cost of composite airframe structures.

Low maintenance. This is related to the minimum cost described above, but merits special mention. Over the long life of an aircraft, maintenance cost (including inspection, disposition

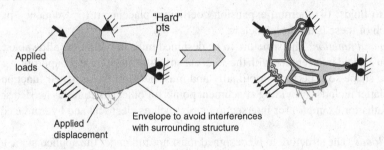

Figure 5.2 Designing a part to meet specific requirements

of problems found and associated repair) can become a very significant portion of the life-cycle cost of the aircraft. Designing structure that minimizes this cost is very desirable and attractive to customers.

Replaceability across assemblies. This is also (indirectly) related to the cost and low maintenance items mentioned earlier. Depending on the part geometry, adjacent structure and fabrication and assembly methods selected, the accuracy of the part geometry and how closely it mates with adjacent structure can vary widely, to the point that exchanging nominally the same parts between two different assemblies can be almost impossible without significant rework to eliminate interferences or fill in gaps (shimming). If, however, the design and fabrication process yield parts of acceptable cost and high accuracy, then replacing a part will only require simple disassembly of the part to be replaced and assembly of the replacement part. This drastically reduces repair and maintenance costs and minimizes turn-around times so that aircraft grounded for inspection and repairs can be returned to service very quickly.

The design process applied to a specific geometry is shown schematically in Figure 5.2.

Given the applied loads and the available space (the shaded area on the left of Figure 5.2), the structural analyst/designer has to come up with a shape that fits the given space, provides hard points for load applications and attachments and includes cutouts for any equipment that passes through. Some of the cutouts may be included as 'lightening holes' to reduce weight. In addition, the designer uses local reinforcements, doublers or flanges, around the cutouts and the attachment points for better load transfer across the part and for increased stability. The geometry (thicknesses, widths, heights, etc.) is selected so that the weight is minimized (for most cases minimum weight is one of the desirable attributes). This results in the structure shown on the right of the figure.

In terms of the sequence of steps and decision flow, the design process can be summarized in the chart of Figure 5.3. The analyst obtains the applied loads and local design requirements and uses the available materials to select the preliminary design. The use of simple analysis methods and experience (if available) with similar parts in the past firms up the geometry and this becomes the structural configuration. The structural configuration is a combination of geometry, material and fabrication process. Typically, at this point the requirements are met or are close to being met (e.g. the applied loads may cause failure, but the reserve factor is close to 1). A series of iterations, as shown by the loops in Figure 5.3, follows in order to fine-tune the design. They consist of more detailed analysis to minimize the weight (or meet other desirable attributes) without failing under the applied loads and, if needed, fabrication or producibility trials to verify that the design is manufacturable at an acceptable cost. Tests

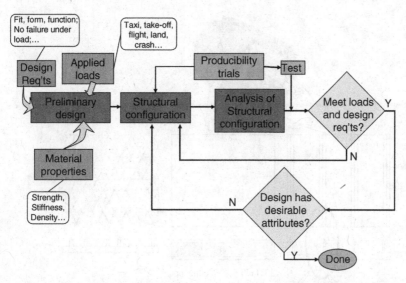

Figure 5.3   Simplified flow diagram of the design/structural analysis process

may also be used to verify the analysis predictions and to check that there are no issues that were not satisfactorily addressed by the design process.

At this point, it is worthwhile to go through some order of magnitude calculations to see what this process implies in terms of time required. A typical aircraft is designed for a series of flight manoeuvres (takeoff, climb, turn, approach, land, etc.) and taxi manoeuvres. These are done for a variety of speeds, accelerations, load factors, etc. and correspond to a large number of static and fatigue load combinations for which each part of the aircraft must be designed. In addition, there are crash load cases that must be included in the design of each part (or at least the parts that see substantial loads during a crash). Nowadays, with the advanced simulation software available and our improved understanding of structural behaviour during different manoeuvres, the total number of load cases (static and fatigue) that have to be analysed is of the order of 1000.

Assume now that the structural analyst has, on average, three design concepts to consider for each part to be designed. For example, for a skin structure, the three design concepts can be: (a) stiffened panel; (b) sandwich panel; and (c) isogrid panel as shown in Figure 5.4a. Also assume that for each of the three design concepts there are, on the average, three fabrication processes/material combinations. For the case of the stiffened panel for example, these could be co-cured, fastened or bonded as shown in Figure 5.4b. Note that all these options, so far, assume one choice of lay-up for each of the components.

In order to determine the optimum solution, that is the solution that meets the design requirements and optimizes the desired attribute(s) such as weight, cost, etc., a certain optimization algorithm must be used. Genetic algorithms are one of the most effective optimization schemes because they are very efficient in dealing with multiple optima and discontinuous variables such as the laminate thickness [1]. A genetic algorithm optimization scheme works by generating a certain number of designs during each iteration (or generation) and evaluating each design against the constraints and objective function. This evaluation implies detailed analysis

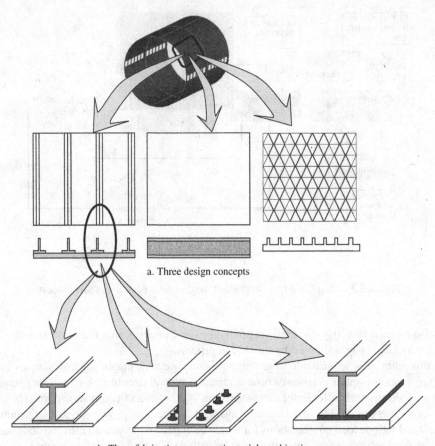

a. Three design concepts

b. Three fabrication processes/material combinations

Figure 5.4   Options to be considered during design/analysis of a part (See Plate 13 for the colour figure)

of each design to determine if it meets the applied loads. The worst performers are eliminated from the design pool while the best performers are recombined to create new designs for the next generation. Typically, to converge to an optimum solution a genetic algorithm will need approximately 1000 iterations (or generations) with approximately 15 designs per generation. With these assumptions, the number of analyses needed to optimize a single part is of the order of:

1000 (load cases or manoeuvres) × 3 design concepts × 3 process/material
combinations per concept × 1000 generations × 15 designs analyzed per generation
= 135 million analyses!

Of course there are shortcuts one can use by eliminating less critical load cases for example, but if one considers: (a) additional analyses that are needed for convergence checks if the finite element method is used for each analysis; (b) load redistribution runs to account for the fact that as the design changes load transfer through the part and around it changes and thus the applied loads change; and (c) applied load changes (mostly increases) that invariably occur during the design effort, the above estimate of 135 million analyses is probably representative of what would be needed.

Clearly, this number of analyses for each part to be designed is prohibitive if the analysis method is time-consuming such as the finite element or finite difference method. For example, to finish 135 million analyses in 1 year working for 365 days 24 h/day, one would have to complete more than four analyses per second. In practice, the number of analyses is reduced by reducing the design concepts and process/material combinations per part and limiting the number of parts to be optimized to a subset of the entire structure. But this, in turn, means that the structure is by necessity suboptimal since not all options are considered nor are all parts optimized. Even with those shortcuts, the number of finite element or finite difference analyses required is still prohibitive.

Therefore, until the computation time for the more accurate analysis methods such as finite elements improves by at least a factor of 50, extensive optimization of large quantities of parts or assemblies is not economically feasible. For this reason, simpler, reasonably accurate and much faster methods of analysis are necessary. In the following chapters, some of these simpler analysis methods are presented. In general, they lend themselves to automation and can be combined with efficient optimization schemes to optimize large quantities of parts. However, it is important to note and it will be stressed time and again throughout this text, that, in order to simplify the analysis, approximations have to be made which lead to results that are not as accurate as more detailed methods would generate and do not apply to all cases. If used judiciously, they can help hone in on the final design (or close to it) in terms of finalizing process, material, design concept and most of the geometry so that more detailed (and more time consuming) methods need only be used once (or few times) per part to firm up the final design.

5.1.2 Basic Design Concept and Process/Material Considerations for Aircraft Parts

This section gives some of the top-level alternatives a designer/analyst has to consider when designing a composite part. This is summarized in Table 5.2 and is by no means an exhaustive discussion, but helps in understanding the process one has to go through before even analyzing a part. The relation of some of these decisions to the analysis methodology that must be done is highlighted. The different types of analyses are only mentioned here and presented in detail in subsequent chapters. The types of parts discussed here follow the listing of Table 5.1.

As shown in Table 5.2, the different options for design concepts for each type of part make it difficult to know a priori the optimum configuration for each application and type of loading. At times, a compromise is necessary in order to better blend the structure to be designed with adjacent structure where some geometry, for example the stiffener spacing, is fixed. Also, knowledge of the fabrication options and corresponding process capabilities is necessary in order to fully exploit the potential of a design concept. For example, maintaining fibre continuity in all three directions in a three-dimensional structure such as a fitting may not be possible, thereby creating interfaces where only resin is available to carry loads if fasteners are not used. This is the case in Figure 5.5 where there are no fibres across planes a–a and b–b.

Table 5.2 Design considerations, alternatives and implications for analysis

Part	Configuration	Alternatives to be considered	Implications for analysis
Skins	Monolithic with stiffeners	• Stiffeners co-cured, fastened, secondarily bonded? • Cutouts moulded in or cut afterwards? Reinforced with doublers or flanged? Reinforcement co-cured, fastened or secondarily bonded?	Material strength Notched strength[a] Buckling Delamination
	Sandwich	• Full-depth core everywhere or with rampdown for attachments? • Assembly via co-curing, fastened or secondarily bonded?	Material strength (facesheet, core, adhesive), notched strength[a] Buckling Wrinkling (symmetric, antisymmetric) Shear crimping Intracellular buckling Delamination, disbond
Stringers, Stiffeners, panel breakers	Cross-sectional shape: L, C, Z, T, I, J, Hat	• Confine buckling pattern between stiffeners (panel breakers)?	Material strength Notched strength[a] Column buckling Crippling Skin/stiffener separation Inter-rivet buckling
Frames and bulkheads		• Co-cured with skin, fastened or secondarily bonded? • Single piece or multiple pieces? • Cutouts flanged or with doublers? • Cutouts moulded-in or cut afterwards?	Material strength Notched strength[a] Buckling of webs Crippling of stiffeners or caps Crippling of reinforcements

Component	Questions	Failure modes
Beams, spars, ribs, intercostals	• Co-cured with skin, fastened or secondarily bonded? • Single piece or multiple pieces? • Cutouts flanged or with doublers? • Cutouts moulded-in or cut afterwards?	Material strength, notched strength[a] Web buckling Crippling of flanges Crippling of reinforcements around cutouts
Fittings	• How to mould a 3-D piece with continuous fibres in all directions?	Material strength, notched strength[a] Lug failure[b] Bearing failure Delamination
Decks and floors	• Stiffened, grid-stiffened or sandwich?	Material strength, notched strength[a] Stiffened panel failure modes Sandwich failure modes
Doors and fairings	• Stiffened or sandwich? • How does compound curvature change fibre orientation locally?	Material strength, notched strength[a] Stiffened panel failure modes Sandwich failure modes

[a]Notched strength: OHT = open hole tension; OHC = open hole compression; TAI = tension after impact; CAI = compression after impact; SAI = shear after impact.

[b]Net section failure, shear-out failure, bearing failure: see also Chapter 11.2.2.

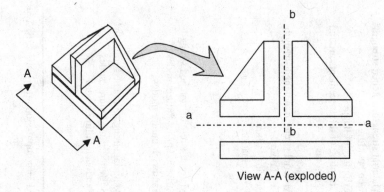

View A-A (exploded)

Figure 5.5 Schematic of a three-dimensional connection of parts without fibre continuity in all primary load directions

As summarized in Table 5.2, apart from the basic strength and notched strength failure modes, different parts may have different failure modes that must be analysed separately. It is important to note that if these failure modes are not anticipated in advance, they cannot be picked up by analysis methods that are not set up to accurately capture them. For example, a finite element method may not pick up failure of a lug if the mesh is not fine enough at the three different locations where net tension, shear-out or bearing failure may occur (Figure 5.6). Or, without proper mesh size and boundary conditions, long-wave (global) and short-wave (e.g. crippling or wrinkling) buckling modes cannot be accurately quantified.

5.1.3   Sources of Uncertainty: Applied Loads, Usage and Material Scatter

It should be recognized that in any large-scale design problem, such as that of an airframe, there are sources of uncertainty. As a result, several input quantities in the design process are not accurately known and the design/analysis process must take these uncertainties into account to make sure that the worst case scenario, however improbable, if it were to occur, would not lead to failure. The three most important sources of uncertainty are: (1) knowledge of applied loads; (2) variability in usage; and (3) material scatter. These are examined briefly below.

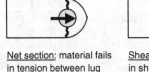

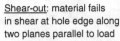

Net section: material fails   Shear-out: material fails   Bearing: Lug hole
in tension between lug   in shear at hole edge along   elongates and material
hole and edge of part   two planes parallel to load   fails in bearing/compr.
                                                                       ahead of hole

Figure 5.6   Three of the failure modes in a lug

5.1.3.1 Knowledge of Applied Loads

As mentioned in Section 5.1.1 the structure of an aircraft must be designed for a large variety of manoeuvres. For each of those manoeuvres, the externally applied loads (e.g. aerodynamic loads) must be known accurately. However, it is difficult to determine exactly these applied loads because of the complexity of the phenomena involved (e.g. flow separation), the complexity of the structure (e.g. wing–fuselage interaction) and the limitations in computational power available. Typically, some approximations are necessary in the computer simulation and it is not uncommon to introduce safety factors to provide a degree of conservatism in determining the applied loads. This is one of the reasons for the use of the 1.5 multiplicative factor between limit and ultimate load. The structure is not designed to the highest expected load during its service life, which is called the limit load, but to that load multiplied by a safety factor of 1.5, which is the ultimate load.

5.1.3.2 Variability in Usage

Even if the applied loads were accurately known for a certain manoeuvre, there is uncertainty in practice in performing the manoeuvre. Nominally the same manoeuvre (e.g. $3g$ turn) will have differences in the transient loads exerted on the aircraft from one operator to the next. For this reason, each manoeuvre is simulated many times while varying different parameters (rate of control action for example) but staying within the parameters defining the manoeuvre and the peak load(s) calculated during the simulation are recorded. Then, the loads corresponding to this manoeuvre are selected so as to cover most loads recorded (for example, the 95th percentile may be selected). This process is shown schematically in Figure 5.7.

Depending on the manoeuvre, there are, in general more than one load that may be of interest, corresponding to different load types or load directions and different times during the manoeuvre, such as maximum power, maximum or minimum control stick input, etc. The situation shown in Figure 5.7 is simplified in that it isolates one load type and shows one maximum load of interest, the peak load recorded during the simulation. Each of the peak loads can be plotted in a frequency plot as shown at the bottom of the figure. Standard statistical methods are then used to determine the percentile of interest. Note that the statistical distribution of the peak load is not necessarily a normal distribution and the one shown in Figure 5.7 is just an example.

5.1.3.3 Material Scatter

The strength of the material used in fabricating a specific design is not a single well-defined number. Inherent variability in the microstructure of the material, material variability from one material batch to another, fluctuations in the fabrication method (e.g. curing cycle), variations in geometry within tolerances (e.g. thickness variation within the same specimen) lead to a range of strength values when the same nominal geometry and lay-up are tested. This variability is shown for typical unidirectional graphite/epoxy in tension and compression in Figure 5.8.

A design must account for this variation and protect against situations where the strength of the material used may be at the low end of the corresponding statistical strength distribution.

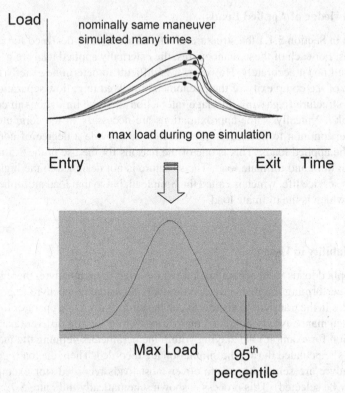

Figure 5.7 Selection of applied load (95th percentile used here as an example) to be used in designing for a specific manoeuvre

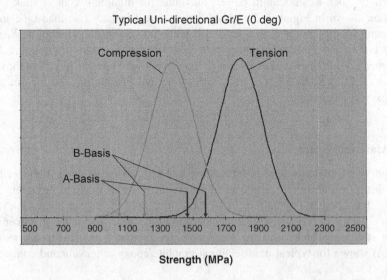

Figure 5.8 Typical ranges for tension and compression strength values for 0° unidirectional graphite/epoxy

For this purpose, specific statistically meaningful values are selected that are guaranteed to be lower than most of the strength population. The two most commonly used values are the B-Basis and the A-Basis strength values [2]. The A-Basis value is the one percentile of the population: 99% of the tests performed will have strength greater than or equal to the A-Basis value. The B-Basis value is the tenth percentile of the population: 90% of the tests performed will have strength greater than or equal to the B-Basis value.

In general, the A-Basis value is used with single load path primary structure, where failure may lead to loss of structural integrity of a component. The B-Basis is used with secondary structure or structure with multiple load paths, where loss of one load path does not lead to loss of structural integrity of the component. The A- and B-Basis values are calculated on the basis of statistical methods accounting for batch-to-batch variation, the type of statistical strength distribution and the number of data points [2, 3].

Stiffness has a similar variability to that of strength. However, one should be careful in using low percentile values for stiffness because they may not represent a conservative scenario. If the material used in a structure has stiffness at the low end of the stiffness statistical distribution, this means that the surrounding structure, being stiffer, will absorb more load. This would require appropriate adjustment of applied loads and it opens up a series of scenarios that may or may not be realistic. Instead, using the average or mean stiffness everywhere in the structure would not unduly transfer load from one part to its neighbours and is more representative. So stiffness-sensitive calculations such as buckling do not, usually, require the lowest stiffness values (B- or A-Basis) but the mean values.

5.1.4 Environmental Effects

Composites are susceptible to environmental effects. In general, as the temperature and/or the moisture content increase beyond room temperature ambient (RTA) conditions, the strength and stiffness properties degrade. Also, at temperatures lower than room temperature, most strength properties are also lower than at room temperature. An example for a typical graphite/epoxy material is shown in Figure 5.9.

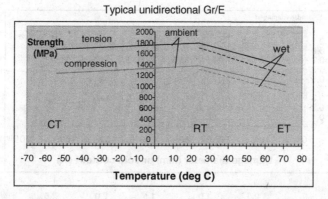

Figure 5.9 Variation of tension and compression strength as a function of temperature and moisture content

Two sets of curves are shown in Figure 5.9. The continuous lines correspond to the 'dry' or ambient condition. The exposure of the specimens to moisture has been minimal. The dashed lines correspond to the wet condition where the specimens are fully saturated with moisture. Increasing the moisture level decreases the strength. Increasing the temperature beyond room temperature decreases the strength. Depending on the property and the material, decreasing the temperature below room temperature may increase or decrease the strength. Typically, a decrease is observed, as shown in Figure 5.9.

Given the type of behaviour shown in Figure 5.9, complete characterization of a material would require knowledge of its properties over the entire range of anticipated temperature and moisture environments during service. In general this is accomplished by doing tests at representative conditions that identify extreme points of the trends and interpolating in between. The number of such key conditions depends on material loading and application (e.g. civilian versus military application). The minimum number is three. These are: (1) the cold temperature (CT in Figure 5.9) condition (usually CTA for cold temperature ambient); (2) the room temperature (RT) condition which is split into RTA, the room temperature ambient, and RTW, the room temperature wet condition where the specimens are fully saturated; and (3) the elevated temperature (ET) condition which is split into elevated temperature ambient (ETA) and elevated temperature wet (ETW) condition.

For design purposes, the most conservative strength properties across all conditions are used. It is important to keep in mind, however, that when trying to match specific test results, the properties corresponding to the test environment and material condition at the time during test should be used. Stiffness also shows a similar sensitivity to environment and it is customary to perform preliminary design using the lowest stiffness across environments.

5.1.5 Effect of Damage

Composites exhibit notch sensitivity. A notch can be any form of damage, such as impact or crack or cutout. The strength in the presence of damage is significantly lower and varies with the damage size and type. Typical trends of compression strength in the presence of damage are shown in Figure 5.10, adjusted from reference [4].

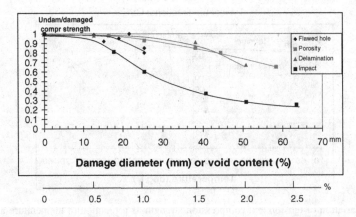

Figure 5.10 Sensitivity of compression strength of composites to various types of damage

While the trends shown in Figure 5.10 are representative of the behaviour of most composites, the specific values shown are only applicable to specific lay-ups and materials. For other materials and/or lay-ups, specific analysis supported by tests must be carried out for each type of damage in order to accurately quantify the residual strength in the presence of damage.

The types of damage shown in Figure 5.10 are the most common types encountered in practice. Of those, the most critical is impact damage. Impact damage is caused by a large variety of sources, ranging from tool drops and foot traffic to impact with large objects (e.g. luggage) and hail damage.

This strength reduction in the presence of damage must be taken into account in the design process. The approach is dependent on the inspection method used and its reliability. First, the type and size of damage or flaw that the chosen inspection method can find consistently and reliably must be determined. Then, the threshold of detectability is defined as the damage size above which all damage can be found by the inspection method with a certain confidence (e.g. 99% of the time). This threshold of detectability divides the damage that may occur during manufacturing or service in two categories: (a) nondetectable damage and (b) detectable damage. These are then tied to specific load levels the structure must withstand. As already mentioned in Section 5.1.3 the two main load levels of interest in structural design of airframe structures are the limit and ultimate load. The limit load is the highest load the structure is ever expected to encounter during service. The ultimate load is the limit load multiplied by 1.5.

A structure with damage below the threshold of detectability of the selected inspection method must be capable of withstanding ultimate load without failure. A structure with damage above the threshold of detectability level of the selected inspection method must be capable of withstanding limit load without failure.

In practice, the most common inspection method used is visual inspection. This is because it combines low cost with ease of implementation. This does not mean that more accurate and more reliable inspection techniques such as ultrasound, X-rays, etc. are not used at different times in the life of an aircraft. Usually, however, these methods are applied during planned detailed inspections at the depot level where an aircraft is taken out of service and specially trained personnel with appropriate equipment conduct a thorough inspection of the structure. On a more regular basis, the structure is inspected visually.

With visual inspection the preferred method of inspection during service today, the structural requirements in the presence of damage become:

- Structure with damage up to barely visible impact damage (BVID) must withstand ultimate load without failure
- Structure with damage greater than BVID, that is structure with visible damage (VD) must withstand limit load without failure

The VD is usually defined as damage that is clearly visible from a distance of 1.5 m under ambient light conditions. Then BVID is damage just below the VD. It is recognized that the definition of BVID is subjective and dependent on the inspector and his/her experience level. For this reason attempts to more accurately define BVID have been made by tying the BVID to a specific indentation size. Usually, 1 mm deep indentation is considered to correspond to BVID.

It should be emphasized that besides limit and ultimate load, other load levels may be used in practice, albeit less frequently. One example is the 'safe return to base load', which is usually a fraction of the limit load (typically 80%) and limits the structure to loads that will not cause catastrophic failure in the presence of larger damage levels such as those caused by bird or lightning strike, etc.

In view of the experimentally measured strength reductions shown in Figure 5.10, the design/analysis process must use analytical methods that allow determination of the reduced strength in the presence of various types of damage. Usually, a conservative approach is selected and the structure is designed for the worst type of damage (impact) since this will cover all other cases. Due to the complexity of the analysis for determining the amount of damage caused by a specific threat and the subsequent complexity of the analysis for determining the strength of the structure in the presence of damage, simplified methods are commonly used for preliminary design [5–11].

A conservative approach is usually followed that avoids computationally intensive analysis methods that model damage creation and its evolution under load. The method consists of designing the structure to meet (a) limit load in the presence of a 6 mm diameter hole (VD) and (b) ultimate load in the presence of low-speed impact damage (BVID). It is important to note, however, that this approach has its limitations because it is not applicable to all threat scenarios. For example, it can be extremely conservative in cases of thick composite structures. The typical damage scenarios based on common threats during manufacturing and service should not include a 6 mm through hole for example because it is a very unlikely event. Designing for damage must be done with care on a case-by-case basis after careful examination of threats and requirements. And, most importantly, it should be supported by tests that verify the analysis method and its applicability to the loading, lay-ups and configurations under consideration.

5.1.6   Design Values and Allowables

The discussion in Sections 5.1.3–5.1.5, 5.2.4 indicated that the strength of a composite structure covers a range of values as a result of material variability, environmental effects and sensitivity to damage. As a result, the strength value used in a design must be such that if the 'worst of all situations' is combined in service, the resulting structure will still meet the load requirements without failure. The 'worst of all situations' combines material at the low end of the strength distribution (Figure 5.8) operating at the worst environment (Figure 5.9) with the worst type of damage present (Figure 5.10). Therefore, sufficiently conservative strength values must be used. A procedure that leads to such design values for strength is shown schematically in Figure 5.11.

The mean RTA strength at the far right of the figure is reduced by a 'knockdown' factor representing the worst environment for the loading and material selected. This is further reduced by another factor that represents the worst type of damage (usually impact damage). This value is treated as the mean with the effect of damage and environment already included. Around this mean value the statistical distribution representing the material scatter for the property in question (tension, compression, shear, etc.) is created. The design value is determined as a value to the left end of the statistical distribution (e.g. A- or B-Basis value as described in Section 5.1.3), which is expected to be lower than a certain high percentage (90% for B-Basis and 99% for A-Basis) of all test results for the property of interest at the most degrading environment and with the highest permissible amount of damage.

This approach can be done rigorously by determining the worst type of damage, which usually is BVID and the worst type of environment, which, usually, is ETW for the strength

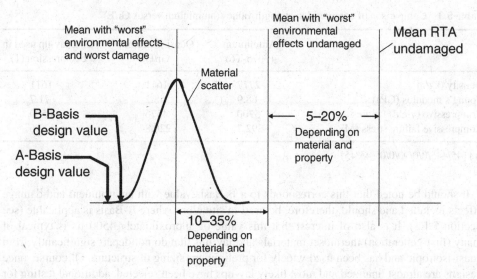

Mean with "worst"
environmental effects
and worst damage

Mean with "worst"
environmental
effects undamaged

Mean RTA
undamaged

Material
scatter

B-Basis
design value

5–20%

Depending on
material and
property

A-Basis
design value

10–35%

Depending on
material and
property

Figure 5.11 Determination of design strength values accounting for effects of damage, environment and material scatter

property of interest. Then, a sufficient number of specimens with this damage are conditioned in that environment and strength tests are carried out. The number of specimens is selected so expected batch-to-batch variation is reproduced and the results give sufficient statistical confidence in the conclusions. Statistical analysis methods [3] are then used to determine the design values which, when they are statistically significant values, are called allowables.

To save time and reduce cost during preliminary design is customary to test a limited number of specimens at various environments to obtain a percentage reduction between mean strengths at various environments. This is done with and without damage to determine the reduction due to the presence of damage at different environments. Finally, a sufficient number of tests (six per batch) from at least three batches in one of the environments give the material scatter. This gives design values that can be used in the preliminary design.

An example follows. Assume that the compression failure strain for undamaged quasi-isotropic lay-up of a material at RTA environment is 11,000 microstrain (µs). And tests have shown that the environment with the biggest reduction in strength is ETW with a mean failure strain equal to 80% of the RTA mean strain. Also, tests at RTA have shown that the mean failure strain with BVID is 65% of the mean RTA strain. Finally, tests of undamaged specimens at RTA have shown a B-Basis value that is 80% of the mean RTA value (this corresponds to a normal distribution with coefficient of variation, i.e. standard deviation divided by the mean, of about 11%). Following the procedure described above and shown in Figure 5.11, a design value that can be used for preliminary design is

$$\varepsilon_{des} = 11000 \times \underset{\substack{\text{worst} \\ \text{envir.} \\ \text{(ETW)}}}{0.8} \times \underset{\substack{\text{BVID} \\ \text{effect}}}{0.65} \times \underset{\substack{\text{mat'l} \\ \text{scatter} \\ \text{effect}}}{0.8} = 4576\,\mu s \tag{5.1}$$

mean
RTA

Table 5.3 Comparison of compression strength values (aluminium versus Gr/E)

	Aluminium (7075-T6)	Quasi-isotropic Gr/E	Gr/E lay-up used in compression (1)
Density (kg/m^3)	2777	1611	1611
Young's modulus (GPa)	68.9	48.2	71.7
Compressive (yield) failure strain (μs)	5700	4576	~4500
Compressive failure stress (MPa)	392.7	220.8	~322.6

(1) [45/–45/0/0/90/0/0/–45/45]

It should be noted that this corresponds to a B-Basis value with environment and damage effects included and should, therefore, be used in situations where B-Basis is applicable (see Section 5.1.3). It is also of interest that this value of approximately 4500 μs is typical of many (first-generation) thermoset materials and lay-ups that do not depart significantly from quasi-isotropic and has been used widely for preliminary sizing of structure. Of course, once designs are almost finalized and most likely lay-ups have been selected, additional testing for spot-checking the validity of this value and more rigorous statistical analysis are necessary to verify or update this value.

Equation (5.1) demonstrates that the design values for composite materials may be less than half the RTA mean undamaged values. This is an important consideration in anticipating the weight savings that can result from the use of composites because the significant reduction in strength offsets a lot of the weight savings one would expect on the basis of density difference alone. A simple comparison between aluminium and graphite/epoxy (Gr/E) composites is shown in Table 5.3. This is a simple comparison to show the relative differences. For aluminium the yield strain is used as the failure strain. The failure stress (yield stress in the case of aluminium) is approximated in Table 5.3 as the product of the Young's modulus times the failure strain. The failure strain and stress for the second Gr/E lay-up (last column) are approximate.

It can be seen from Table 5.3 that the strength of aluminium can be significantly higher than that of Gr/E. This means that in order to carry the same load with Gr/E, as with aluminium, one has to use higher thickness. For a plate-type application, the weight is calculated as

$$W = \rho t \, (\text{Area})$$

where ρ is the density, t, the thickness and Area, the planform area of the plate.

If the structure fails exactly when the required applied load is reached, the thickness needed is calculated from

$$\sigma_{\text{fail}} = \frac{F_a}{wt} \Rightarrow t = \frac{F_a}{w\sigma_{\text{fail}}}$$

where F_a is the applied load, σ_{fail} is the failure strength of the material and w is the width of the cross-section over which F_a acts.

Using this expression for the thickness t to substitute in the weight expression,

$$W = \rho \frac{F_a}{w\sigma_{\text{fail}}} (\text{Area})$$

Table 5.4   Weight comparison for plate application based on strength

	Quasi-isotropic design/Al	$[45/-45/0_2/\overline{90}]_s/Al$
$\dfrac{W_{Gr}}{W_{Al}}$	1.03	0.706

Note: The overbar over the 90° ply in the stacking sequence in the third column denotes a ply that does not repeat symmetrically with respect to mid-plane, that is 90° is the mid-ply in this case.

Then, the ratio of the weight W_{Gr} of a graphite/epoxy panel to that of aluminium W_{Al} for the same applied load F_a and the same planform area, can be found after some rearranging to be:

$$\frac{W_{Gr}}{W_{Al}} = \frac{\left(\dfrac{\rho}{\sigma_{fail}}\right)_{Gr}}{\left(\dfrac{\rho}{\sigma_{fail}}\right)_{Al}}$$

Using the values from Table 5.3 gives the weight ratios in Table 5.4.

As can be seen from Table 5.4, the quasi-isotropic composite design (column 2) is approximately the same weight as the aluminium counterpart, in fact it is 3% heavier. The more tailored $[45/-45/0_2/\overline{90}]_s$ lay-up is approximately 30% lighter than the aluminium counterpart. This case serves as an example that shows the advantages of tailoring designs for the maximum use of composite capabilities instead of using quasi-isotropic lay-ups that lead to the so-called 'black aluminium' designs.

5.1.7   Additional Considerations of the Design Process

The analysis methods used are always a tradeoff between accuracy and cost and ease of use. In a preliminary design stage where many candidate designs must be traded quickly, especially if formal optimization is introduced early in the design process, using conservative 'reasonably accurate' methods is preferred over very accurate computationally intensive approaches. This allows the examination of many more options that would not be possible with more detailed methods. The term 'reasonably accurate' is, of course, subjective and, usually, is tied to how conservative one can afford to be before the design weight starts increasing beyond acceptable levels. Often, approximate analytical methods are modified based on test results and adjusted accordingly to give accurate predictions over a limited range of applicability. In addition, test methods are often used to circumvent problems with analytical modelling of structural details present in the structure, the detailed modelling of which would make the entire analysis very expensive. Two such examples are: (a) modelling of fasteners in bolted structures; and (b) knowing the exact type of boundary conditions provided by the edge supports or intermediate structure present.

Typical airframe structures have a large variety of failure modes. Which failure mode starts failure and which one eventually leads to catastrophic failure of the structure is a function of the material, lay-up and geometry used. Changing any of these can alter the failure mode scenario. One example is the sandwich structure where each of the skins may fail in:(1) material strength, (2) wrinkling (symmetric or antisymmetric), (3) dimpling or intracellular buckling, (4) shear crimping (precipitated by core failure). In addition, the adhesive connecting core and facesheets may fail in (5) adhesive strength (tension, compression or shear) or the core itself may fail in (6) core strength (tension, compression or shear) and finally the entire sandwich may fail in (7) sandwich buckling. And these do not include additional failure modes specific to sandwich rampdown if there is one present. See Chapter 10 for a detailed discussion of sandwich structures.

In general, a priori knowledge of the possible failure modes is necessary for a good design. Different failure modes may interact which makes their analytical simulation without the use of extensive very detailed analysis tools, such as finite elements, very difficult. This is a case where tests are used to adjust the simpler analysis methods or suggest how the existing methods must be modified to more accurately match test results.

The most efficient design is the one that just fails when the applied design load (ultimate or limit depending on the requirement) is reached. Trying to implement this during preliminary design may not be advisable since the analysis methods may not be sufficiently accurate, test results with allowables may not be completed, loads may increase, etc. So there may be a difference between the failure load of the design and the applied load. The relative magnitudes of failure load and applied loads are related through the loading index, the reserve factor or the margin of safety. All three refer to the same thing in a slightly different way. The loading index is the ratio of the applied load to the failure load. If less than one, there is no failure. The reserve factor is the inverse of the loading index and equals the ratio of the failure load to the applied load. If greater than one it implies the structure does not fail and the applied load must be increased by a factor equal to the reserve factor for failure to occur. Finally, the margin of safety is the reserve factor minus one. Expressed in percent, if it is positive it implies no failure and denotes by what percentage the applied load must be increased to cause failure. If negative, it implies failure and denotes by what percentage the applied load must be decreased to prevent failure. It is customary to maintain positive (but not very high) margins of safety during preliminary design and, later on, as the design is finalized, detailed analysis supported by testing increases confidence in the design and the applied loads are 'frozen', can be driven as close to zero as possible by fine-tuning the design to increase its efficiency.

5.2   Governing Equations

The starting point is the governing equations for a composite plate. These are: (a) the equilibrium equations; (b) the stress–strain equations; and (c) the strain–displacement equations. Versions of the stress–strain and strain–displacement equations have been used already in Sections 3.2 and 3.3. The reader is referred to the literature for the detailed derivation of these equations [12–14]. Only the final form of these equations is given here.

5.2.1   Equilibrium Equations

With reference to Figure 5.12, the equilibrium equations (no body forces) have the form,

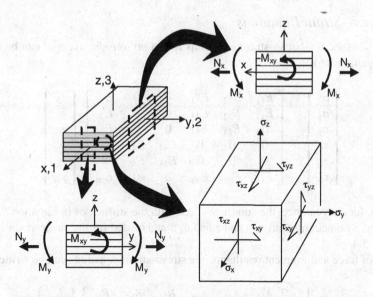

Figure 5.12   Coordinate system and force and moment sign convention

$$\frac{\partial \sigma_x}{\partial x} + \frac{\partial \tau_{xy}}{\partial y} + \frac{\partial \tau_{xz}}{\partial z} = 0$$

$$\frac{\partial \tau_{xy}}{\partial x} + \frac{\partial \sigma_y}{\partial y} + \frac{\partial \tau_{yz}}{\partial z} = 0 \qquad (5.2)$$

$$\frac{\partial \tau_{xz}}{\partial x} + \frac{\partial \tau_{yz}}{\partial y} + \frac{\partial \sigma_z}{\partial z} = 0$$

or in terms of force and moment resultants,

$$\frac{\partial N_x}{\partial x} + \frac{\partial N_{xy}}{\partial y} = 0$$

$$\frac{\partial N_{xy}}{\partial x} + \frac{\partial N_y}{\partial y} = 0 \qquad (5.3a–c)$$

$$\frac{\partial Q_x}{\partial x} + \frac{\partial Q_y}{\partial y} = 0$$

with

$$Q_x = \frac{\partial M_x}{\partial x} + \frac{\partial M_{xy}}{\partial y}$$

$$Q_y = \frac{\partial M_{xy}}{\partial x} + \frac{\partial M_y}{\partial y} \qquad (5.3d–e)$$

5.2.2   Stress–Strain Equations

In terms of stresses, the stress–strain equations for an orthotropic material can be written as (see also Equation 3.5),

$$
\begin{Bmatrix} \sigma_x \\ \sigma_y \\ \sigma_z \\ \tau_{yz} \\ \tau_{xz} \\ \tau_{xy} \end{Bmatrix} =
\begin{bmatrix}
E_{11} & E_{12} & E_{13} & 0 & 0 & E_{16} \\
E_{12} & E_{22} & E_{23} & 0 & 0 & E_{26} \\
E_{13} & E_{23} & E_{33} & 0 & 0 & E_{36} \\
0 & 0 & 0 & E_{44} & E_{45} & 0 \\
0 & 0 & 0 & E_{45} & E_{55} & 0 \\
E_{16} & E_{26} & E_{36} & 0 & 0 & E_{66}
\end{bmatrix}
\begin{Bmatrix} \varepsilon_x \\ \varepsilon_y \\ \varepsilon_z \\ \gamma_{yz} \\ \gamma_{xz} \\ \gamma_{xy} \end{Bmatrix}
\tag{5.4}
$$

Note that, for convenience, the subscripts used with the stiffnesses in Equation 5.4 are 1–6 with 1, 2 and 3 coinciding with x, y and z and 4, 5 and 6 used for the shear moduli as shown in Equation 5.4.

In terms of force and moment resultants, the stress–strain equations can be written as,

$$
\begin{Bmatrix} N_x \\ N_y \\ N_{xy} \\ M_x \\ M_y \\ M_{xy} \end{Bmatrix} =
\begin{bmatrix}
A_{11} & A_{12} & A_{16} & B_{11} & B_{12} & B_{16} \\
A_{12} & A_{22} & A_{26} & B_{12} & B_{22} & B_{26} \\
A_{16} & A_{26} & A_{66} & B_{16} & B_{26} & B_{66} \\
B_{11} & B_{12} & B_{16} & D_{11} & D_{12} & D_{16} \\
B_{12} & B_{22} & B_{26} & D_{12} & D_{22} & D_{26} \\
B_{16} & B_{26} & B_{66} & D_{16} & D_{26} & D_{66}
\end{bmatrix}
\begin{Bmatrix} \varepsilon_{xo} \\ \varepsilon_{yo} \\ \gamma_{xyo} \\ \kappa_x \\ \kappa_y \\ \kappa_{xy} \end{Bmatrix}
\tag{5.5}
$$

where A_{ij} are the elements of the membrane stiffness matrix for a laminate, B_{ij} are the elements of the membrane–bending coupling matrix for a laminate and D_{ij} are the elements of the bending stiffness matrix of the laminate (see also Chapter 3).

The vector multiplying the stiffness matrix in the right-hand side of Equation (5.5) consists of the mid-plane strains and curvatures of the laminate. The curvatures κ_x, κ_y and κ_{xy} are given by

$$
\kappa_x = -\frac{\partial^2 w}{\partial x^2}
$$

$$
\kappa_y = -\frac{\partial^2 w}{\partial y^2}
\tag{5.6}
$$

$$
\kappa_{xy} = -2\frac{\partial^2 w}{\partial x \partial y}
$$

The strains at any through-the-thickness location of a laminate are obtained assuming the standard linear variation with the out-of-plane coordinate z as

$$
\begin{aligned}
\varepsilon_x &= \varepsilon_{xo} + z\kappa_x \\
\varepsilon_y &= \varepsilon_{yo} + z\kappa_y \\
\gamma_{xy} &= \gamma_{xyo} + z\kappa_{xy}
\end{aligned}
\tag{5.7}
$$

5.2.3   Strain–Displacement Equations

For small displacements and rotations, the equations relating mid-plane strains to displacements are,

$$\varepsilon_{xo} = \frac{\partial u}{\partial x}$$

$$\varepsilon_{yo} = \frac{\partial v}{\partial y} \tag{5.8}$$

$$\gamma_{xyo} = \frac{\partial u}{\partial y} + \frac{\partial v}{\partial x}$$

Similarly, the out-of-plane strains are given by,

$$\varepsilon_z = \frac{\partial w}{\partial z}$$

$$\gamma_{yz} = \frac{\partial v}{\partial z} + \frac{\partial w}{\partial y} \tag{5.9}$$

$$\gamma_{xz} = \frac{\partial u}{\partial z} + \frac{\partial w}{\partial x}$$

Since the three mid-plane strains in Equation (5.8) are expressed in terms of only two displacements, a strain compatibility condition can be derived by eliminating the displacements from Equations (5.8). Differentiate the first of Equation (5.8) twice with respect to y and the second of Equation (5.8) twice with respect to x. Finally differentiate the last of Equation (5.8) once with respect to x and once with respect to y. Combining the results leads to

$$\frac{\partial^2 \varepsilon_{xo}}{\partial y^2} + \frac{\partial^2 \varepsilon_{yo}}{\partial x^2} - \frac{\partial^2 \gamma_{xyo}}{\partial x \partial y} = 0 \tag{5.10}$$

Similarly, two more compatibility relations can be obtained by combining corresponding equations from Equations (5.8) and (5.9) or using cyclic symmetry:

$$\frac{\partial^2 \varepsilon_{yo}}{\partial z^2} + \frac{\partial^2 \varepsilon_z}{\partial y^2} - \frac{\partial^2 \gamma_{yz}}{\partial y \partial z} = 0 \tag{5.11}$$

$$\frac{\partial^2 \varepsilon_z}{\partial x^2} + \frac{\partial^2 \varepsilon_{xo}}{\partial z^2} - \frac{\partial^2 \gamma_{xz}}{\partial x \partial z} = 0 \tag{5.12}$$

Depending on which quantities are used as variables, Equations (5.2) to (5.12) form a system of equations in these unknown variables. For example, if stresses, strains and displacements are used as unknowns, Equations (5.2), (5.4), (5.8) and (5.9) form a system of 15 equations in the 15 unknowns: σ_x, σ_y, σ_z, τ_{yz}, τ_{xz}, τ_{xy}, ε_{xo}, ε_{yo}, ε_z, γ_{yz}, γ_{xz}, γ_{xyo}, u, v and w. Alternatively, for a plate problem, if forces, moments, strains and displacements are used, Equations (5.3), (5.5), (5.6) and (5.8) form a system of 17 equations in the 17 unknowns, N_x, N_y, N_{xy}, M_x, M_y, M_{xy}, Q_x, Q_y, ε_{xo}, ε_{yo}, γ_{xyo}, κ_x, κ_y, κ_{xy}, u, v and w.

These systems of equations can be reduced all the way to one equation in some cases, by eliminating appropriate variables according to the needs of specific problems. Some of these reductions will be shown in later chapters.

5.2.4 von Karman Anisotropic Plate Equations for Large Deflections

The case of large deflections merits special attention since they become important in some problems such as post-buckling of composite plates. The von Karman equations for large deflections are derived in this section. Consider the case of a plate undergoing large deflections with distributed loads p_x, p_y and p_z (units of force/area). The basic assumptions that: (a) the out-of-plane stress σ_z is negligible compared with the in-plane stresses; and (b) plane sections remain plane and normal to the mid-plane after deformation (leading to zero out of plane shear strains γ_{yz} and γ_{xz}) are still valid.

To keep the resulting equations relatively simple (and still covering a wide variety of applications) it is also assumed that: (a) the laminate is symmetric (coupling matrix $B = 0$); (b) the coupling terms D_{16} and D_{26} of the bending matrix D are zero; and (c) the laminate is balanced (shearing–stretching coupling terms $A_{16} = A_{26} = 0$). The deformed and undeformed states of a plate element dx in the xz plane is shown in Figure 5.13.

With reference to Figure 5.13 the coordinates of any point A' on the left edge of element dx, are given by:

$$A'_x = x_o + u - \zeta \frac{\partial w}{\partial x}$$

$$A'_z = w + \zeta$$

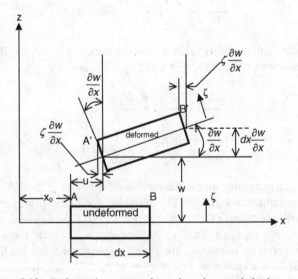

Figure 5.13   Deformation pattern for a plate element dx in the xz plane

The x coordinate of any point B' on the right edge of element dx is given by a Taylor series expansion (truncated after the second term) of the coordinate at the left end shifted by the length of the element dx:

$$B'_x = x_o + u - \zeta \frac{\partial w}{\partial x} + \frac{\partial}{\partial x}\left[x_o + u - \zeta \frac{\partial w}{\partial x}\right]dx + dx$$

$$= x_o + u - \zeta \frac{\partial w}{\partial x} + \frac{\partial}{\partial x}\left[u - \zeta \frac{\partial w}{\partial x}\right]dx + dx$$

The z coordinate of any point B' on the right edge of element dx is given by

$$B'_z = w + dx\frac{\partial w}{\partial x} + \zeta$$

The x and z components of the deformed element $A'B'$ are then given by

$$A'B'_x = B'_x - A'_x = dx + \frac{\partial}{\partial x}\left[u - \zeta \frac{\partial w}{\partial x}\right]dx$$

$$A'B'_z = B'_z - A'_z = \frac{\partial w}{\partial x}dx$$

Therefore, the length of the deformed element $A'B'$ is given by

$$A'B' = \sqrt{\left(A'B'_x\right)^2 + \left(A'B'_z\right)^2} = \sqrt{\left(dx + \left(\frac{\partial u}{\partial x} - \zeta \frac{\partial^2 w}{\partial x^2}\right)dx\right)^2 + \left(\frac{\partial w}{\partial x}\right)^2 dx^2}$$

$$= dx\sqrt{1 + \left(\frac{\partial u}{\partial x} - \zeta \frac{\partial^2 w}{\partial x^2}\right)^2 + 2\left(\frac{\partial u}{\partial x} - \zeta \frac{\partial^2 w}{\partial x^2}\right) + \left(\frac{\partial w}{\partial x}\right)^2}$$

The second term under the square root is small compared with the remaining terms and is neglected. The remaining expression is expanded using the binomial theorem:

$$(a + b)^r = a^r + ra^{r-1}b + \dots$$

and letting $a = 1$ in the expression for $A'B'$ and keeping only leading terms:

$$A'B' = \left\{1 + \frac{1}{2}\left[2\left(\frac{\partial u}{\partial x} - \zeta \frac{\partial^2 w}{\partial x^2}\right) + \left(\frac{\partial w}{\partial x}\right)^2\right]\right\}dx \Rightarrow$$

$$A'B' = \left[1 + \frac{\partial u}{\partial x} - \zeta \frac{\partial^2 w}{\partial x^2} + \frac{1}{2}\left(\frac{\partial w}{\partial x}\right)^2\right]dx$$

Then, the axial strain ε_x is given by

$$\varepsilon_x = \frac{A'B' - AB}{AB} = \frac{\left[1 + \frac{\partial u}{\partial x} - \zeta \frac{\partial^2 w}{\partial x^2} + \frac{1}{2}\left(\frac{\partial w}{\partial x}\right)^2\right]dx - dx}{dx} = \frac{\partial u}{\partial x} - \zeta \frac{\partial^2 w}{\partial x^2} + \frac{1}{2}\left(\frac{\partial w}{\partial x}\right)^2$$

Using now the first of Equations (5.7) and (5.6) and noting that in this case $z \to \zeta$

$$\varepsilon_x = \varepsilon_{xo} + \zeta\left(-\frac{\partial^2 w}{\partial x^2}\right)$$

Comparing the two expressions for ε_x it follows that,

$$\varepsilon_{xo} = \frac{\partial u}{\partial x} + \frac{1}{2}\left(\frac{\partial w}{\partial x}\right)^2 \tag{5.13a}$$

which is a nonlinear strain displacement equation because of the square of the slope $\partial w/\partial x$.

In a similar fashion, it can be shown that the other two mid-plane strains are given by:

$$\varepsilon_{yo} = \frac{\partial v}{\partial y} + \frac{1}{2}\left(\frac{\partial w}{\partial y}\right)^2$$

$$\gamma_{xyo} = \frac{\partial u}{\partial y} + \frac{\partial v}{\partial x} + \left(\frac{\partial w}{\partial x}\right)\left(\frac{\partial w}{\partial y}\right) \tag{5.13b, c}$$

The curvatures κ_x, κ_y and κ_{xy} are still given by Equations (5.6).

Now the first two of Equations (5.3), which represent force equilibrium along the x and y axes, are the same as before, with the addition of the distributed loads p_x and p_y:

$$\frac{\partial N_x}{\partial x} + \frac{\partial N_{xy}}{\partial y} + p_x = 0$$

$$\frac{\partial N_{xy}}{\partial x} + \frac{\partial N_y}{\partial y} + p_y = 0 \tag{5.3a, b}$$

For force equilibrium along the z axis, the situation is as shown in Figure 5.14. Angles are sufficiently small so that

$$\tan \varphi \approx \varphi$$
$$\sin \varphi \approx \varphi$$
$$\cos \varphi \approx 1$$

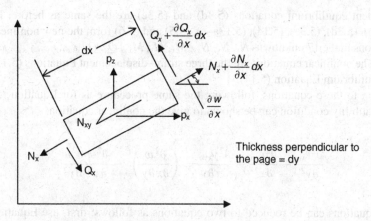

Figure 5.14 Force equilibrium of plate element in x and z directions

With $\varphi = \partial w/\partial x$ or $\partial w/\partial y$, (respectively), summation of forces in the z direction gives

$$-Q_x dy + \left(Q_x + \frac{\partial Q_x}{\partial x}dx\right)dy - Q_y dx + \left(Q_y + \frac{\partial Q_y}{\partial y}dy\right)dx - N_x\frac{\partial w}{\partial x}dy$$

$$+ \left(N_x\frac{\partial w}{\partial x} + \frac{\partial}{\partial x}\left(N_x\frac{\partial w}{\partial x}\right)dx\right)dy - N_y\frac{\partial w}{\partial y}dx$$

$$+ \left(N_y\frac{\partial w}{\partial y} + \frac{\partial}{\partial y}\left(N_y\frac{\partial w}{\partial y}\right)dy\right)dx - N_{xy}\frac{\partial w}{\partial x}dx$$

$$+ \left(N_{xy}\frac{\partial w}{\partial x} + \frac{\partial}{\partial y}\left(N_{xy}\frac{\partial w}{\partial x}\right)dy\right)dx + N_{xy}\frac{\partial w}{\partial y}dy$$

$$+ \left(N_{xy}\frac{\partial w}{\partial y} + \frac{\partial}{\partial x}\left(N_{xy}\frac{\partial w}{\partial y}\right)dx\right)dy + p_z dxdy = 0$$

Canceling and collecting terms,

$$\frac{\partial Q_x}{\partial x} + \frac{\partial Q_y}{\partial y} + N_x\frac{\partial^2 w}{\partial x^2} + 2N_{xy}\frac{\partial^2 w}{\partial x\partial y} + N_y\frac{\partial^2 w}{\partial y^2}$$

$$+ \frac{\partial w}{\partial x}\left(\frac{\partial N_x}{\partial x} + \frac{\partial N_{xy}}{\partial y}\right) + \frac{\partial w}{\partial y}\left(\frac{\partial N_{xy}}{\partial x} + \frac{\partial N_y}{\partial y}\right) + p_z = 0$$

But, from Equations (5.3a) and (5.3b) the quantities in parentheses in the equation above are equal to $-p_x$ and $-p_y$, respectively. Substituting leads to the nonlinear equation:

$$\frac{\partial Q_x}{\partial x} + \frac{\partial Q_y}{\partial y} + N_x\frac{\partial^2 w}{\partial x^2} + 2N_{xy}\frac{\partial^2 w}{\partial x\partial y} + N_y\frac{\partial^2 w}{\partial y^2} - p_x\frac{\partial w}{\partial x} - p_y\frac{\partial w}{\partial y} + p_z = 0 \qquad (5.14)$$

The moment equilibrium equations (5.3d) and (5.3e) are the same as before. Equations (5.3a), (5.3b), (5.3d), (5.3e), (5.14), (5.13a–c), (5.5) and (5.6) form the new nonlinear system of 17 equations in the 17 unknowns N_x, N_y, N_{xy}, M_x, M_y, M_{xy}, Q_x, Q_y, ε_{xo}, ε_{yo}, γ_{xyo}, κ_x, κ_y, κ_{xy}, u, v and w. The nonlinear equations are the three strain–displacement Equations (5.13a–c) and the force equilibrium Equation (5.14).

In addition to these equations, following the same procedure as for Equation (5.10), the strain compatibility condition can be shown to give the nonlinear equation:

$$\frac{\partial^2 \varepsilon_{xo}}{\partial y^2} + \frac{\partial^2 \varepsilon_{yo}}{\partial x^2} - \frac{\partial^2 \gamma_{xyo}}{\partial x \partial y} = \left(\frac{\partial^2 w}{\partial x \partial y} \right)^2 - \frac{\partial^2 w}{\partial x^2} \frac{\partial^2 w}{\partial y^2} \qquad (5.10a)$$

The 17 equations can be reduced to two equations as follows: first, use Equations (5.3d, 5.3e) to substitute in Equation (5.14). This gives

$$\frac{\partial^2 M_x}{\partial x^2} + 2\frac{\partial^2 M_{xy}}{\partial x \partial y} + \frac{\partial^2 M_y}{\partial y^2} + N_x \frac{\partial^2 w}{\partial x^2} + 2N_{xy} \frac{\partial^2 w}{\partial x \partial y} + N_y \frac{\partial^2 w}{\partial y^2} - p_x \frac{\partial w}{\partial x} - p_y \frac{\partial w}{\partial y} + p_z = 0$$

$$(5.15)$$

Then, use the moment–curvature relations from Equation (5.5) and recall that $B_{ij} = D_{16} = D_{26} = 0$:

$$M_x = -D_{11} \frac{\partial^2 w}{\partial x^2} - D_{12} \frac{\partial^2 w}{\partial y^2}$$

$$M_y = -D_{12} \frac{\partial^2 w}{\partial x^2} - D_{22} \frac{\partial^2 w}{\partial y^2}$$

$$M_{xy} = -2D_{66} \frac{\partial^2 w}{\partial x \partial y}$$

to substitute in Equation (5.15):

$$D_{11} \frac{\partial^4 w}{\partial x^4} + 2(D_{12} + 2D_{66})\frac{\partial^4 w}{\partial x^2 \partial y^2} + D_{22} \frac{\partial^4 w}{\partial y^4}$$

$$= N_x \frac{\partial^2 w}{\partial x^2} + 2N_{xy} \frac{\partial^2 w}{\partial x \partial y} + N_y \frac{\partial^2 w}{\partial y^2} - p_x \frac{\partial w}{\partial x} - p_y \frac{\partial w}{\partial y} + p_z$$

$$(5.16)$$

Equation (5.16) is the first von Karman equation, describing the bending behaviour of the plate (left-hand side) and how it couples with stretching (right-hand side). As can be seen from the first three terms in the right-hand side, it is nonlinear.

For the second von Karman equation, the Airy stress function F is introduced so that the equilibrium equations (5.3a) and (5.3b) are satisfied:

$$N_x = \frac{\partial^2 F}{\partial y^2} + V$$

$$N_y = \frac{\partial^2 F}{\partial x^2} + V \tag{5.17}$$

$$N_{xy} = -\frac{\partial^2 F}{\partial x \partial y}$$

with V the potential function for the distributed loads p_x and p_y,

$$p_x = -\frac{\partial V}{\partial x}$$

$$\tag{5.18}$$

$$p_y = -\frac{\partial V}{\partial y}$$

From Equation (5.5), the in-plane portion

$$N_x = A_{11}\varepsilon_{xo} + A_{12}\varepsilon_{yo}$$

$$N_y = A_{12}\varepsilon_{xo} + A_{22}\varepsilon_{yo}$$

$$N_{xy} = A_{66}\gamma_{xyo}$$

can be solved for the mid-plane strains,

$$\varepsilon_{xo} = \frac{A_{22}}{A_{11}A_{22} - A_{12}^2} N_x - \frac{A_{12}}{A_{11}A_{22} - A_{12}^2} N_y$$

$$\varepsilon_{yo} = -\frac{A_{12}}{A_{11}A_{22} - A_{12}^2} N_x + \frac{A_{11}}{A_{11}A_{22} - A_{12}^2} N_y \tag{5.19}$$

$$\gamma_{xyo} = \frac{1}{A_{66}} N_{xy}$$

which, in turn, can be substituted in the strain compatibility relation (5.10a) to give,

$$\frac{1}{A_{11}A_{22} - A_{12}^2} \left(A_{22}\frac{\partial^2 N_x}{\partial y^2} - A_{12}\frac{\partial^2 N_y}{\partial y^2} + A_{11}\frac{\partial^2 N_y}{\partial x^2} - A_{12}\frac{\partial^2 N_x}{\partial x^2} \right) - \frac{1}{A_{66}}\frac{\partial^2 N_{xy}}{\partial x \partial y}$$

$$= \left(\frac{\partial^2 w}{\partial x \partial y} \right)^2 - \frac{\partial^2 w}{\partial x^2}\frac{\partial^2 w}{\partial y^2}$$

Now Equation (5.17) is used to express N_x, N_y and N_{xy}, in terms of F and V:

$$\frac{1}{A_{11}A_{22} - A_{12}^2}\left(A_{22}\frac{\partial^4 F}{\partial y^4} - 2A_{12}\frac{\partial^4 F}{\partial x^2 \partial y^2} + A_{11}\frac{\partial^4 F}{\partial x^4} + (A_{22} - A_{12})\frac{\partial^2 V}{\partial y^2} + (A_{11} - A_{12})\frac{\partial^2 V}{\partial x^2}\right)$$

$$+ \frac{1}{A_{66}}\frac{\partial^4 F}{\partial x^2 \partial y^2} = \left(\frac{\partial^2 w}{\partial x \partial y}\right)^2 - \frac{\partial^2 w}{\partial x^2}\frac{\partial^2 w}{\partial y^2} \tag{5.20}$$

This is the second von Karman equation, relating the membrane behaviour of the plate (left-hand side) with the out-of-plane curvatures (right-hand side). The terms in the right-hand side are nonlinear.

5.3   Reductions of Governing Equations: Applications to Specific Problems

This section shows two examples where the governing equations are solved exactly and the results are used in the design of specific applications.

5.3.1   Composite Plate under Localized In-Plane Load [15]

The situation is shown in Figure 5.15. In practice, besides the obvious case where an in-plane point load is applied on a plate, this case arises when a stiffener is terminated. This happens when the axial load applied to a stiffened panel is reduced to the point that a monolithic panel may be sufficient to take the load or in cases with moderate loads where there is not enough room to accommodate the stiffeners.

The situation shown in Figure 5.15 represents the load from a single stiffener introduced in a rectangular panel and reacted by a uniform load at the other end. It is assumed that the

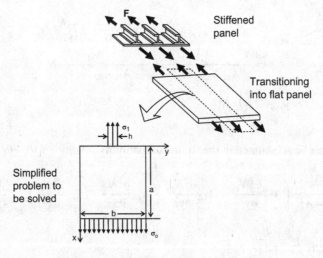

Figure 5.15   Model for a stiffener termination

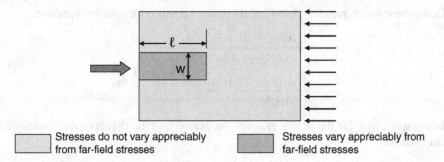

Figure 5.16   Doubler dimensions

stiffener spacing is such that there is no interaction between stiffeners (which covers most cases of realistic stiffener spacing, which is at least 6 cm). Also, to simplify the derivation, the length of the panel a is assumed to be sufficiently long that the details of the concentrated load introduction at one end have died down before the other end is reached. This is also a realistic assumption since the typical panel length, such as that corresponding to the frame spacing in a fuselage, is much longer than the distance required for the transient effects to die out.

It is exactly these transients that the designer is required to design for. In the vicinity of the point of introduction of the concentrated load, high normal and shear stresses develop that converge to their far-field (uniform stress) values fairly quickly. The size of this transition region both along the x axis and along the y axis defines the size of reinforcement or doubler that must be added to help transition the local load without failure. Determining the stresses in the vicinity of the load application will help determine the dimensions $\ell \times w$ of the required reinforcement as shown in Figure 5.16.

In addition to the assumptions already mentioned, the following conditions are imposed:

- Plate is homogeneous and orthotropic
- Lay-up is symmetric (B matrix $= 0$) and balanced ($A_{16} = A_{26} = 0$)
- No bending/twisting coupling ($D_{16} = D_{26} = 0$)

Under these assumptions, the stress–strain equations (5.5) can be solved for the mid-plane strains to give the relations (5.19). The average stresses through the plate thickness are given by

$$\sigma_x = \frac{N_x}{H}$$

$$\sigma_y = \frac{N_y}{H}$$

$$\tau_{xy} = \frac{N_{xy}}{H}$$

where H is the plate thickness.

Placing these into Equations (5.19) and dropping the subscript o for convenience,

$$\varepsilon_x = \frac{HA_{22}\sigma_x - HA_{12}\sigma_y}{A_{11}A_{22} - A_{12}^2}$$

$$\gamma_{xy} = H\frac{\tau_{xy}}{A_{66}} \qquad (5.21)$$

$$\varepsilon_y = \frac{HA_{11}\sigma_y - HA_{12}\sigma_x}{A_{11}A_{22} - A_{12}^2}$$

These expressions for the strains can now be placed into the (linear) strain compatibility condition (5.10):

$$\frac{A_{11}A_{22} - A_{12}^2}{A_{66}}\frac{\partial^2\tau_{xy}}{\partial x\partial y} = A_{22}\frac{\partial^2\sigma_x}{\partial y^2} - A_{12}\frac{\partial^2\sigma_y}{\partial y^2} + A_{11}\frac{\partial^2\sigma_y}{\partial x^2} - A_{12}\frac{\partial^2\sigma_x}{\partial x^2} \qquad (5.22)$$

Now, for a plane stress problem the out-of-plane stresses σ_z, τ_{xz} and τ_{yz} are zero. Then, the stress equilibrium condition (5.2a) gives,

$$\frac{\partial^2\tau_{xy}}{\partial x\partial y} = -\frac{\partial^2\sigma_x}{\partial x^2} \qquad (5.23)$$

Similarly, from (5.2b),

$$\frac{\partial^2\sigma_y}{\partial y^2} = -\frac{\partial^2\tau_{xy}}{\partial x\partial y} \qquad (5.24)$$

which, in view of Equation (5.23), gives

$$\frac{\partial^2\sigma_y}{\partial y^2} = \frac{\partial^2\sigma_x}{\partial x^2} \qquad (5.25)$$

Substituting in Equation (5.22) gives

$$-\frac{A_{11}A_{22} - A_{12}^2}{A_{66}}\frac{\partial^2\sigma_x}{\partial x^2} = A_{22}\frac{\partial^2\sigma_x}{\partial y^2} - A_{12}\frac{\partial^2\sigma_x}{\partial x^2} + A_{11}\frac{\partial^2\sigma_y}{\partial x^2} - A_{12}\frac{\partial^2\sigma_x}{\partial x^2} \qquad (5.26)$$

Now differentiate Equation (5.25) twice with respect to x and Equation (5.26) twice with respect to y to obtain

$$\frac{\partial^4\sigma_y}{\partial x^2\partial y^2} = \frac{\partial^4\sigma_x}{\partial x^4} \qquad (5.27)$$

and

$$-\frac{A_{11}A_{22} - A_{12}^2}{A_{66}}\frac{\partial^4\sigma_x}{\partial x^2\partial y^2} = A_{22}\frac{\partial^4\sigma_x}{\partial y^4} - A_{12}\frac{\partial^4\sigma_x}{\partial x^2\partial y^2} + A_{11}\frac{\partial^4\sigma_y}{\partial x^2\partial y^2} - A_{12}\frac{\partial^4\sigma_x}{\partial x^2\partial y^2} \qquad (5.28)$$

The stress σ_y can be eliminated from Equation (5.28) with the use of Equation (5.27). Then, collecting terms gives the governing equation for σ_x;

$$\frac{\partial^4\sigma_x}{\partial x^4} + \left[\frac{A_{11}A_{22} - A_{12}^2}{A_{11}A_{66}} - 2\frac{A_{12}}{A_{11}}\right]\frac{\partial^4\sigma_x}{\partial x^2\partial y^2} + \frac{A_{22}}{A_{11}}\frac{\partial^4\sigma_x}{\partial y^4} = 0$$

or defining

$$\beta = \frac{A_{11}A_{22} - A_{12}^2}{A_{11}A_{66}} - 2\frac{A_{12}}{A_{11}}$$

$$\gamma = \frac{A_{22}}{A_{11}}$$

$$\frac{\partial^4 \sigma_x}{\partial x^4} + \beta \frac{\partial^4 \sigma_x}{\partial x^2 \partial y^2} + \gamma \frac{\partial^4 \sigma_x}{\partial y^4} = 0 \qquad (5.29)$$

Equation (5.29) must be solved subject to the following boundary conditions:

$$\sigma_x(x = 0) = 0 \quad 0 \le y \le \frac{b-h}{2} \quad \text{and} \quad \frac{b+h}{2} \le y \le b$$

$$\sigma_x(x = 0) = \sigma_1 = \frac{F}{Hh} \quad \text{for} \quad \frac{b-h}{2} \le y \le \frac{b+h}{2}$$

$$\sigma_x(x = a) = \sigma_o = \frac{F}{bH} \qquad (5.30a\text{–}e)$$

$$\sigma_y(y = 0) = \sigma_y(y = b) = 0$$

$$\tau_{xy}(x = 0) = \tau_{xy}(x = a) = \tau_{xy}(y = 0) = \tau_{xy}(y = b) = 0$$

Conditions (5.30a) and (5.30b) define the applied concentrated load on one end ($x = 0$) of the plate. The stress σ_x is zero there except for the narrow region of width h at the centre where it equals $F/(Hh)$. Condition (5.30c) defines the uniform stress applied at the other end of the plate (at $x = a$). Finally, conditions (5.30d) and (5.30e) state that the transverse stress σ_y and the shear stress τ_{xy} are zero at the corresponding plate edges.

The solution of Equation (5.29) can be obtained using separation of variables [16]. Following this procedure, it is expedient to assume a solution of the form,

$$\sigma_x \approx f_n(x) \cos \frac{n\pi y}{b} \qquad (5.31)$$

Substituting in the governing Equation (5.29), the y dependence cancels out and the following ordinary differential equation for f_n is obtained:

$$\frac{d^4 f_n}{dx^4} - \beta \left(\frac{n\pi}{b}\right)^2 \frac{d^2 f_n}{dx^2} + \gamma \left(\frac{n\pi}{b}\right)^4 f_n = 0 \qquad (5.32)$$

From the theory of linear ordinary differential equations with constant coefficients, the solution to Equation (5.32) is found as

$$f_n = C e^{\varphi x} \qquad (5.33)$$

with

$$\varphi = \pm \frac{1}{\sqrt{2}} \left(\frac{n\pi}{b}\right) \sqrt{\beta \pm \sqrt{\beta^2 - 4\gamma}} \tag{5.34}$$

Note that Equation (5.34) implies four different values of φ to be used in Equation (5.33) yielding four different solutions for f_n as should be expected from the fourth order differential equation (5.32). It is also important to note that the quantities under the square roots in Equation (5.34) can be negative, leading to complex values for φ. In such a case the four different right-hand sides of Equation (5.34) appear in pairs of complex conjugates leading to a real solution for the stress σ_x.

If the real part of φ given by Equation (5.34) is positive, the stress σ_x will increase with increasing x. And for a long plate (value of a in Figure 5.15 is large) this would lead to unbounded stresses. So, if the plate is long enough for the effect of the load introduction at $x = 0$ to have died down, the two solutions for φ with positive real parts must be neglected. The remaining two solutions (with negative real parts) are denoted by φ_1 and φ_2 and can be combined with Equations (5.31) and (5.33) to give the most general expression for σ_x as a linear combination of all the possible solutions (all possible values of n in Equation (5.31)):

$$\sigma_x = K_o + \sum_{n=1}^{\infty} A_n \left[e^{\varphi_1 x} + C_n e^{\varphi_2 x}\right] \cos \frac{n\pi y}{b} \tag{5.35}$$

A constant K_o, which is also a solution to Equation (5.29) has been added in Equation (5.35) to obtain the most general form of the solution. The right-hand side of Equation (5.35) is a Fourier cosine series.

Now, as mentioned earlier, the out-of-plane stresses σ_z, τ_{xz}, τ_{yz} are assumed to be zero, which eliminates the last term in each of the equilibrium equations (5.2a) and (5.2b) and identically satisfies Equation (5.2c). Then, from Equation (5.2a),

$$\frac{\partial \tau_{xy}}{\partial y} = -\frac{\partial \sigma_x}{\partial x} \tag{5.36}$$

Differentiating Equation (5.35) with respect to x and then integrating the result with respect to y to substitute in Equation (5.36) leads to

$$\tau_{xy} = -\sum_{n=1}^{\infty} A_n \left[\varphi_1 e^{\varphi_1 x} + C_n \varphi_2 e^{\varphi_2 x}\right] \frac{b}{n\pi} \sin \frac{n\pi y}{b} + G_1(z) \tag{5.37}$$

Applying now the boundary condition (5.30e) at $y = 0$ leads to the following condition

$$\tau_{xy}(y = 0) = 0 \Rightarrow G_1(z) = 0$$

Then, Equation (5.30e) at $x = 0$ leads to

$$\tau_{xy}(x = 0) = 0 \Rightarrow \varphi_1 + C_n \varphi_2 = 0 \Rightarrow C_n = -\frac{\varphi_1}{\varphi_2}$$

Note that the condition (5.30e) at $x = a$ is satisfied as long as a is large enough and the exponentials in Equation (5.37) have died out. Incorporating these results in Equation (5.37), τ_{xy} is obtained as:

$$\tau_{xy} = -\sum_{n=1}^{\infty} \varphi_1 A_n \left[e^{\varphi_1 x} - e^{\varphi_2 x} \right] \frac{b}{n\pi} \sin \frac{n\pi y}{b}$$

The last of the conditions (5.30e) is at $y = b$ and it leads to

$$\tau_{xy}(y = b) = 0 \Rightarrow \sin n\pi = 0 \Rightarrow \text{ satisfied for any } n$$

The final expression for τ_{xy} is, therefore,

$$\tau_{xy} = -\sum_{n=1}^{\infty} \varphi_1 A_n \left[e^{\varphi_1 x} - e^{\varphi_2 x} \right] \frac{b}{2n\pi} \sin \frac{2n\pi y}{b} \tag{5.38}$$

Note that $2n$ is substituted for n; this is needed in order to satisfy Equation (5.30d). This, in turn implies that Equation (5.35) for σ_x has the form,

$$\sigma_x = Ko + \sum_{n=1}^{\infty} A_n \left[e^{\varphi_1 x} + C_n e^{\varphi_2 x} \right] \cos \frac{2n\pi y}{b} \tag{5.35a}$$

In an analogous manner, σ_y is determined from Equation (5.2b), with $\tau_{yz} = 0$

$$\frac{\partial \sigma_y}{\partial y} = -\frac{\partial \tau_{xy}}{\partial x} \tag{5.39}$$

which, combined with Equation (5.38) and condition (5.30d) leads to

$$\sigma_y = \sum_{n=1}^{\infty} \left(\frac{b}{2n\pi} \right)^2 \varphi_1 A_n \left(\varphi_1 e^{\varphi_1 x} - \varphi_2 e^{\varphi_2 x} \right) \left(1 - \cos \frac{2n\pi y}{b} \right) \tag{5.40}$$

At this point all unknowns in the stress expressions (5.35a), (5.38) and (5.40) have been determined except for K_o and A_n. These are determined as Fourier cosine series coefficients using conditions (5.30a) and (5.30b). The constant K_o is the average of stress σ_x at any x value,

$$K_o = \frac{F}{bH} \tag{5.41}$$

For the A_n coefficients, multiplying both sides of Equation (5.30a) by $\cos 2q\pi y/b$ and integrating from 0 to b leads to

$$\int_0^b \sigma_x(x = 0) \cos \frac{2q\pi y}{b} dy = \int_0^b \left(K_o + \sum A_n \left(e^{\varphi_1 x} - \frac{\varphi_1}{\varphi_2} e^{\varphi_2 x} \right)_{x=0} \cos \frac{2n\pi y}{b} \right) \cos \frac{2q\pi y}{b} dy \tag{5.42}$$

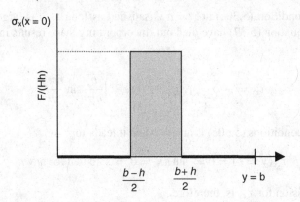

Figure 5.17   Applied normal stress σ_x at one end of the plate

Now $\sigma_x(x = 0)$ is zero everywhere except at the centre of the plate where it equals the applied load F divided by the area over which F acts. This is shown in Figure 5.17.

Substituting in Equation (5.42) and carrying out the integrations leads to the final expression for A_n:

$$A_n = \frac{F}{hH} \frac{\varphi_2}{\varphi_2 - \varphi_1} \frac{2}{n\pi} \cos n\pi \, \sin \frac{n\pi h}{b} \tag{5.43}$$

This completes the determination of the stresses in the plate. It is in closed form and exact within the assumptions made during the derivation. Since the solution is in terms of infinite series, see Equations (5.35), (5.36) and (5.40), some guidelines on selecting the number of terms after which they can be truncated and still give sufficient accuracy in the results are needed. One way to do that is to evaluate Equation (5.35) at $x = 0$ and compare it with the applied load. This is shown in Figure 5.18.

The laminate selected in Figure 5.18 has the lay-up $(\pm 45)_4$ consisting of four plies of plain weave fabric material each at $45°$ with the load direction. The geometry and applied loading information are shown in Figure 5.18. The basic material properties are as follows:

$$E_x = E_y = 73 \text{ GPa}$$

$$G_{xy} = 5.3 \text{ GPa}$$

$$v_{xy} = 0.05$$

$$\text{ply thickness } (t_{\text{ply}}) = 0.19 \text{ mm}$$

It can be seen from Figure 5.18 that, even with 160 terms in the series in Equation (5.35a), the step function behaviour of the applied load is not exactly reproduced. In addition, outside the region of the applied load, that is $y < (b - h)/2$ and $y > (b + h)/2$, the σ_x stress at $x = 0$ is very small, but not exactly zero as it should be. More terms would be necessary for even better accuracy. In what follows, predictions of the method are compared with finite element results (obtained with ANSYS) using $n = 80$.

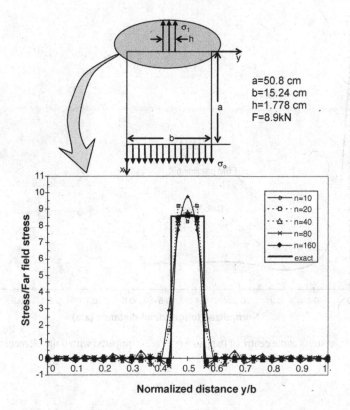

Figure 5.18 Applied stress at panel edge ($x = 0$) and approximation as a function of the number of terms in the series

The axial stress σ_x as a function of x obtained from Equation (5.35a) is compared with the finite element prediction in Figure 5.19. Very good agreement between the two methods is observed. The shear stress τ_{xy} as a function of y is compared with the finite element results at $x/a = 0.0075$ in Figure 5.20. Excellent agreement between the two methods is observed. Finally the transverse stress σ_y is compared with the finite element predictions in Figure 5.21 where the stress is plotted as a function of y at $x/a = 0.0075$. Again, very good agreement is observed.

It appears from the results in Figures 5.19–5.21 that, even though the applied σ_x is not exactly reproduced at $x = 0$ (see Figure 5.18), $n = 80$ gives sufficient accuracy for predicting the in-plane stresses in this problem. The good agreement of the method with the finite element results gives confidence in its use for the design of reinforcements in composite plates with localized loads, such as those coming from stiffener terminations shown in Figure 5.15.

There are two main issues that need to be addressed. The first is which lay-up minimizes the peak stresses that develop in the vicinity of the point-load introduction. It should be noted that the peak σ_x stress is $F/(Hh)$ where F is the applied load, H is the laminate thickness and h is the width over which the concentrated load is applied.

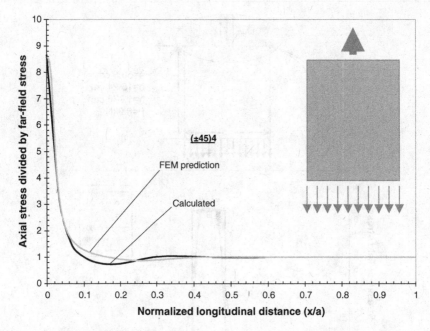

Figure 5.19   Axial stress at the centre of the panel ($y = b/2$) compared with finite element results (FEM)

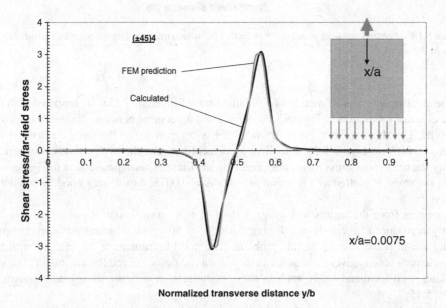

Figure 5.20   Shear stress as a function of the transverse coordinate y at $x/a = 0.0075$

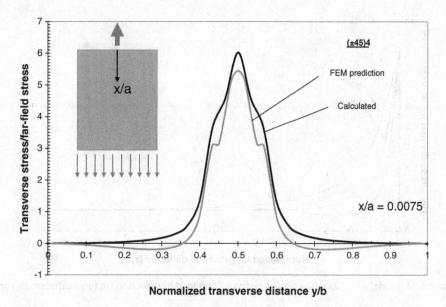

Figure 5.21   Transverse stress as a function of y at $x/a = 0.0075$

The peak σ_y and τ_{xy} stresses are not as obvious and only through methods like the one presented here can they be calculated and their potentially deleterious effect on panel performance be mitigated. The second issue is the size (in terms of length ℓ and width w) (Figure 5.16) of the reinforcement needed to transition the applied concentrated load to the far-field uniform load without failure. Some results showing how the method can be applied to specific problems are shown in Figures 5.22–5.24.

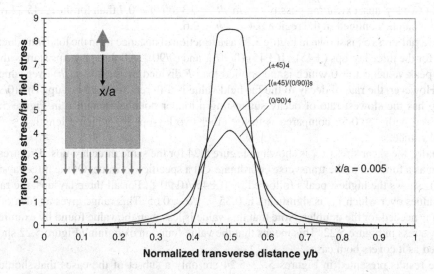

Figure 5.22   Transverse stress σ_y as a function of y for different lay-ups ($x/a = 0.005$)

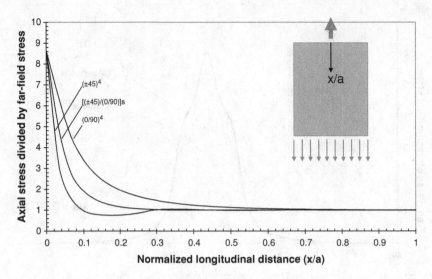

Figure 5.23   Axial stress as a function of distance from load introduction for three different lay-ups

The effect of lay-up on the peak stress σ_y is shown in Figure 5.22. Only a few representative lay-ups are used here to show trends. The material used is the same plain weave fabric mentioned earlier. The geometry is the same as that shown in Figure 5.18.

Of the three lay-ups shown in Figure 5.22, the softest, $(\pm 45)_4$ has the highest peak stress and is, therefore, to be avoided in such applications. The remaining two lay-ups, the quasi-isotropic $[(\pm 45)/(0/90)]_s$ and the orthotropic $(0/90)_4$ have much lower peak stresses. Also of interest is the fact that the region over which the σ_y stress is appreciable in the y direction is much narrower for the $(\pm 45)_4$ lay-up than for the $[(\pm 45)/(0/90)]_s$ or the $(0/90)_4$ lay-up. This means that the width w of the reinforcement or doubler needed (see Figure 5.16) must be wider for the last two lay-ups, extending possibly from $y/b = 0.3$ to $y/b = 0.7$ than for the $(\pm 45)_4$ lay-up where it can be confined in the region $0.4 < y/b < 0.6$.

The axial stress σ_x is shown in Figure 5.23 as a function of distance from the load introduction point for the three lay-ups $(\pm 45)_4$, $[(\pm 45)/(0/90)]_s$ and $(0/90)_4$. All three lay-ups start with the same peak value at $x = 0$ which is the applied load F divided by the area (Hh) over which it acts. However, the rate of decay to the far-field value is different for each lay-up. The $(0/90)_4$ lay-up has the slowest rate of decay suggesting a longer doubler (length ℓ in Figure 5.16) is needed with $\ell \approx 0.5a$ compared with the other two lay-ups for which a length $\ell \approx 0.3a$ would suffice.

Finally, the shear stress τ_{xy} is shown in Figure 5.24 for the same three lay-ups. The stress is shown as a function of the transverse coordinate y at a specific x/a value. Of the three lay-ups, $(\pm 45)_4$ shows the highest peaks followed by $[(\pm 45)/(0/90)]_s$. For all three lay-ups, the range of y values over which τ_{xy} is significant is $0.35 < y/b < 0.65$. This range gives an idea of the width w needed for the doubler. Note that this value is less than the value found by examining the σ_y stress in Figure 5.22. This means that the value found in examining Figure 5.22 should be used as it covers both cases.

The results presented in Figures 5.22–5.24 are only a subset of the cases that should be examined for a complete assessment of the doubler requirements. Other locations in the panel should also be checked so that the extreme values and locations of all three stresses σ_x,

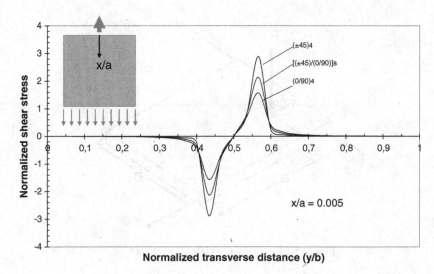

Figure 5.24 Shear stress versus y for three different lay-ups

σ_y and τ_{xy} can be determined so the doubler characteristics can be defined. The results in Figures 5.22–5.24 give a good idea of the basic trends. Based on these results, the basic characteristics of the doubler needed from the analysis so far are as follows:

- Axial stresses in $(0/90)_4$ panels decay more slowly (require longer doublers) than $(\pm45)_4$ or $[(\pm45)/(0/90)]_s$ panels.
- On the other hand, transverse and shear stresses in $(\pm45)_4$ or $[(\pm45)/(0/90)]_s$ panels are more critical than in $(0/90)_4$ panels.
- Preliminary doubler (reinforcement dimensions): $\ell = 0.5a$ for $(0/90)_4$ and $0.3a$ for $(\pm45)_4$ or $[(\pm45)/(0/90)]_s$ panels and $w = 0.3b$ for all panels.

While the analysis so far is very helpful in giving basic design guidelines, it is by no means complete. It should be borne in mind that once a doubler is added to the panel, the load distribution changes and additional iterations are necessary. The previous discussion is a good starting point for a robust design. Finally, for the specific case of a stiffener termination, the analysis presented assumed that the concentrated load acts at the mid-plane of the plate. For a terminating stiffener, the load in the stiffener acts at the stiffener neutral axis and is, therefore, offset from the centre of the plate. This means that, in addition to the axial load examined here, a moment equal to the axial load times the offset from the stiffener neutral axis to the plate mid-plane should be added.

5.3.2 Composite Plate under Out-of-Plane Point Load

The situation is shown in Figure 5.25. The plate of dimensions $a \times b$ is loaded by a vertical force F. The coordinates of the point where the load is applied are x_o and y_o. Besides the obvious application of a point load on a plate, this problem can be used to obtain the basic trends in structural response of a plate under low-speed impact damage (Figure 5.26).

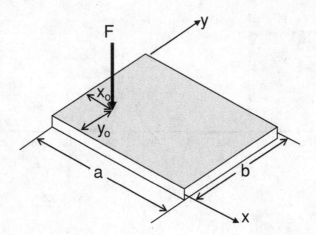

Figure 5.25   Composite plate under point load

The plate is assumed to have zero out-of-plane deflection w all around its boundary (simply supported). It is also assumed that the plate is symmetric (B matrix $= 0$) and there is no bending–twisting coupling ($D_{16} = D_{26} = 0$). Finally, the out-of-plane stresses σ_z, τ_{xz} and τ_{yz} are neglected.

The goal is to determine the out-of-plane displacement w of the plate as a function of location. Since the B matrix of the lay-up of the plate is zero, the out-of-plane behaviour of the plate decouples from the in-plane behaviour. Then, the governing equation is Equation (5.16) with $p_x = p_y = 0$ and the nonlinear terms neglected since we are interested in a linear (small deflections) solution:

$$D_{11}\frac{\partial^4 w}{\partial x^4} + 2(D_{12} + 2D_{66})\frac{\partial^4 w}{\partial x^2 \partial y^2} + D_{22}\frac{\partial^4 w}{\partial y^4} = p_z \qquad (5.16a)$$

The out-of-plane applied load p_z in this case can be expressed with the use of delta functions as

$$p_z = F\delta(x - x_o)\delta(y - y_o) \qquad (5.44)$$

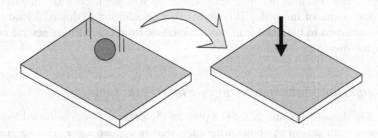

Figure 5.26   Low-speed impact modelled as point load

where

$$\delta(x - x_o) = 1 \quad \text{when} \quad x = x_o$$

$$= 0 \quad \text{otherwise}$$

Then, the governing equation is

$$D_{11}\frac{\partial^4 w}{\partial x^4} + 2(D_{12} + 2D_{66})\frac{\partial^4 w}{\partial x^2 \partial y^2} + D_{22}\frac{\partial^4 w}{\partial y^4} = F\delta(x - x_o)\delta(y - y_o) \tag{5.45}$$

with D_{ij} the bending stiffness terms for the plate lay-up.

Since $w = 0$ at the plate boundary, a solution to Equation (5.45) is sought in the form

$$w = \sum\sum A_{mn} \sin\frac{m\pi x}{a} \sin\frac{n\pi y}{b} \tag{5.46}$$

with A_{mn} unknown coefficients.

It can be seen that if Equation (5.46) is placed in Equation (5.45) the left-hand side will contain terms multiplied by $\sin(m\pi x/a)\sin(n\pi y/b)$. In order to proceed, the right-hand side of Equation (5.45) must also be expanded in a double Fourier series in order to be able to match terms in the left- and right-hand sides. Setting,

$$F\delta(x - x_o)\delta(y - y_o) = \sum\sum B_{mn} \sin\frac{m\pi x}{a} \sin\frac{n\pi y}{b} \tag{5.47}$$

where B_{mn} are unknown coefficients, one can multiply both sides of Equation (5.47) by $\sin(p\pi x/a)\sin(q\pi y/b)$ and integrate over the plate domain (x from 0 to a and y from 0 to b) to obtain,

$$\iint F\delta(x - x_o)\delta(y - y_o) \sin\frac{p\pi x}{a} \sin\frac{q\pi y}{b} dxdy$$

$$= \iint \sum\sum B_{mn} \sin\frac{m\pi x}{a} \sin\frac{p\pi x}{a} \sin\frac{n\pi y}{b} \sin\frac{q\pi y}{b} dxdy \tag{5.48}$$

Now the integral of a function multiplied by the delta function is equal to the function evaluated at the location where the delta function is nonzero. So, carrying out the integrations in Equation (5.48),

$$F\sin\frac{m\pi x_o}{a} \sin\frac{n\pi y_o}{b} = B_{mn}\frac{ab}{4} \tag{5.49}$$

from which

$$B_{mn} = \frac{4F}{ab} \sin\frac{m\pi x_o}{a} \sin\frac{n\pi y_o}{b} \tag{5.50}$$

Equation (5.50) can be placed in Equation (5.47) which, along with Equation (5.46), can be placed in Equation (5.45) to give

$$
\sum_{m=1}^{\infty}\sum_{n=1}^{\infty} A_{mn}\left[D_{11}\left(\frac{m\pi}{a}\right)^4 + 2(D_{12}+2D_{66})\frac{m^2n^2\pi^4}{a^2b^2} + D_{22}\left(\frac{n\pi}{b}\right)^4 \right]\sin\frac{m\pi x}{a}\sin\frac{n\pi y}{b}
$$

$$
= \sum_{m=1}^{\infty}\sum_{n=1}^{\infty}\frac{4F}{ab}\sin\frac{m\pi x_o}{a}\sin\frac{n\pi y_o}{b}\sin\frac{m\pi x}{a}\sin\frac{n\pi y}{b} \tag{5.51}
$$

and matching coefficients of $\sin(m\pi x/a)\sin(n\pi y/b)$ the coefficients A_{mn} are determined as

$$
A_{mn} = \frac{\dfrac{4F}{ab}\sin\dfrac{m\pi x_o}{a}\sin\dfrac{n\pi y_o}{b}}{D_{11}\left(\dfrac{m\pi}{a}\right)^4 + 2(D_{12}+2D_{66})\dfrac{m^2n^2\pi^4}{a^2b^2} + D_{22}\left(\dfrac{n\pi}{b}\right)^4} \tag{5.52}
$$

Combining Equations (5.52) with (5.46) gives the complete expression for w for this case:

$$
w = \sum\sum \frac{\dfrac{4F}{ab}\sin\dfrac{m\pi x_o}{a}\sin\dfrac{n\pi y_o}{b}\sin\dfrac{m\pi x}{a}\sin\dfrac{n\pi y}{b}}{D_{11}\left(\dfrac{m\pi}{a}\right)^4 + 2(D_{12}+2D_{66})\dfrac{m^2n^2\pi^4}{a^2b^2} + D_{22}\left(\dfrac{n\pi}{b}\right)^4} \tag{5.53}
$$

For the case where F acts at the centre of the plate, the maximum out-of-plane deflection δ at the plate centre is obtained by substituting $x = a/2$ and $y = b/2$ in Equation (5.53),

$$
\delta = w_{max} = \sum\sum \frac{\dfrac{4F}{ab}\sin^2\dfrac{m\pi}{2}\sin^2\dfrac{n\pi}{2}}{D_{11}\left(\dfrac{m\pi}{a}\right)^4 + 2(D_{12}+2D_{66})\dfrac{m^2n^2\pi^4}{a^2b^2} + D_{22}\left(\dfrac{n\pi}{b}\right)^4} \tag{5.54}
$$

Once the deflections are determined, classical laminated-plate theory can be used to obtain bending moments and, in turn, strains and stresses to check the plate for failure.

As in the previous section, this is an exact solution to the problem within the context of the assumptions made. It should also be kept in mind that because of the linearization in Equation (5.16a), the solution is only valid for small out-of-plane deflections w.

5.4　Energy Methods

For most practical problems, the governing equations described in the previous section cannot be solved exactly and, in some cases, approximate solutions are hard to obtain. As a powerful alternative, energy methods can be used. Minimizing the energy stored in the system or structure can yield useful, approximate and reasonably accurate solutions.

Two energy minimization principles are of interest here: (1) minimum potential energy; and (2) minimum complementary energy.

In both cases, some of the governing equations are satisfied exactly and some approximately through energy minimization. They both derive from the following two theorems [17]:

Minimum potential energy: Of all *geometrically compatible displacement states*, those which also satisfy the *force balance conditions* give stationary values to the potential energy.

Minimum complementary energy: Of all *self-balancing force states*, those which also satisfy the requirements of *geometric compatibility* give stationary values to *the complementary energy*.

The governing equations, given in the previous section can be split into: (a) equilibrium equations; (b) compatibility equations (which are the strain compatibility equations obtained once the displacements are eliminated from the strain–displacement equations); and (c) the constitutive law or stress–strain equations.

In the case of the principle of minimum potential energy, if the strain compatibility relations and displacement boundary conditions are exactly satisfied, then minimization of the potential energy results in a solution that satisfies the equilibrium equations in an average sense. In the case of the principle of minimum complementary energy, if the stress equilibrium equations and force boundary conditions are exactly satisfied, then minimization of the complementary energy results in a solution that satisfies strain compatibility in an average sense. Both approaches yield approximate solutions whose accuracy depends on the number of terms assumed in the displacement (minimum potential energy) or stress expressions (minimum complementary energy) and how well the assumed functions approximate the sought-for response.

The two situations as well as the situation corresponding to the exact solution are shown in Table 5.5.

The energy methods are not limited to the two approaches just described. Hybrid approaches where combinations of some stresses and displacements are assumed are also possible [18].

5.4.1 Energy Expressions for Composite Plates

According to the principle of virtual work for linear elasticity, the incremental internal energy stored in a body equals the incremental work done by external forces:

$$\delta U = \delta W_s + \delta W_b$$

where W_s is the work done by surface forces and W_b is the work done by body forces.

Table 5.5 Approximate and exact evaluations of field equations during energy minimization

Equilibrium equations	Strain compatibility condition	Force boundary conditions	Displacement boundary conditions	Energy	Solution is
Exactly	Exactly	Exactly	Exactly	Minimized	Exact
Approximately	Exactly	(In an average sense)	Exactly	Minimize potential (displacement-based)	Approximate
Exactly	Approximately	Exactly	(In an average sense)	Minimize complementary	Approximate

Then, if we define the total incremental energy $\delta\Pi$ as the difference between internal energy and external work,

$$\delta\Pi = \delta U - \delta W_s - \delta W_b \tag{5.55}$$

the exact solution would make the energy variation $\delta\Pi$ zero or would minimize the total energy Π:

$$\Pi = U - W_s - W_b = U - W \tag{5.56}$$

5.4.1.1 Internal Strain Energy U

The increment in the internal potential energy δU is obtained by integrating all contributions of products of stresses and incremental strains

$$\delta U = \iiint_V \left\{\sigma_x \delta\varepsilon_x + \sigma_y \delta\varepsilon_y + \sigma_z \delta\varepsilon_z + \tau_{yz}\delta\gamma_{yz} + \tau_{xz}\delta\gamma_{xz} + \tau_{xy}\delta\gamma_{xy}\right\} \mathrm{d}x\mathrm{d}y\mathrm{d}z \tag{5.57}$$

where the integration is over the entire volume V of the body in question.

For a plate, Equation (5.57) reduces to

$$\delta U = \iiint_V \left\{\sigma_x \delta\varepsilon_x + \sigma_y \delta\varepsilon_y + \tau_{xy}\delta\gamma_{xy}\right\} \mathrm{d}x\mathrm{d}y\mathrm{d}z \tag{5.58}$$

Using Equations (5.7) to substitute for the strains in terms of curvatures and mid-plane strains gives

$$\delta U = \iiint_V \left\{\sigma_x \left(\delta\varepsilon_{xo} + z\delta\kappa_x\right) + \sigma_y \left(\delta\varepsilon_{yo} + z\delta\kappa_y\right) + \tau_{xy} \left(\delta\gamma_{xyo} + z\delta\kappa_{xy}\right)\right\} \mathrm{d}x\mathrm{d}y\mathrm{d}z \tag{5.59}$$

For a plate of constant thickness h, the z integration in Equation (5.59) can be carried out using

$$\int_{-\frac{h}{2}}^{\frac{h}{2}} \begin{bmatrix} \sigma_x \\ \sigma_y \\ \tau_{xy} \end{bmatrix} \mathrm{d}z = \begin{bmatrix} N_x \\ N_y \\ N_{xy} \end{bmatrix} \quad \text{and} \quad \int_{-\frac{h}{2}}^{\frac{h}{2}} \begin{bmatrix} z\sigma_x \\ z\sigma_y \\ z\tau_{xy} \end{bmatrix} \mathrm{d}z = \begin{bmatrix} M_x \\ M_y \\ M_{xy} \end{bmatrix}$$

to give

$$\delta U = \iint_A N_x \delta\varepsilon_{xo} + N_y \delta\varepsilon_{yo} + N_{xy}\delta\gamma_{xyo} + M_x \delta\kappa_x + M_y \delta\kappa_y + M_{xy}\delta\kappa_{xy}\mathrm{d}x\mathrm{d}y \tag{5.60}$$

where A is the area of the plate.

At this point, several options are available depending on which version of energy minimization principle (e.g. displacement-based or stress-based) is to be used.

For a displacement-based formulation, Equations (5.5) can be used to express forces and moments in terms of mid-plane strains and curvatures.

$$
\begin{aligned}
\delta U = \iint_A \Big\{ & \left(A_{11}\varepsilon_{xo} + A_{12}\varepsilon_{yo} + A_{16}\gamma_{xyo} + B_{11}\kappa_x + \ldots \right) \delta\varepsilon_{xo} \\
& + \left(A_{12}\varepsilon_{xo} + A_{22}\varepsilon_{yo} + A_{26}\gamma_{xyo} + B_{12}\kappa_x + \ldots \right) \delta\varepsilon_y \\
& + \left(A_{16}\varepsilon_{xo} + A_{26}\varepsilon_{yo} + A_{66}\gamma_{xyo} + B_{16}\kappa_x + \ldots \right) \delta\gamma_{xy} \\
& + \left(B_{11}\varepsilon_{xo} + B_{12}\varepsilon_{yo} + B_{16}\gamma_{xyo} + D_{11}\kappa_x + \ldots \right) \delta\kappa_x \\
& + \left(B_{12}\varepsilon_{xo} + B_{22}\varepsilon_{yo} + B_{26}\gamma_{xyo} + D_{12}\kappa_x + \ldots \right) \delta\kappa_y \\
& + \left(B_{16}\varepsilon_{xo} + B_{26}\varepsilon_{yo} + B_{66}\gamma_{xyo} + D_{16}\kappa_x + \ldots \right) \delta\kappa_{xy} \Big\} \, dxdy
\end{aligned}
$$

(5.61)

It is now observed that

$$
\varepsilon_{xo}\delta\varepsilon_{xo} = \frac{1}{2}\delta\left(\varepsilon_{xo}\right)^2
$$

$$
\varepsilon_{xo}\delta\varepsilon_{yo} + \varepsilon_{yo}\delta\varepsilon_{xo} = \delta(\varepsilon_{xo}\varepsilon_{yo})
$$

$$
\varepsilon_{xo}\delta\kappa_x + \kappa_x\delta\varepsilon_{xo} = \delta\left(\varepsilon_{xo}\kappa_x\right)
$$

with analogous expressions for the other mid-plane strains and curvatures.

These expressions are substituted in Equation (5.61) and integrated term by term. For example, the first term of Equation (5.61) becomes

$$
\iint_A A_{11}\left(\varepsilon_{xo}\right)\delta\varepsilon_{xo}dxdy = \frac{1}{2}\iint_A A_{11}\delta\left(\varepsilon_{xo}\right)^2 dxdy \rightarrow \frac{1}{2}\iint_A A_{11}\left(\varepsilon_{xo}\right)^2 dxdy
$$

This substitution leads to the following expression for the internal strain energy U:

$$
\begin{aligned}
U = \frac{1}{2}\iint_A & \left\{ \begin{array}{l} A_{11}\left(\varepsilon_{xo}\right)^2 + 2A_{12}\left(\varepsilon_{xo}\right)\left(\varepsilon_{yo}\right) + 2A_{16}\left(\varepsilon_{xo}\right)\left(\gamma_{xyo}\right) + A_{22}(\varepsilon_{yo})^2 + \\ 2A_{26}(\varepsilon_{yo})(\gamma_{xyo}) + A_{66}(\gamma_{xyo})^2 \end{array} \right\} dxdy \\
& + \iint_A \left\{ \begin{array}{l} B_{11}\left(\varepsilon_{xo}\right)\kappa_x + B_{12}\left((\varepsilon_{yo})\kappa_x + (\varepsilon_{xo})\kappa_y\right) + B_{16}\left((\gamma_{xyo})\kappa_x + (\varepsilon_{xo})\kappa_{xy}\right) + \\ B_{22}(\varepsilon_{yo})\kappa_y + B_{26}\left((\gamma_{xyo})\kappa_y + (\varepsilon_{yo})\kappa_{xy}\right) + B_{66}(\gamma_{xyo})\kappa_{xy} \end{array} \right\} dxdy \\
& + \frac{1}{2}\iint_A \left\{ D_{11}\kappa_x^2 + 2D_{12}\kappa_x\kappa_y + 2D_{16}\kappa_x\kappa_{xy} + D_{22}\kappa_y^2 + 2D_{26}\kappa_y\kappa_{xy} + D_{66}\kappa_{xy}^2 \right\} dxdy
\end{aligned}
$$

(5.62)

Finally, to express the internal strain energy in terms of displacements u, v and w, the strain–displacement Equations (5.6) and (5.8) are used to obtain:

$$
U = \frac{1}{2} \iint_A \left\{ \begin{array}{l} A_{11}\left(\dfrac{\partial u}{\partial x}\right)^2 + 2A_{12}\dfrac{\partial u}{\partial x}\dfrac{\partial v}{\partial y} + 2A_{16}\dfrac{\partial u}{\partial x}\left(\dfrac{\partial u}{\partial y}+\dfrac{\partial v}{\partial x}\right) + A_{22}\left(\dfrac{\partial u}{\partial y}\right)^2 \\[3mm] +2A_{26}\dfrac{\partial v}{\partial y}\left(\dfrac{\partial u}{\partial y}+\dfrac{\partial v}{\partial x}\right) + A_{66}\left(\dfrac{\partial u}{\partial y}+\dfrac{\partial v}{\partial x}\right)^2 \end{array} \right\} dxdy
$$

$$
- \iint_A \left\{ \begin{array}{l} B_{11}\left(\dfrac{\partial u}{\partial x}\dfrac{\partial^2 w}{\partial x^2}\right) + B_{12}\left(\dfrac{\partial v}{\partial y}\dfrac{\partial^2 w}{\partial x^2}+\dfrac{\partial u}{\partial x}\dfrac{\partial^2 w}{\partial y^2}\right) + B_{16}\left[\left(\dfrac{\partial u}{\partial y}+\dfrac{\partial v}{\partial x}\right)\dfrac{\partial^2 w}{\partial x^2}+2\dfrac{\partial u}{\partial x}\dfrac{\partial^2 w}{\partial x\partial y^2}\right] \\[3mm] +B_{22}\dfrac{\partial v}{\partial y}\dfrac{\partial^2 w}{\partial y^2} + B_{26}\left[\left(\dfrac{\partial u}{\partial y}+\dfrac{\partial v}{\partial x}\right)\dfrac{\partial^2 w}{\partial y^2}+2\dfrac{\partial v}{\partial y}\dfrac{\partial^2 w}{\partial x\partial y}\right] + 2B_{66}\left(\dfrac{\partial u}{\partial y}+\dfrac{\partial v}{\partial x}\right)\dfrac{\partial^2 w}{\partial x\partial y} \end{array} \right\} dxdy
$$

$$
+ \frac{1}{2} \iint_A \left\{ \begin{array}{l} D_{11}\left(\dfrac{\partial^2 w}{\partial x^2}\right)^2 + 2D_{12}\dfrac{\partial^2 w}{\partial x^2}\dfrac{\partial^2 w}{\partial y^2} + 4D_{16}\dfrac{\partial^2 w}{\partial x^2}\dfrac{\partial^2 w}{\partial x\partial y} + D_{22}\left(\dfrac{\partial^2 w}{\partial y^2}\right)^2 \\[3mm] +4D_{26}\dfrac{\partial^2 w}{\partial y^2}\dfrac{\partial^2 w}{\partial x\partial y} + 4D_{66}\left(\dfrac{\partial^2 w}{\partial x\partial y}\right)^2 \; dxdy \end{array} \right\} dxdy \qquad (5.63)
$$

The first set of terms in Equation (5.63) involves the membrane stiffnesses A_{ij} ($i,j = 1,2,6$) and represents stretching (or membrane) energy. The last set, involving bending stiffnesses D_{ij} ($i,j = 1,2,6$) represents energy stored in bending of the plate. The remaining terms involving B_{ij} ($i,j = 1,2,6$) represent energy stored through bending–membrane coupling. If the plate has symmetric lay-up, $B_{ij} = 0$ and Equation (5.63) decouples in two parts, the membrane (involving the A matrix) and the bending (involving the D matrix) portion.

At the other extreme, a stress-based energy formulation starts with Equation (5.60) and uses the inverse of the stress–strain Equations (5.5) to substitute for the strains. For simplicity, only the case of a symmetric lay-up is shown here. The inverted stress–strain equations,

$$
\left\{ \begin{array}{c} \varepsilon_x^o \\ \varepsilon_y^o \\ \gamma_{xy}^o \\ \kappa_x \\ \kappa_y \\ \kappa_{xy} \end{array} \right\} = \left[\begin{array}{cccccc} a_{11} & a_{12} & a_{16} & 0 & 0 & 0 \\ a_{12} & a_{22} & a_{26} & 0 & 0 & 0 \\ a_{16} & a_{26} & a_{66} & 0 & 0 & 0 \\ 0 & 0 & 0 & d_{11} & d_{12} & d_{16} \\ 0 & 0 & 0 & d_{12} & d_{22} & d_{26} \\ 0 & 0 & 0 & d_{16} & d_{26} & d_{66} \end{array} \right] \left\{ \begin{array}{c} N_x \\ N_y \\ N_{xy} \\ M_x \\ M_y \\ M_{xy} \end{array} \right\} \qquad (5.64)
$$

where $[a]$ and $[d]$ are the inverses of the laminate $[A]$ and $[D]$ matrices can be used to substitute in Equation (5.60):

$$
\begin{aligned}
\delta U = \iint_A \; & N_x\delta\left(a_{11}N_x + a_{12}N_y + a_{16}N_{xy}\right) + N_y\delta\left(a_{12}N_x + a_{22}N_y + a_{26}N_{xy}\right) \\
& + N_{xy}\delta\left(a_{16}N_x + a_{26}N_y + a_{66}N_{xy}\right) + M_x\delta\left(d_{11}M_x + d_{12}M_y + d_{16}M_{xy}\right) \\
& + M_y\delta\left(d_{12}M_x + d_{22}M_y + d_{26}M_{xy}\right) + M_{xy}\delta\left(d_{16}M_x + d_{26}M_y + d_{66}M_{xy}\right) dxdy
\end{aligned} \qquad (5.65)
$$

A completely analogous procedure as in deriving Equation (5.63) from Equation (5.61) leads to the final expression for the stress-based (complementary) energy:

$$U = \frac{1}{2} \iint_A \left\{ a_{11} N_x^2 + 2a_{12} N_x N_y + 2a_{16} N_x N_{xy} + a_{22} N_y^2 + 2a_{26} N_y N_{xy} + a_{66} N_{xy}^2 \right\} dxdy$$

$$+ \frac{1}{2} \iint_A \left\{ d_{11} M_x^2 + 2d_{12} M_x M_y + 2d_{16} M_x M_{xy} + d_{22} M_y^2 + 2d_{26} M_y M_{xy} + d_{66} M_{xy}^2 \right\} dxdy$$

$$(5.66)$$

Equation (5.66) has the stretching and bending portions already decoupled because the laminate was assumed symmetric.

5.4.1.2 External Work W

The derivation for the external work does not have any difference between composite and noncomposite plates. It is derived for a general plate and included here for completeness. With reference to Equation (5.55), the incremental work δW_b done by applied body forces on a body is given by

$$\delta W_b = \iiint_V \left\{ f_x \delta u + f_y \delta v + f_z \delta w \right\} dxdydz$$

where V is the volume of the body, f_x, f_y and f_z are forces per unit volume in the x, y and z directions, respectively and δu, δv and δw are incremental displacements in the x, y and z directions.

For a plate, the body forces can be integrated through the thickness

$$\int f_x dz = p_{xb}$$

$$\int f_y dz = p_{yb}$$

$$\int f_z dz = p_{zb}$$

with the subscript b denoting that these contributions to surface forces come from integrating the body forces.

Combining these with any surface forces applied over the plate surface and the contribution from any forces or moments applied on the plate edges, gives

$$\delta W = \delta W_b + \delta W_s = \iint_{A_p} \left\{ p_x \delta u + p_y \delta v + p_z \delta w \right\} dxdy$$

$$+ \int_0^a \left[N_x \delta u + N_{xy} \delta v + Q_x \delta w - M_x \delta \left(\frac{\partial w}{\partial x} \right) \right]_{x=0}^{x=a} dy$$

$$+ \int_0^a \left[N_{xy} \delta u + N_y \delta v + Q_y \delta w - M_y \delta \left(\frac{\partial w}{\partial y} \right) \right]_{y=0}^{y=b} dx$$

where a and b are the plate dimensions and A_p is the plate area. The contributions from p_{xb}, p_{yb} and p_{zb} are included in the first term within p_x, p_y and p_z, respectively. The second and third terms in the right-hand side of the above expression include contributions from applied forces N_x, N_y and N_{xy} (in-plane) or (transverse shear) forces Q_x and Q_y (out-of-plane) or bending moments M_x and M_y at the plate edges ($x = 0,a$ and/or $y = 0,b$).

Integrating the incremental contributions on left- and right-hand sides gives

$$
W = \iint_{A_p} \{p_x u + p_y v + p_z w\}\, dxdy + \int_0^b \left[N_x u + N_{xy} v + Q_x w - M_x \frac{\partial w}{\partial x} \right]_{x=0}^{x=a} dy
$$

$$
+ \int_0^a \left[N_{xy} u + N_y v + Q_y w - M_y \frac{\partial w}{\partial y} \right]_{y=0}^{y=b} dx \qquad\qquad (5.67)
$$

For the case of plate buckling problems, p_x and p_y in Equation (5.57) can be evaluated further. Assuming there is no stretching or shearing of the plate mid-plane during buckling, the mid-plane strains ε_{xo}, ε_{yo} and γ_{xyo} are zero. Then, for large deflections, Equations (5.13a–c) imply

$$
\frac{\partial u}{\partial x} + \frac{1}{2}\left(\frac{\partial w}{\partial x}\right)^2 = 0
$$

$$
\frac{\partial v}{\partial y} + \frac{1}{2}\left(\frac{\partial w}{\partial y}\right)^2 = 0
$$

$$
\frac{\partial u}{\partial y} + \frac{\partial v}{\partial x} + \left(\frac{\partial w}{\partial x}\right)\left(\frac{\partial w}{\partial y}\right) = 0
$$

Consider now the first term of Equation (5.67) with $p_z = 0$ for a buckling problem. Using Equations (5.3a,b), to substitute for p_x and p_y,

$$
\iint_A (p_x u + p_y v)\, dxdy = \iint_A \left\{ \left(-\frac{\partial N_x}{\partial x} - \frac{\partial N_{xy}}{\partial y}\right) u + \left(-\frac{\partial N_{xy}}{\partial x} - \frac{\partial N_y}{\partial y}\right) v \right\} dxdy
$$

Integrating by parts, for a rectangular plate of dimensions a and b, gives

$$
\iint_A (p_x u + p_y v)\, dxdy = \int_0^b \left\{ \left[-N_x u - N_{xy} v\right]_{x=0}^{x=a} + \int_0^a \left[N_x \frac{\partial u}{\partial x} + N_{xy} \frac{\partial v}{\partial x} \right] dx \right\} dy
$$

$$
+ \int_0^a \left\{ \left[-N_{xy} u - N_y v\right]_{y=0}^{y=b} + \int_0^b \left[N_{xy} \frac{\partial u}{\partial y} + N_y \frac{\partial v}{\partial y} \right] dy \right\} dx
$$

$$= \int\limits_0^a [-N_{xy}u - N_y v]_{y=0}^{y=b} \, dx + \int\limits_0^b [-N_x u - N_{xy} v]_{x=0}^{=a} \, dy$$

$$+ \int\limits_0^a \int\limits_0^a \left[N_x \frac{\partial u}{\partial x} + N_{xy} \frac{\partial v}{\partial x} + N_{xy} \frac{\partial u}{\partial y} + N_y \frac{\partial v}{\partial y} \right] dxdy$$

The derivatives $\partial u/\partial x$, $\partial v/\partial y$ and the sum $\partial u/\partial y + \partial v/\partial x$ can be substituted by derivatives of w, as shown in the large deflection equations above. Then, combining everything in Equation (5.67), canceling terms and noting that for a typical buckling problem $Q_x = Q_y = M_x = M_y = 0$ leads to

$$W = \int\limits_0^a \int\limits_0^b \left\{ -\frac{1}{2} N_x \left(\frac{\partial w}{\partial x} \right)^2 - \frac{1}{2} N_y \left(\frac{\partial w}{\partial y} \right)^2 - N_{xy} \left(\frac{\partial w}{\partial x} \right) \left(\frac{\partial w}{\partial y} \right) \right\} dxdy$$

or

$$W = -\frac{1}{2} \int\limits_0^a \int\limits_0^b \left\{ N_x \left(\frac{\partial w}{\partial x} \right)^2 + N_y \left(\frac{\partial w}{\partial y} \right)^2 + 2N_{xy} \left(\frac{\partial w}{\partial x} \right) \left(\frac{\partial w}{\partial y} \right) \right\} dxdy \qquad (5.68)$$

valid for plate buckling problems.

Exercises

5.1 A certain composite material is proposed for use at two different locations of the same application. Location 1 is designed by tension with a design (ultimate) load of 1750 N/mm. Location 2 is designed by shear with a design (ultimate) load of 2450 N/mm. The proposed material has been tested in tension and shear with the results shown in Table E5.1.

Table E5.1 Test data for proposed composite material

Specimen	Tension (Pa)	Shear (Pa)
1		3.0918E + 08
2	6.7217E + 08	4.0789E + 08
3	6.1025E + 08	3.2922E + 08
4	6.3263E + 08	2.9084E + 08
5	6.5498E + 08	3.6868E + 08
6	5.3391E + 08	3.3140E + 08
7	6.5647E + 08	3.6039E + 08
Mean	6.2673E + 08	3.4251E + 08

Originally, the two parts at the locations of interest were made with aluminium with the following properties shown in Table E5.2:

Table E5.2   Aluminium properties (7075 Al from [19])

	Tension (Pa)	Shear (Pa)
Mean	5.1016E + 08	3.2815E + 08
B-Basis	4.9637E + 08	3.2402E + 08
A-Basis	4.7569E + 08	3.1712E + 08

An aspiring engineer looks at the two tables of properties, in particular the mean values, and claims that he/she can save at least 30% of the weight at both locations by switching from aluminium to composite. You are to check if the engineer is right in his/her claim for both locations considering: (a) a single load path application and (b) a multiple load path application.

You are to assume that the test data in Table E5.1 follows a normal distribution for both tension and shear. Note that for a normal distribution the B- and A-Basis values are given by

$$B = \text{Mean} - k_B \sigma$$
$$A = \text{Mean} - k_A \sigma$$

where σ is the standard deviation of the test results and k_A, k_B are the so-called one-sided tolerance limit factors given in Table E5.3 (see for example [2] Chapter 9).

Table E5.3   One-sided tolerance limit factors for normal distribution

Number of specimens	$k90(1)$	$k99(1)$
6	3.006	5.741
7	2.755	4.642

(1) 90 and 99 refer to the % of tests that will be stronger than the corresponding basis value

In your calculations consider *only* the material scatter. Do not include effects of damage and environment.

5.2   The tension strength data for a specific composite material and lay-up at RTA conditions is given in the table below:

Specimen	Value (MPa)
1	538.8
2	475.6
3	447.9
4	461.7
5	495.4
6	483
7	479.3
8	442.5
9	471.6
10	525.5

Assuming that the experimental data are normally distributed, it can be shown that the fraction of the population with strength less than any given value (cumulative probability) is given by the following graph:

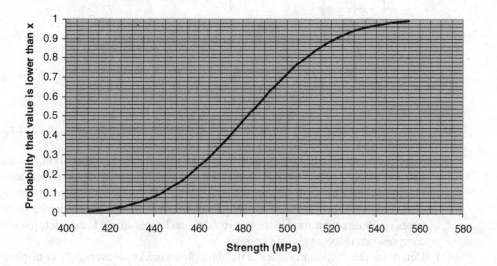

Assume that at any other environmental condition, the tension strength is given as a fraction of its corresponding value at RTA condition and that fraction can be obtained from the material covered in this chapter.

This composite material and lay-up are to be used in a wing-box (single load path primary structure). The best available aluminium is 7075-T6 with the following properties (from reference [19]):

	RTA	ETW
Mean (MPa)	586.0	562.5
B-Basis (MPa)	551.5	529.4
A-Basis (MPa)	537.7	516.2
Density (kg/m³)	2773.8	2773.8

Determine the weight savings of using the composite instead of aluminium if: (a) the design is based on RTA properties and (b) the design is based on ETW properties. Looking at your results, are the weight savings resulting from using composites in this application worth the extra material and processing cost associated with composites?

5.3 A simply supported rectangular composite plate (Figure E5.1) with dimensions 152.4×508 mm is loaded at $x = 127$ mm and $y = 38.1$ mm by a force F perpendicular to the plate. The lay-up of the plate is $(\pm 45)/(0/90)_3/(\pm 45)$ and the basic material properties are as shown in Figure E5.1.

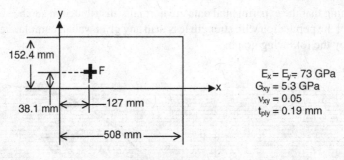

$E_x = E_y = 73$ GPa
$G_{xy} = 5.3$ GPa
$v_{xy} = 0.05$
$t_{ply} = 0.19$ mm

Figure E5.1

(a) Determine the location in the plate where each of the three stresses σ_x, σ_y and τ_{xy} is maximized.

(b) Since the three stresses do not reach their peak values at the same location, discuss how one would go about predicting the load F at which the plate would fail (assume that the ultimate strength values such as X^t, X^c, Y^t, Y^c and S with X strength along fibres or warp direction for a plain weave fabric and Y strength perpendicular to the fibres or fill direction for a plain weave fabric and superscripts t and c tension and compression, respectively, are known).

(c) Determine the maximum values of the through the thickness averaged out-of-plane shear stresses τ_{xz} and τ_{yz} and their locations for a unit load $F = 1$ N. Compare these values with the maximum values for σ_x, σ_y and τ_{xy} from part (a) and comment on whether the assumption made in Section 5.3.2 that τ_{xz} and τ_{yz} can be neglected is valid.

5.4 For a composite rectangular panel simply supported all around under pressure loading, determine if the linear solution for the out-of-plane deflections is sufficient to use in design. The applied pressure corresponds to an overload pressure case of a pressurized composite fuselage of almost 1.4 atmospheres or 20 psi. (Note: the units are British (Imperial) in this problem because you are to use the ESDU data sheets which have charts in these units). The situation is shown in Figure E5.2.

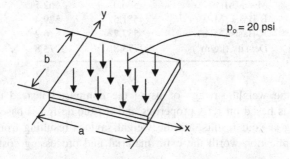

$p_o = 20$ psi

Figure E5.2

(a) Derive an expression for the deflection at the centre of the plate δ.

(b) For the case $a = b = 50$ in, $D_{11} = D_{22} = 347,000$ in lb, $D_{12} = 110,000$ in lb, $D_{66} = 120,000$ in lb and $t = 0.5$ in, the solution for δ as a function of p_o can be found in the ESDU data sheets. The ESDU solution is a large-deflection, moderate-rotation solution that will be more accurate as the applied pressure increases. Find the ESDU solution and plot δ versus applied pressure for pressures from 0 to 20 psi for your solution and the ESDU solution. Compare the two solutions and determine when your (linear) solution departs significantly from the ESDU (nonlinear) solution. Can your linear solution be used for the overpressure case of 20 psi? Before you give your final answer on this, keep in mind that this is a design problem so you do not always have to be accurate as long as you are conservative (and can afford the associated increase in weight).

(c) In view of your comparison in Exercise 5.4 and the ESDU curves you found, what exactly does 'simply supported plate' mean in this case? (discuss in-plane and out-of-plane boundary conditions that your linear solution satisfies versus the cases that ESDU provides).

5.5 Three different materials are considered for use at a specific location of the lower skin of the wing of an aircraft. The designing load condition is wing up-bending which causes a limit load N_x at that location of 580 N/mm. The materials considered are 7075-T6 aluminium, Ti-6-4 Titanium and quasi-isotropic Graphite/Epoxy. The aircraft must operate in the standard range of environments ($-54°C$ to $82°C$ and from dry to fully saturated with moisture). Neglecting the effects of damage and assuming that moisture has no discernible effect on the metal properties, determine the ratios of weights Graphite/Aluminium and Graphite/Titanium for the skin at the location in question. Which material gives the lightest design for this application? Is this the material mostly used on aircraft today? If not why not? Note that if the skin fails at this location, the surrounding structure can still take the loads and there is no catastrophic failure. For metal properties, refer to Tables E5.4 and E5.5 and Figures E5.3 and E5.4 from [19, 20] (note that the density is denoted by ω). For the Graphite/Epoxy properties note that at RTA condition, a [45/−45/0/90]s lay-up has a strength of 482.58 MPa. The ply thickness is 0.1524 mm and the density is 1636 kg/m^3.

Table E5.4 Basic properties of 7075 aluminium

Material: 7075 aluminium sheet							
Temper: T6 and T62							
Thickness (in)	0.008–0.011	0.012–0.039		0.040–0.125		0.126–0.249	
Basis	S	A	B	A	B	A	B
Ftu (ksi)							
L		76	78	78	80	78	80
LT	74	76	78	78	80	78	80
Fty (ksi)							
L		69	72	70	72	71	73
LT		67	70	68	70	69	71

(*Continued*)

Table E5.4   (*Continued*)

	Material: 7075 aluminium sheet						
	Temper: T6 and T62						
Thickness (in)	0.008–0.011	0.012–0.039		0.040–0.125		0.126–0.249	
Basis	S	A	B	A	B	A	B
Fcy (ksi)							
L		68	71	69	71	70	72
LT		71	74	72	74	73	75
E (msi)	10.3	10.3	10.3	10.3	10.3	10.3	10.3
ω (lb/in^3)	0.101	0.101	0.101	0.101	0.101	0.101	0.101

Table E5.5   Basic properties of Ti-6-4 Titanium

	Material: Ti-6Al-4V						
	Annealed						
Thickness (in)	>0.5	0.5–1.0		1.0–2.0		2.0–3.0	
Basis	S	A	B	A	B	A	B
Ftu (ksi)							
L	130	130	142	130	140	130	138
LT	130	130	144	130	143	130	142
Fty (ksi)							
L	120	120	134	120	131	120	128
LT	120	120	134	120	132	120	131
Fcy (ksi)							
L	124	124	138	124	125		
L							
E (msi)	16.9	16.9	16.9	16.9	16.9	16.9	16.9
ω (lb/in^3)	0.16	0.16	0.16	0.16	0.16	0.16	0.16

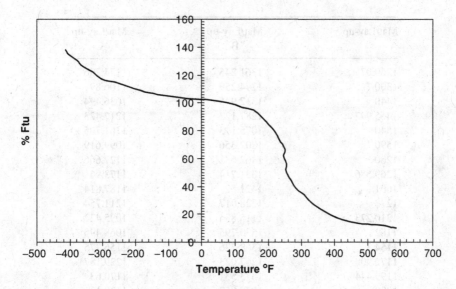

Figure E5.3

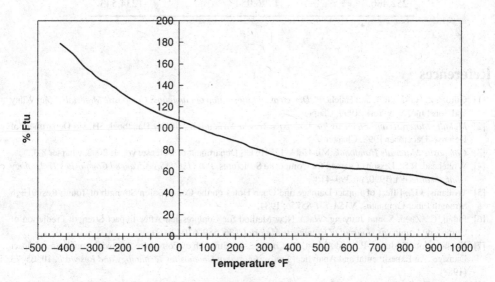

Figure E5.4

5.6 Compression strength (in MPa) data for three candidate materials and lay-ups are given in the table below.

Which one is better to use for design in aerospace applications and why?

Mat/Lay-up A	Mat/Lay-up B	Mat/Lay-up C
1260.97	1361.245	1174.796
1300	1274.259	1106.89
1440	1312.945	1036.494
1465.942	1287.122	1217.423
1540	1075.113	1161.705
1590	1202.356	1099.619
1260	1462.632	1127.666
1269.556	1331.714	1173.96
1601	1421.52	1187.814
1278	1228.017	1211.754
1210.273	1115.834	1095.422
1182	1530.795	1068.495
1154.3	1368.726	1156.775
1172.908	1267.023	1229.187
1199.444	1373.531	1170.63
1230	1078.057	1128.131
1226.17	1509.975	1264.322
1247.474	1024.368	1279.4
1380	1238.391	1122.747
1252.486	1796.954	1234.543

References

[1] Gürdal, Z, Haftka, R.T. and Hajela, P., *Design and Optimization of Laminated Composite Materials*, John Wiley and Sons, Inc, New York, 1999, Chapter 5.4.

[2] *Metallic Materials and Elements for Aerospace Vehicle Structures*, Military Handbook 5H, US Department of Defense, December 1998, Chapter 9.

[3] *Composite Materials Handbook*, Mil-Hdbk-17–1F, US Department of Defense, vol. 1, 2002, Chapter 8.3.

[4] Whitehead, R.S., Lessons Learned for Composite Structures, *Proc 1st NASA Advanced Composites Technology Conf., Seattle WA*, 1990, pp 399–415.

[5] Williams, J.C., Effect of Impact Damage and Open Holes on the Compression Strength of Tough Resin/High Strength Fiber Laminates, *NASA-TM-85756*, 1984.

[6] Puhui, C., Zhen, S. and Junyang, W., A New Method for Compression After Impact Strength Prediction of Composite Laminates, *Journal of Composite Materials*, **36**, 589–610 (2002).

[7] Kassapoglou, C., Jonas, P.J. and Abbott, R., Compressive Strength of Composite Sandwich Panels after Impact Damage: An Experimental and Analytical Study, *Journal of Composites Technology and Research*, **10**, 65–73 (1988).

[8] Nyman, T., Bredberg, A. and Schoen, J., Equivalent Damage and Residual Strength of Impact Damaged Composite Structures, *Journal of Reinforced Plastics and Composites*, **19**, 428–448 (2000).

[9] Dost, E.F., Ilcewicz, L.B., Avery, W.B. and Coxon, B.R., Effect of Stacking Sequence on Impact Damage Resistance and Residual Strength for Quasi-isotropic Laminates, *ASTM STP 1110*, 1991, pp 476–500.

[10] Xiong, Y., Poon, C., Straznicky, P.V. and Vietinghoff, H., A Prediction Method for the Compressive Strength of Impact Damaged Composite Laminates, *Composite Structures*, **30**, 357–367 (1993).

[11] Kassapoglou, C., Compression Strength of Composite Sandwich Structures After Barely Visible Impact Damage, *Journal of Composites Technology and Research*, New York, St Louis, 1996, pp 274–284.

[12] Rivello, R.M., *Theory and Analysis of Flight Structures*, McGraw-Hill Book Co., 1969, Chapter 2.

[13] Whitney, J.M., *Structural Analysis of Laminated Anisotropic Plates*, Technomic Publishing Co., Lancaster, PA, 1987, Chapters 1 and 2.

[14] Herakovich, C.T., *Mechanics of Fibrous Composites*, John Wiley and Sons, Inc, New York, 1998, Chapter 5.

[15] Kassapoglou, C. and Bauer, G., Composite Plates Under Concentrated Load on One End and Distributed Load on the Opposite End, *Mechanics of Advanced Materials and Structures*, **17**, 196–203 (2010).

[16] Hildebrand, F.B., *Advanced Calculus for Applications*, Prentice Hall, Englewood Cliffs, NJ, 1976, Chapter 9.

[17] Crandall, S.H. *Engineering Analysis*, McGraw-Hill, New York, 1956, Section 4.4.

[18] Pian, T.H.H., Variational principles for incremental finite element methods, *Journal of the Franklin Institute*, **302**, 473–488 (1976).

[19] *Metallic Materials and Elements for Aerospace Vehicle Structures*, Military Handbook 5H, US Department of Defense, December 1998, Chapter 3.

[20] *Metallic Materials and Elements for Aerospace Vehicle Structures*, Military Handbook 5H, US Department of Defense, December 1998, Chapter 5.

6

Buckling of Composite Plates

Composite plates under compression and/or shear loading are sensitive to buckling failures. A typical situation where a stiffened composite plate has buckled between the stiffeners is shown in Figure 6.1.

Unlike beams, where buckling is, typically, very close to final failure, plates may have significant post-buckling ability (see Chapter 7). However, post-buckling of composite plates requires accurate knowledge of the possible failure modes and their potential interaction. For example, in a stiffened panel such as that of Figure 6.1, the portion of the skin buckling away from the reader tends to peel off the stiffeners. The skin–stiffener separation mode is fairly common in post-buckled stiffened panels and may lead to premature failure. Depending on the application, designing for buckling and using any post-buckling capability as an extra degree of conservatism is one of the possible approaches. Even in post-buckled panels, accurate calculation of the buckling load for different loading combinations and boundary conditions is paramount in the design.

6.1 Buckling of Rectangular Composite Plate under Biaxial Loading

The derivation of the buckling equation follows the approach described by Whitney [1]. A rectangular composite plate under biaxial loading is shown in Figure 6.2.

The governing equation is obtained from Equation (5.16) by setting $N_{xy} = p_x = p_y = p_z = 0$:

$$D_{11}\frac{\partial^4 w}{\partial x^4} + 2(D_{12} + 2D_{66})\frac{\partial^4 w}{\partial x^2 \partial y^2} + D_{22}\frac{\partial^4 w}{\partial y^4} = N_x\frac{\partial^2 w}{\partial x^2} + N_y\frac{\partial^2 w}{\partial y^2} \tag{6.1}$$

where w is the out-of-plane displacement of the plate.

Note that the governing equation (6.1) assumes that the bending–twisting coupling terms D_{16} and D_{26} are negligible compared with the remaining terms D_{11}, D_{12}, D_{22} and D_{66}. The

Design and Analysis of Composite Structures: With Applications to Aerospace Structures, Second Edition. Christos Kassapoglou.
© 2013 John Wiley & Sons, Ltd. Published 2013 by John Wiley & Sons, Ltd.

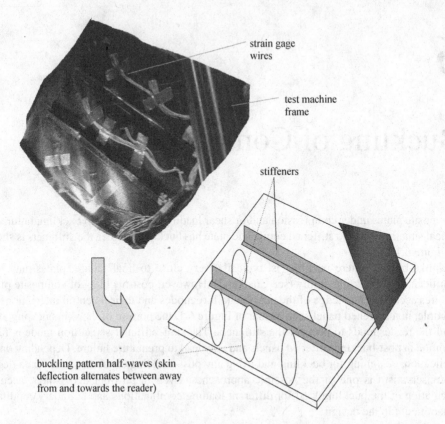

strain gage
wires

test machine
frame

stiffeners

buckling pattern half-waves (skin
deflection alternates between away
from and towards the reader)

Figure 6.1   Composite stiffened panel buckling under shear (see Plate 14 for the colour figure)

plate is assumed simply supported all around its boundary and the only loads applied are N_x and N_y as shown in Figure 6.2. Then, the boundary conditions are,

$$w = M_x = -D_{11}\frac{\partial^2 w}{dx^2} - D_{12}\frac{\partial^2 w}{dy^2} = 0 \quad \text{at} \quad x = 0 \quad \text{and} \quad x = a$$

$$w = M_y = -D_{12}\frac{\partial^2 w}{dx^2} - D_{22}\frac{\partial^2 w}{dy^2} = 0 \quad \text{at} \quad y = 0 \quad \text{and} \quad y = b$$

$$(6.2)$$

An expression for w that satisfies all boundary conditions (Equations 6.2) is,

$$w = \sum\sum A_{mn} \sin\frac{m\pi x}{a} \sin\frac{n\pi y}{b}$$

$$(6.3)$$

Substituting in Equation (6.1) and rearranging and defining the plate aspect ratio AR = a/b gives,

$$\pi^2 A_{mn}[D_{11}m^4 + 2(D_{12} + 2D_{66})m^2 n^2 (AR)^2 + D_{22}n^4 (AR)^4] = -A_{mn}a^2[N_x m^2 + N_y n^2 (AR)^2]$$

$$(6.4)$$

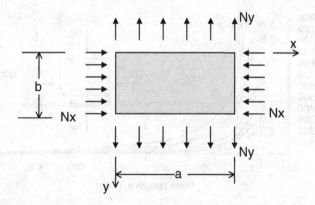

Figure 6.2 Rectangular composite panel under biaxial loading

When buckling occurs, the out-of-plane deflection w of the plate is nonzero. This means that the coefficients A_{mn} of Equation (6.3) are nonzero and cancel out in Equation (6.4). It is convenient to let $k = N_y/N_x$ and to let the buckling load N_x be denoted by $-N_o$ (minus sign to indicate compression). Then, from Equation (6.4),

$$N_o = \frac{\pi^2[D_{11}m^4 + 2(D_{12} + 2D_{66})m^2n^2(AR)^2 + D_{22}n^4(AR)^4]}{a^2(m^2 + kn^2(AR)^2)} \tag{6.5}$$

The buckling load N_o is a function of the number of half-waves m in the x direction and n in the y direction and thus, changes as m and n, which define the buckling mode, change. The sought-for buckling load is the lowest value of Equation (6.5) so the right-hand side of Equation (6.5) must be minimized with respect to m and n.

As an application of Equation (6.5), consider a square plate with quasi-isotropic lay-up $[(45/-45)_2/0_2/90_2]_s$ with basic ply properties (x parallel to fibres):

$$E_x = 137.9 \text{ GPa}$$

$$E_y = 11.7 \text{ GPa}$$

$$\nu_{xy} = 0.31$$

$$G_{xy} = 4.82 \text{ GPa}$$

$$t_{ply} = 0.1524 \text{ mm}$$

where t_{ply} is the (cured) ply thickness.

Determine the compressive buckling load N_o for various values of k.

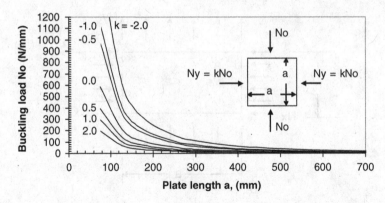

Figure 6.3 Buckling load of a square quasi-isotropic plate as a function of plate size and biaxial loading ratio N_y/N_x

Using classical laminated-plate theory (CLPT), the bending stiffness terms are found to be:

$$D_{11} = 65.4 \text{ kNmm}$$

$$D_{12} = 37.2 \text{ kNmm}$$

$$D_{22} = 51.1 \text{ kNmm}$$

$$D_{66} = 38.6 \text{ kNmm}$$

$$D_{16} = 5.40 \text{ kNmm}$$

$$D_{26} = 5.40 \text{ kNmm}$$

The bending–twisting coupling terms D_{16} and D_{26} are less than 15% of the next larger term so using Equation (6.1) will give accurate trends and reasonable buckling predictions.

For a given value of k, Equation (6.5) is evaluated for successive values of n and m until the combination that minimizes the buckling load N_o is obtained. This load is shown in Figure 6.3 as a function of the plate size and different ratios k.

As expected, increasing the plate size decreases the buckling load, which varies with the inverse of the square of the plate size. Both positive and negative values of k are shown in Figure 6.3. Positive values mean that the sign of N_y is the same as N_o and since N_o is compressive, $k > 0$ implies biaxial compression. Then, negative values of k correspond to tensile N_y values. As is seen from Figure 6.3, a tensile N_y ($k < 0$) tends to stabilize the plate and increase its buckling load. Compressive N_y ($k > 0$) tends to precipitate buckling earlier (material is pushed from both x and y directions) and decreases the buckling load. The case of $k = 0$ corresponds to uniaxial compression (see below).

It is interesting to note that the minimum buckling load was obtained for $n = 1$ in all cases. It can be shown [2, 3] that for a rectangular plate under biaxial loading the number of half-waves n in one of the two directions will always be 1.

Finally, Equation (6.5) also gives negative values of N_o when $k < 0$. This means that N_o is tensile and, since $k < 0$, N_y is compressive. So the plate still buckles, but now the compressive load is in the y direction while the load in the x direction is tensile.

6.2 Buckling of Rectangular Composite Plate under Uniaxial Compression

This case was derived as a special case in the previous section when $k = 0$. The buckling load when the plate is under compression is given by Equation (6.5) with k set to zero:

$$N_o = \frac{\pi^2 \left[D_{11}m^4 + 2(D_{12} + 2D_{66})m^2n^2(AR)^2 + D_{22}n^4(AR)^4 \right]}{a^2 m^2} \tag{6.6}$$

The right-hand side is minimized when $n = 1$, that is only one half-wave is present in the direction transverse to the applied load. Setting $n = 1$ and rearranging,

$$N_o = \frac{\pi^2}{a^2} \left[D_{11}m^2 + 2(D_{12} + 2D_{66})(AR)^2 + D_{22}\frac{(AR)^4}{m^2} \right] \tag{6.7}$$

The value of m that minimizes the right-hand side of Equation (6.7) gives the buckling load of a simply supported rectangular composite plate under compression.

As can be seen from Equation (6.7), in addition to the bending stiffnesses D_{11}, D_{12}, D_{22} and D_{66}, the buckling load is also dependent on the aspect ratio (AR = length/width) of the plate. This dependence is shown in Figure 6.4 for a plate with fixed length 508 mm.

As is seen from Figure 6.4, as the aspect ratio increases, the number of half-waves m in the direction of the load increases. Typically, for each m value, there is a value of AR that minimizes the buckling load. Points of intersection of curves corresponding to successive m values indicate that the plate may buckle in either of the two modes (differing by one half-wave) and have the same buckling load. In practice, due to eccentricities and inaccuracies due to fabrication, these cusps cannot be reproduced. The plate will tend to buckle in one of the two modes and will not switch to the other.

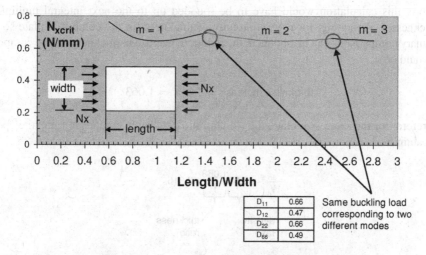

Figure 6.4 Dependence of buckling load on plate aspect ratio

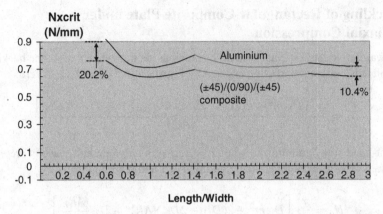

Figure 6.5 Comparison of buckling loads for equal thickness aluminium and quasi-isotropic composite plates (length = 508 mm)

The results in Figure 6.4 correspond to a quasi-isotropic lay-up (±45)/(0/90)/(±45) with D matrix values as shown in the same figure. The laminate thickness for this laminate is 0.5715 mm. It is of interest to compare with an aluminium plate of the same thickness, length and aspect ratio. This is done in Figure 6.5. Note that the buckling loads for aluminium can be obtained using the same Equation (6.7) with proper redefinition of the D matrix terms.

As is seen from Figure 6.5, the buckling load of an aluminium plate of the same thickness can be as much as 20% higher (for AR ≈ 0.5) than that of an equal thickness quasi-isotropic composite plate. Based on this result, to match the buckling load of the aluminium plate at the worst case (AR = 0.5) the quasi-isotropic plate thickness must be increased by a factor of $(1.2)^{1/3}$. The one-third power is because the D matrix terms are proportional to thickness to the third power (see also Equation 3.47). It is recognized here that typical composite materials are not available at any desired thickness, but only in multiples of specific ply thicknesses. Therefore, this calculation would have to be rounded up to the next integral multiple of ply thicknesses. Assuming, for now, continuity of thickness for the composite plate so that preliminary comparisons can be obtained, the required increase in thickness for the composite plate would be

$$\text{thickness increase} = 1.2^{1/3} = 1.063$$

Therefore, for the same size plate, the weight ratio between a composite (graphite/epoxy) and an aluminium panel is

$$\frac{W_{Gr/E}}{W_{Al}} = 0.58\underbrace{\frac{1.063}{1}}_{\substack{\text{density} \\ \text{ratio}}} \ \underbrace{}_{\substack{\text{thickness} \\ \text{ratio}}} = 0.616 \tag{6.8}$$

Equation (6.8) implies that a quasi-isotropic composite with the same buckling load under compression as an aluminium plate, is approximately 62% of the aluminium weight or results

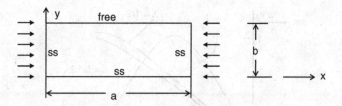

Figure 6.6 Plate under compression with one (unloaded) edge free and three edges simply supported

in, approximately, 38% weight savings. It is important to keep in mind that this result assumes that any thickness is achievable with a composite material (which is not true, as mentioned above) and that there are no other factors that may affect the design such as material scatter, environmental effects and sensitivity to damage. Accounting for these effects tends to decrease the weight savings.

6.2.1   Uniaxial Compression, Three Sides Simply Supported, One Side Free

The discussion so far in this section has been confined to a simply supported plate. The effect of the boundary conditions can be very important. As a special case, of interest in future discussion (Section 8.5 on stiffener crippling) the case of a rectangular composite plate under compression with three sides simply supported and one (not loaded) side free, is discussed here. The situation is shown in Figure 6.6.

An approximate solution is obtained following the same steps as for the plate simply supported all around. Analogous to Equation (6.3) an expression for w is assumed in the form,

$$w = \sum\sum A_{mn} \sin\frac{m\pi x}{a} \sin\frac{\lambda n\pi y}{b} \tag{6.9}$$

where λ is a parameter appropriately selected to satisfy the boundary conditions of the problem.

The governing equation is the same as Equation (6.1) with $N_y = 0$:

$$D_{11}\frac{\partial^4 w}{\partial x^4} + 2(D_{12} + 2D_{66})\frac{\partial^4 w}{\partial x^2 \partial y^2} + D_{22}\frac{\partial^4 w}{\partial y^4} = N_x\frac{\partial^2 w}{\partial x^2} \tag{6.10}$$

The boundary conditions for Equation (6.10) are

$$w(x = 0) = w(x = a) = 0$$

$$w(y = 0) = 0$$

$$M_x = -D_{11}\frac{\partial^2 w}{\partial x^2} - D_{12}\frac{\partial^2 w}{\partial y^2} = 0 \quad \text{at} \quad x = 0, a \tag{6.11}$$

$$M_y = -D_{12}\frac{\partial^2 w}{\partial x^2} - D_{22}\frac{\partial^2 w}{\partial y^2} = 0 \quad \text{at} \quad y = 0, b$$

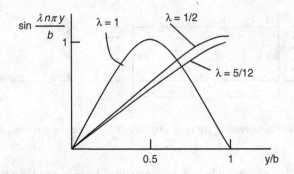

Figure 6.7 Shape of w deflection for various values of the parameter λ

The value of λ must be chosen such that w given by Equation (6.9) is free to attain any value at the free edge $y = b$. For example, if $\lambda = 1$, w at $y = b$ is zero and the simply supported case discussed earlier is recovered. A plot of w as a function of y for different λ values is shown in Figure 6.7.

It appears from Figure 6.7, that λ values in the vicinity of 1/2 would give a reasonable representation of w. It should be noted that $\lambda = 1/2$ gives a slope of w at $y = b$ that equals zero which is unlikely to be the case since w is arbitrary at $y = b$ and there is no reason for its slope to be equal to zero all along the edge $y = b$.

The results obtained with this expression for w are approximate for another reason: the last of the boundary conditions (Equation 6.11) is not satisfied. The moment M_y at $y = b$ will not be zero and its value will depend on λ.

Following the same procedure as for the simply supported case above, the expression for the buckling load corresponding to Equation (6.7) is,

$$N_o = \frac{\pi^2}{a^2}\left[D_{11}m^2 + 2(D_{12} + 2D_{66})\lambda^2(AR)^2 + D_{22}\frac{(AR)^4}{m^2}\lambda^4\right] \tag{6.12}$$

The exact solution to this problem is [4]

$$N_o = 12\frac{D_{66}}{b^2} + \frac{1}{(AR)^2}\sqrt{\frac{D_{11}}{D_{22}}} \tag{6.13}$$

The approximation of Equation (6.12) and the exact solution (6.13) are compared in Figure 6.8 for the same quasi-isotropic lay-up $(\pm45)/(0/90)/(\pm45)$ of Figure 6.4.

The approximate solution is very close to the exact answer especially for $\lambda = 5/12$. In particular, for an infinitely long plate, the exact solution (6.13) becomes

$$N_{xcrit} = \frac{12D_{66}}{b^2} \tag{6.13a}$$

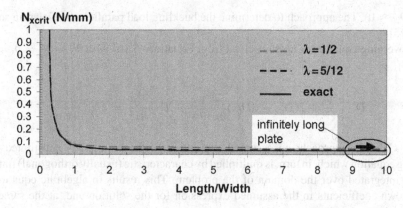

Figure 6.8 Comparison of approximate (two λ values) and exact solutions for buckling load of a rectangular composite panel under compression with three simply supported edges and one (unloaded) edge free

and the approximate solution becomes

$$N_{xcrit} = \frac{4\pi^2}{b^2}\lambda^2 D_{66} + \frac{2\pi^2}{b^2}D_{12} \qquad (6.12a)$$

Note that since the plate is infinitely long, the two expressions are only dependent on the plate width b now.

Setting $b = 508$ mm, for $\lambda = 1/2$ the two answers differ by 46.9%, but for $\lambda = 5/12$ the two differ by only 12.5%. Obviously, if an exact solution to a problem such as the one under discussion exists and does not require expensive computation (e.g. solution of a large eigenvalue problem), it will be preferred over an approximate solution. Unfortunately, in most cases, approximate solutions may be all that is available during design and preliminary analysis. The example given is meant to show the potential and the drawbacks of approximate methods.

6.3 Buckling of Rectangular Composite Plate under Shear

A rectangular composite plate under shear is shown in Figure 6.9. As before, the lay-up of the plate is assumed symmetric (B matrix $= 0$) and with negligible bending–twisting coupling

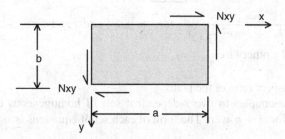

Figure 6.9 Rectangular composite plate under shear

$(D_{16} \approx D_{26} \approx 0)$. The approach to determine the buckling load parallels the Galerkin solution given in [5].

The governing equation is again derived from Equation (5.16) with $N_x = N_y = p_x = p_y = p_z = 0$:

$$D_{11}\frac{\partial^4 w}{\partial x^4} + 2(D_{12} + 2D_{66})\frac{\partial^4 w}{\partial x^2 \partial y^2} + D_{22}\frac{\partial^4 w}{\partial y^4} = 2N_{xy}\frac{\partial^2 w}{\partial x \partial y} \tag{6.14}$$

In the Galerkin approach, an assumed expression of the solution is substituted in the governing equation which, in turn, is multiplied by characteristic (usually orthogonal) functions and then integrated over the domain of the problem. This results in algebraic equations for the unknown coefficients in the assumed expression for the solution and, at the same time, minimizes the error [6].

To solve Equation (6.14) by the Galerkin method, the following expression for w is used which is the same as Equation (6.3):

$$w = \sum\sum A_{mn} \sin\frac{m\pi x}{a} \sin\frac{n\pi y}{b} \tag{6.3}$$

where A_{mn} are unknowns to be determined.

As the terms in Equation (6.3) comprise orthogonal sine functions, the same characteristic functions are used. Multiplying Equation (6.14) by the characteristic functions $\sin(m\pi x/a)\sin(n\pi y/b)$ and integrating gives

$$\iint\left[D_{11}\frac{\partial^4 w}{\partial x^4} + 2(D_{12} + 2D_{66})\frac{\partial^4 w}{\partial x^2 \partial y^2} + D_{22}\frac{\partial^4 w}{\partial y^4} - 2N_{xy}\frac{\partial^2 w}{\partial x \partial y}\right]$$
$$\times \sin\frac{m\pi x}{a} \sin\frac{n\pi y}{b}dxdy = 0 \tag{6.15}$$

where the integrations are carried over the entire plate ($0 \leq x \leq a$ and $0 \leq y \leq b$).

Note that each set of m,n values gives a different equation to be solved where all unknowns A_{mn} appear. Substituting for w from Equation (6.3) and carrying out the integrations give:

$$\pi^4\left[D_{11}m^4 + 2(D_{12} + 2D_{66})m^2n^2(AR)^2 + D_{22}n^4(AR)^4\right]A_{mn}$$
$$-32mn(AR)^3b^2N_{xy}\sum\sum T_{ij}A_{ij} = 0$$

$$T_{ij} = \frac{ij}{(m^2 - i^2)(n^2 - j^2)} \quad \text{for} \quad m \pm i \quad \text{odd} \quad \text{and} \quad n \pm j \quad \text{odd} \tag{6.16}$$

$$T_{ij} = 0 \quad \text{otherwise}$$

with $AR = a/b$ the aspect ratio of the plate.

Equation (6.16) uncouples to two independent sets of homogeneous equations, one for $m + n$ odd and one for $m + n$ even. The form of each set of equations is:

$$[E]\{A_{mn}\} = 0 \tag{6.17}$$

with $[E]$ a coefficient matrix with ijth entry given by

$$E_{ij} = -32mn(AR)^3b^2N_{xy}T_{ij}+$$
$$\pi^4\left[D_{11}m^4 + 2(D_{12} + 2D_{66})m^2n^2(AR)^2 + D_{22}n^4(AR)^4\right]\delta(m-i)\delta(n-j) \quad (6.18)$$

where $\delta(m-i) = 1$ when $m = i$ and 0 otherwise and $\delta(n-j) = 1$ when $n = j$ and zero otherwise.

Equations (6.17) have coefficients A_{mn} that are a function of the shear load N_{xy} as shown in Equation (6.18). For each of the independent sets of Equations (6.17), a nontrivial solution ($A_{mn} \neq 0$) is obtained when the determinant of the coefficient matrix is set equal to zero,

$$\det[E]_{m+n=odd} = 0$$
$$\det[E]_{m+n=even} = 0$$

Each of these two equations results in an eigenvalue problem where the eigenvalue is the buckling load N_{xy} and the eigenvector gives the buckling mode. The lowest eigenvalue across both problems is the sought-for buckling load. For symmetric and balanced (specially orthotropic) plates, the eigenvalues appear in pairs of positive and negative values, indicating that if the load direction changes the plate will buckle when the applied load reaches the same magnitude.

The approach just described gives very accurate buckling loads, provided sufficient terms in Equation (6.3) are used and an accurate eigenvalue solver is available. The following is a less involved, approximate method to obtain the buckling load under shear.

For $0.5 \leq a/b < 1$, the buckling load is given by

$$N_{xyEcr} = \cfrac{\dfrac{\pi^4 b}{a^3}}{\sqrt{\dfrac{14.28}{D1^2} + \dfrac{40.96}{D1D2} + \dfrac{40.96}{D1D3}}} \quad \text{with}$$

$$D1 = D_{11} + D_{22}\left(\frac{a}{b}\right)^4 + 2(D_{12} + 2D_{66})\left(\frac{a}{b}\right)^2 \quad (6.19)$$

$$D2 = D_{11} + 81D_{22}\left(\frac{a}{b}\right)^4 + 18(D_{12} + 2D_{66})\left(\frac{a}{b}\right)^2$$

$$D3 = 81D_{11} + D_{22}\left(\frac{a}{b}\right)^4 + 18(D_{12} + 2D_{66})\left(\frac{a}{b}\right)^2$$

For $a/b = 0$, use the results of the next section for long plates. Finally, for $0 \leq a/b < 0.5$, interpolate linearly between the result for $a/b = 0$ and $a/b = 0.5$. The accuracy of this approach depends on the bending stiffnesses of the plate and its aspect ratio a/b and ranges from less than one percent to 20% for typical lay-ups used in practice.

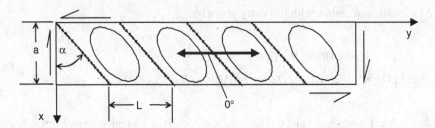

Figure 6.10   Buckling pattern in a long rectangular plate under shear

6.4   Buckling of Long Rectangular Composite Plates under Shear

The Galerkin-based derivation of the previous section can be simplified significantly if one of the plate dimensions is long compared with the other. In such a case, the long dimension does not affect the buckling load and the buckling pattern is confined over a length L, which is significantly lower than the panel long dimension. The situation is shown in Figure 6.10.

Following Thielemann [7] and assuming a simply supported plate, an expression for the out-of-plane displacement w can be assumed in the form:

$$w = w_o \sin \frac{\pi x}{a} \sin \frac{\pi (y - x \tan \alpha)}{L} \qquad (6.20)$$

This expression satisfies the conditions that w is zero along the long sides ($x = 0$ and $x = a$) and along lines inclined by an angle α to the x axis separated by distance L, as shown in Figure 6.10. It should be noted that in the actual buckling pattern these inclined lines of zero w are not perfectly straight as Equation (6.20) implies, but the error in assuming perfectly straight lines is small.

The buckling load is obtained by minimizing the energy stored in the plate. It is assumed that the laminate is symmetric so the internal potential energy (Equation 5.63) decouples in a membrane (in-plane) portion (the terms involving the A matrix) and a bending (out-of-plane) portion (the terms involving the D matrix). For the buckling problem under consideration, only w is of interest and, therefore, only the terms involving the D matrix are used. In addition, it is assumed that $D_{16} = D_{26} = 0$. Then, the internal potential energy has the form:

$$U = \frac{1}{2} \iint_A \left\{ D_{11} \left(\frac{\partial^2 w}{\partial x^2} \right)^2 + 2D_{12} \frac{\partial^2 w}{\partial x^2} \frac{\partial^2 w}{\partial y^2} + 4D_{66} \left(\frac{\partial^2 w}{\partial x \partial y} \right)^2 + D_{22} \left(\frac{\partial^2 w}{\partial y^2} \right)^2 \right\} dxdy \quad (6.21)$$

Using Equation (6.20) to substitute for w in Equation (6.21) and carrying out the integrations give:

$$U = \frac{aL}{2} \left\{ \begin{array}{l} D_{11} \left[\frac{w_o^2 \pi^4}{4} \left(\frac{1}{a^2} + \frac{\tan^2 \alpha}{L^2} \right)^2 + \frac{w_o^2 \pi^4}{a^2 L^2} \tan^2 \alpha \right] + 2D_{12} \frac{w_o^2 \pi^4}{4L^2} \left(\frac{1}{a^2} + \frac{\tan^2 \alpha}{L^2} \right) \\[3mm] + D_{22} \frac{w_o^2 \pi^4}{4L^4} + 4D_{66} \left(\frac{w_o^2 \pi^4}{4a^2 L^2} + \frac{w_o^2 \pi^4}{4L^4} \tan^2 \alpha \right) \end{array} \right\}$$

which, after rearranging and simplifying, becomes:

$$U = \frac{w_o^2 \pi^4 L}{8a^3} \left[\begin{array}{l} D_{11} \left(1 + 6\tan^2 \alpha AR^2 + \tan^4 \alpha AR^4\right) \\ + 2\left(D_{12} + 2D_{66}\right)\left(AR^2 + AR^4 \tan^2 \alpha\right) + D_{22}AR^4 \end{array} \right] \tag{6.22}$$

where $AR = a/L$.

Now the work done by the applied load N_{xy} is given by Equation (5.68) with $N_x = N_y = 0$

$$W = -\frac{1}{2} \int\limits_0^a \int\limits_0^b \left\{ 2N_{xy} \left(\frac{\partial w}{\partial x}\right)\left(\frac{\partial w}{\partial y}\right) \right\} dxdy \tag{6.23}$$

Using Equation (6.20) to substitute for w and carrying out the integrations gives:

$$W = \frac{w_o^2 AR\pi^2}{4} \tan \alpha N_{xy} \tag{6.24}$$

Minimizing the total potential energy

$$\Pi = U - W \tag{6.25}$$

with respect to the unknown coefficient w_o implies,

$$\frac{\partial \Pi}{\partial w_o} = 0 \tag{6.26}$$

which, using Equations (6.22) and (6.24) results in:

$$\frac{2w_o\pi^4 L}{8a^3} \left[\begin{array}{l} D_{11} \left(1 + 6\tan^2 \alpha AR^2 + \tan^4 \alpha AR^4\right) \\ + 2\left(D_{12} + 2D_{66}\right)\left(AR^2 + AR^4 \tan^2 \alpha\right) + D_{22}AR^4 \end{array} \right]$$
$$- \frac{2w_o AR\pi^2 \tan \alpha}{4} N_{xy} = 0 \tag{6.27}$$

The obvious (trivial) solution to Equation (6.27) is $w_o = 0$ which corresponds to the in-plane pre-buckling situation (out-of-plane displacement w is zero). For $w_o \neq 0$, N_{xy} must attain a critical value which corresponds to the buckling load. Therefore, solving Equation (6.27) for $N_{xy} = N_{xycrit}$ gives the buckling load:

$$N_{xycrit} = \frac{\pi^2}{2AR^2 a^2 \tan \alpha} \left[\begin{array}{l} D_{11} \left(1 + 6\tan^2 \alpha AR^2 + \tan^4 \alpha AR^4\right) \\ + 2\left(D_{12} + 2D_{66}\right)\left(AR^2 + AR^4 \tan^2 \alpha\right) + D_{22}AR^4 \end{array} \right] \tag{6.28}$$

Equation (6.28) shows that the buckling load is a function of the angle α and the length L through the aspect ratio AR. Since the buckling load is the lowest load at which out-of-plane displacements w are permissible, the values of $\tan\alpha$ and AR must be determined for which the

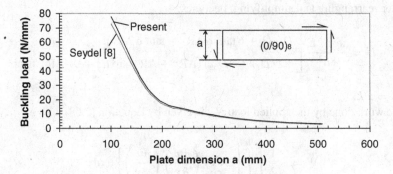

Figure 6.11 Buckling load of a long simply supported plate under shear: comparison of two approaches

right-hand side of Equation (6.28) is minimized. This is done by differentiating with respect to the two parameters $\tan\alpha$ and AR and setting the result equal to zero. Then,

$$\frac{\partial N_{xycrit}}{\partial(\text{AR})} = 0 \Rightarrow \text{AR} = \left[\frac{D_{11}}{D_{11}\tan^4\alpha + 2(D_{12}+2D_{66})\tan^2\alpha + D_{22}}\right]^{1/4} \quad (6.29)$$

and

$$\frac{\partial N_{xycrit}}{\partial(\tan\alpha)} = 0 \Rightarrow 3D_{11}\text{AR}^4\tan^4\alpha + \left(6D_{11}\text{AR}^2 + 2(D_{12}+2D_{66})\text{AR}^4\right)\tan^2\alpha$$
$$- \left(D_{11} + 2(D_{12}+2D_{66})\text{AR}^2 + D_{22}\text{AR}^4\right) = 0 \quad (6.30)$$

Equations (6.29) and (6.30) are solved simultaneously for AR and $\tan\alpha$. The results are substituted in Equation (6.28) to obtain the buckling load N_{xycrit}.

The accuracy of this approach is compared with a solution obtained by Seydel [8] where the governing differential equation (6.14) is solved as a product of an exponential function in y and an unknown function of x. For the comparison, a $(0/90)_8$ laminate with basic ply properties: $E_x = E_y = 68.9$ GPa, $\nu_{xy} = 0.05$, $G_{xy} = 4.83$ GPa and ply thickness $= 0.1905$ mm is selected. The result is shown in Figure 6.11 where the two methods are shown to be in excellent agreement (largest difference is less than 7%).

6.5 Buckling of Rectangular Composite Plates under Combined Loads

A composite plate under compression and shear is shown in Figure 6.12. Its edges are assumed to be simply supported.

The out-of-plane displacement w is assumed to be of the form

$$w = w_1 \sin\frac{\pi x}{a}\sin\frac{\pi y}{b} + w_2\sin\frac{2\pi x}{a}\sin\frac{2\pi y}{b} \quad (6.31)$$

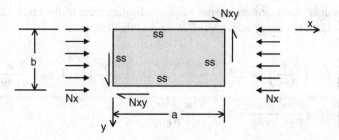

Figure 6.12   Simply supported plate under compression and shear

The two terms in the right-hand side of Equation (6.31) are two of the terms in the w expressions in previous sections (see Equation 6.3). Equation (6.31) satisfies the simply supported boundary conditions on w,

$$w(x = 0) = w(x = a) = 0$$
$$w(y = 0) = w(y = b) = 0$$

and the fact that the bending moments at the plate boundary are also zero

$$M_x = -D_{11}\frac{\partial^2 w}{\partial x^2} - D_{12}\frac{\partial^2 w}{\partial y^2} = 0$$

$$M_y = -D_{12}\frac{\partial^2 w}{\partial x^2} - D_{22}\frac{\partial^2 w}{\partial y^2} = 0$$

However, substituting in the last two of Equations (5.3) shows that Equation (6.31) results in nonzero transverse shear forces V $(= Q_z - \partial M_{xy}/\partial x$ and $Q_y - \partial M_{xy}/\partial y)$ at the plate boundary. This solution is, therefore, an approximation since there are no transverse shear forces applied on the plate boundaries. In an energy minimization approach, which is the method that will be used in this case, it is not necessary to satisfy the force boundary conditions when the problem is formulated in terms of displacements. This was discussed in Section 5.4. The more terms are used in the w expression (6.31) the higher the accuracy and the force boundary conditions will, in the limit, be satisfied in an average sense.

Minimization of the total potential energy of the plate will lead to two equations for the two unknowns w_1 and w_2 in the assumed expression for w. It is important to note that for shear loading cases, assuming a single term for w will not work (see Exercise 6.5). The assumed shape using one term is quite different from the plate deformations caused by the shear loading when the plate buckles. At least two terms are necessary to begin capturing the buckling mode.

For a displacement-based approach, Equations (5.63) and (5.68) can be used. Since the plate is symmetric (B matrix terms are equal to zero) the in-plane and out-of-plane contributions to the energy decouple. To determine the out-of-plane displacement w, therefore, the total energy to be minimized, strain energy minus work done, has the form:

$$\Pi_c = \frac{1}{2} \iint \left\{ \begin{array}{l} D_{11}\left(\dfrac{\partial^2 w}{\partial x^2}\right)^2 + 2D_{12}\dfrac{\partial^2 w}{\partial x^2}\dfrac{\partial^2 w}{\partial y^2} + D_{22}\left(\dfrac{\partial^2 w}{\partial y^2}\right)^2 + 4D_{66}\left(\dfrac{\partial^2 w}{\partial x \partial y}\right)^2 + \\[3mm] 4D_{16}\dfrac{\partial^2 w}{\partial x^2}\dfrac{\partial^2 w}{\partial x \partial y} + 4D_{26}\dfrac{\partial^2 w}{\partial y^2}\dfrac{\partial^2 w}{\partial x \partial y} \end{array} \right\} dxdy$$

$$+ \frac{1}{2} \iint N_x \left(\frac{\partial w}{\partial x}\right)^2 dxdy + \iint N_{xy}\frac{\partial w}{\partial x}\frac{\partial w}{\partial y}dxdy \tag{6.32}$$

It is further assumed that the bending–twisting coupling terms $D_{16} \approx D_{26} \approx 0$.

Equation (6.31) is substituted in the expression (6.32) for Π_c. As an example, the first term is shown below:

$$\left(\frac{\partial^2 w}{\partial x^2}\right)^2 = w_1^2 \frac{\pi^4}{4a^4}\left(1 - \cos\frac{2\pi x}{a}\right)\left(1 - \cos\frac{2\pi y}{b}\right) + w_2^2 \frac{16\pi^4}{4a^4}\left(1 - \cos\frac{4\pi x}{a}\right)\left(1 - \cos\frac{4\pi y}{b}\right)$$

$$+ 2w_1 w_2 \frac{4\pi^4}{a^4}\frac{1}{4}\left(\cos\frac{\pi x}{a} - \cos\frac{3\pi x}{a}\right)\left(\cos\frac{\pi y}{b} - \cos\frac{3\pi y}{b}\right)$$

with similar expressions for the remaining derivatives present in Equation (6.32).

Carrying out the integrations gives

$$\int_0^a \int_0^b \left(\frac{\partial^2 w}{\partial x^2}\right)^2 dxdy = w_1^2 \frac{\pi^4}{4a^4}ab + w_2^2 \frac{4\pi^4}{a^4}ab$$

$$\int_0^a \int_0^b \left(\frac{\partial^2 w}{\partial y^2}\right)^2 dxdy = w_1^2 \frac{\pi^4}{4b^4}ab + w_2^2 \frac{4\pi^4}{b^4}ab$$

$$\int_0^a \int_0^b \left(\frac{\partial^2 w}{\partial x^2}\frac{\partial^2 w}{\partial y^2}\right) dxdy = w_1^2 \frac{\pi^4}{4a^2 b^2}ab + w_2^2 \frac{4\pi^4}{a^2 b^2}ab$$

$$\int_0^a \int_0^b \left(\frac{\partial^2 w}{\partial x \partial y}\right)^2 dxdy = w_1^2 \frac{\pi^4}{4a^2 b^2}ab + w_2^2 \frac{4\pi^4}{a^2 b^2}ab$$

$$\int_0^a \int_0^b \left(\frac{\partial w}{\partial x}\right)^2 dxdy = w_1^2 \frac{\pi^2}{4a^2}ab + w_2^2 \frac{\pi^2}{a^2}ab$$

$$\int_0^a \int_0^b \left(\frac{\partial w}{\partial x}\frac{\partial w}{\partial y}\right) dxdy = \frac{w_1 w_2 \pi^2}{2ab}\left(\frac{2a}{3\pi} + \frac{2a}{\pi}\right)\left(\frac{2b}{3\pi} - \frac{2b}{\pi}\right)$$

$$+ \frac{w_1 w_2 \pi^2}{2ab}\left(\frac{2a}{3\pi} - \frac{2a}{\pi}\right)\left(\frac{2b}{3\pi} + \frac{2b}{\pi}\right)$$

So the final form for Π_c is

$$\Pi_c = \frac{1}{2}\left\{ \begin{array}{c} D_{11}\left[w_1^2\frac{\pi^4}{4a^3}b + w_2^2\frac{4\pi^4}{a^3}b\right] + 2(D_{12}+2D_{66})\left[w_1^2\frac{\pi^4}{4ab} + w_2^2\frac{4\pi^4}{ab}\right] + \\ D_{22}\left[w_1^2\frac{\pi^4}{4b^3}a + w_2^2\frac{4\pi^4}{b^3}a\right] \end{array}\right\}$$

$$-\frac{N_o}{2}\left[w_1^2\frac{\pi^2}{4a}b + w_2^2\frac{\pi^2}{a}b\right] - kN_o w_1 w_2\left(-\frac{32}{9}\right) \qquad (6.32a)$$

where, for simplicity,

$$\frac{N_{xy}}{N_x} = k$$

and $N_o = N_{x\text{crit}}$ the value of N_x which, simultaneously with $N_{xy} = kN_x$ causes buckling of the plate.

The energy expression (6.32a) is minimized with respect to the unknown coefficients w_1 and w_2. This leads to,

$$\frac{\partial \Pi_c}{\partial w_1} = 0$$

$$\frac{\partial \Pi_c}{\partial w_2} = 0$$

and substituting,

$$\frac{1}{2}\left\{D_{11}\frac{w_1\pi^4 b}{2a^3} + 2(D_{12}+2D_{66})\frac{\pi^4 w_1}{2ab} + D_{22}\frac{w_1\pi^4 a}{2b^3}\right\} - N_o\frac{w_1\pi^2 b}{4a} + \frac{32}{9}kN_o w_2 = 0$$

$$\frac{1}{2}\left\{D_{11}\frac{8w_2\pi^4 b}{a^3} + 2(D_{12}+2D_{66})\frac{8\pi^4 w_2}{ab} + D_{22}\frac{8w_2\pi^4 a}{b^3}\right\} - N_o\frac{w_2\pi^2 b}{a} + \frac{32}{9}kN_o w_1 = 0$$

$$(6.33)$$

Setting, for simplicity,

$$K_1 = \frac{1}{4}\left[D_{11}\frac{\pi^4 b}{a^3} + 2(D_{12}+2D_{66})\frac{\pi^4}{ab} + D_{22}\frac{\pi^4 a}{b^3}\right]$$

Equations (6.33) can be recast in the following generalized eigenvalue problem:

$$\begin{bmatrix} K_1 & 0 \\ 0 & 16K_1 \end{bmatrix}\begin{Bmatrix} w_1 \\ w_2 \end{Bmatrix} = N_o\begin{bmatrix} \dfrac{\pi^2 b}{4a} & -\dfrac{32}{9}k \\ -\dfrac{32}{9}k & \dfrac{\pi^2 b}{a} \end{bmatrix}\begin{Bmatrix} w_1 \\ w_2 \end{Bmatrix}$$

which, with terms appropriately defined, is of the form

$$\underline{A}\,\underline{x} = \alpha\underline{B}\,\underline{x}$$

The solution is obtained by premultiplying both sides of the equation by B^{-1}, the inverse of B, to obtain the standard eigenvalue problem,

$$B^{-1}Ax = \alpha I x$$

where I is the identity matrix.
With

$$B^{-1} = \frac{1}{\dfrac{\pi^4 b^2}{4a^2} - \left(\dfrac{32}{9}k\right)^2} \begin{bmatrix} \dfrac{\pi^2 b}{a} & \dfrac{32}{9}k \\ \dfrac{32}{9}k & \dfrac{\pi^2 b}{4a} \end{bmatrix}$$

the standard eigenvalue problem has the form:

$$\begin{bmatrix} \dfrac{\pi^2 b}{a} & 16\dfrac{32}{9}k \\ \dfrac{32}{9}k & \dfrac{\pi^2 b}{4a}16 \end{bmatrix} \begin{Bmatrix} w_1 \\ w_2 \end{Bmatrix} = N_o \underbrace{\left(\frac{\pi^4 b^2}{4a^2} - \left(\frac{32}{9}k\right)^2\right)}_{\alpha} \frac{1}{K_1} \begin{Bmatrix} w_1 \\ w_2 \end{Bmatrix}$$

where the quantity premultiplying the vector $\{w_1\ w_2\}^T$ on the right-hand side is the eigenvalue α.

By bringing the right-hand side to the left of the above equation, a system of homogeneous equations is obtained. For a nontrivial solution, the determinant of the resulting left-hand side must be set equal to zero. Then, the eigenvalues are obtained as solutions to

$$\det\left[B^{-1}A - \alpha I\right] = 0$$

which leads to the following equation for the eigenvalue α:

$$\left(\frac{\pi^2 b}{a} - \alpha\right)\left(\frac{4\pi^2 b}{a} - \alpha\right) - \frac{512(32)}{81}k^2 = 0$$

Solving for α and recovering N_o, leads to

$$N_o = \frac{\pi^2}{a^2} \frac{\left(D_{11} + 2(D_{12} + 2D_{66})\dfrac{a^2}{b^2} + D_{22}\dfrac{a^4}{b^4}\right)}{2 - \dfrac{8192}{81}\dfrac{a^2}{b^2\pi^4}k^2}\left[5 \pm \sqrt{9 + \frac{65536}{81}\frac{a^2}{\pi^4 b^2}k^2}\right] \quad (6.34)$$

Of the two solutions given by Equation (6.34), the one giving the lowest buckling load (in absolute value) is selected.

Before proceeding with the general case where both N_x and N_{xy} are nonzero, two special cases, those of pure compression and pure shear, are examined. This will give insight to how accurate or inaccurate this two-term solution is.

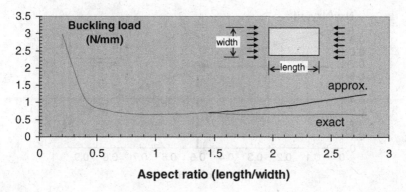

Figure 6.13   Approximate and exact buckling loads as a function of panel aspect ratio

For pure compression, $N_{xy} = 0$ and, therefore, $k = 0$. Substituting in Equation (6.34), the buckling load under compression is given by

$$N_o = \frac{\pi^2}{a^2}\left(D_{11} + 2(D_{12} + 2D_{66})\frac{a^2}{b^2} + D_{22}\frac{a^4}{b^4}\right) \tag{6.35}$$

Comparison of this expression with the general expression (6.7) for buckling under compression shows that the current expression coincides with the exact solution given by that equation when the number of half-waves m parallel to the loading direction equals 1. If the panel aspect ratio is large and/or the difference in bending stiffnesses D_{11} and D_{22} is large, the present approximate solution will depart from the exact solution. The approximate expression just derived and the exact solution are compared in Figure 6.13. In this comparison, the bending stiffness values were taken to be $D_{11} = D_{22} = 0.66$ N m, $D_{12} = 0.47$ N m, $D_{66} = 0.49$ N m and $D_{16} = D_{26} = 0$. As is seen from Figure 6.13, the two solutions are identical up to aspect ratios of approximately 1.5. For greater aspect ratios, the approximate solution gives higher buckling loads than the exact solution.

For pure shear, the ratio $k = N_{xy}/N_x$ is allowed to become large (implying N_x is negligible compared with N_{xy}). Then, Equation (6.34) simplifies to

$$N_o k = \pm\frac{\pi^2}{a^2}\frac{\left(D_{11} + 2(D_{12} + 2D_{66})\frac{a^2}{b^2} + D_{22}\frac{a^4}{b^4}\right)}{\frac{32}{9}\frac{a}{b\pi^2}} \tag{6.36}$$

By recognizing that $N_o k = N_{xy}$ and rearranging,

$$N_{xycrit} = \pm\frac{9\pi^4 b}{32a^3}\left(D_{11} + 2(D_{12} + 2D_{66})\frac{a^2}{b^2} + D_{22}\frac{a^4}{b^4}\right) \tag{6.37}$$

Equation (6.37) gives an approximate expression for the buckling load of a rectangular composite panel under shear. The $\pm$ sign indicates that buckling can be caused by either

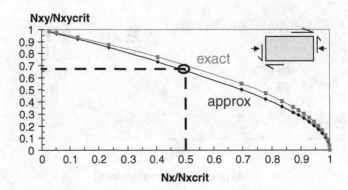

Figure 6.14 Interaction curve for buckling of composite rectangular plate under combined compression and shear

positive or negative shear loads. This expression is, typically, 27–30% higher than the exact solution one can obtain using the procedure in Section 6.3. The accuracy of Equation (6.37) can be improved if more terms are included in Equation (6.31) at considerable increase in algebraic complexity [9].

For the combined load case, Equation (6.34) will provide an approximation to the buckling load. However, for combined loading, the accuracy of this equation is higher than what was obtained for the compression and shear acting alone, as was seen in Equations (6.35) and (6.37). The reason is that, even though the individual buckling loads may be approximate, the interaction between the two loading types is accurately captured by Equation (6.34). A comparison of Equation (6.34) with the interaction curve [10] that has been found to be very accurate for this type of load combination,

$$\frac{N_x}{N_{xcrit}} + \left(\frac{N_{xy}}{N_{xycrit}}\right)^2 = 1 \tag{6.38}$$

is shown in Figure 6.14. The approximate and 'exact' solutions are very close to each other.

Interaction curves such as the one shown in Figure 6.14 can be very useful in design. They provide a means for determining: (a) if a panel fails under combined loads N_x and N_{xy}; or (b) the maximum allowable load in one direction (compression or shear) given the applied load in the other.

Load combinations inside the interaction curve imply that the panel does not buckle. Load combinations corresponding to points outside the interaction curve correspond to a panel that has buckled already. As an example, consider a case where the applied compressive load is half the buckling load of the panel when only compression is applied (N_x/N_{xcrit} = 0.5). This point gives the x coordinate in Figure 6.14. The corresponding y coordinate is (approximately) 0.67. This means that if the applied shear load is less than 67% of the shear buckling load when shear acts alone, the panel will not buckle under this load combination.

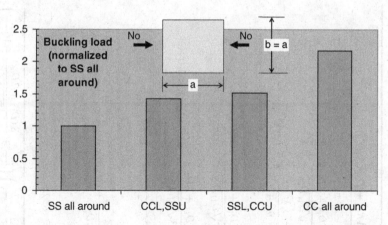

Figure 6.15 Effect of boundary conditions on buckling load of a square composite plate under compression

6.6 Design Equations for Different Boundary Conditions and Load Combinations

Approaches similar to those presented in the three previous sections can be used to obtain expressions for the buckling loads of rectangular composite panels with different boundary conditions and/or applied loads. A brief summary for the most common cases [4, 10–12] is given in Table 6.1. Note that, in all cases, in Table 6.1, the panel is assumed to have no bending–twisting coupling ($D_{16} = D_{26} = 0$).

As an example of using Table 6.1, examine the effect of various boundary conditions on a square composite plate under uniaxial compression. The side of the plate is a and the bending stiffnesses are $D_{11} = D_{22} = 660.5$ N mm, $D_{12} = 467.4$ N mm and $D_{66} = 494.5$ N mm. Normalizing the results to the case of a plate simply supported all around, the results shown in Figure 6.15 are obtained. The notation CCL implies the loaded sides are clamped. The notation CCU implies the unloaded sides are clamped. An analogous notation scheme is used for the simply supported boundary condition.

As is seen from Figure 6.15, the clamped-all-around plate has the highest buckling load. As expected, the simply supported all-around plate has the lowest buckling load and the clamped/simply supported combinations lie in between the two extremes. It should be noted that, unlike beams where the ratio of clamped to simply supported buckling load is 4, for plates, the corresponding ratio is significantly less (less than 2.5 for the case of Figure 6.15).

Exercises

6.1 Consider a composite plate with bending stiffnesses D_{11}, D_{12}, D_{22} and D_{66} ($D_{16} = D_{26} = 0$) and dimensions a, b as shown below (Figure E6.1). Use the derivation shown in Section 6.5 and assume the compression load $N_x = 0$. Verify the approximate expression for the buckling load under shear shown in that section by deriving the new 2×2 eigenvalue problem and solving for all the eigenvalues. What does the sign of the eigenvalue mean?

Table 6.1 Buckling loads for various boundary conditions and load combinations

$$N_o = \frac{\pi^2\left[D_{11}m^4 + 2(D_{12}+2D_{66})m^2(AR)^2 + D_{22}(AR)^4\right]}{a^2 m^2}$$

$$N_o = \frac{\pi^2}{b^2}\sqrt{D_{11}D_{22}}\,(K)$$
$$K = \frac{4}{\lambda^2} + \frac{2(D_{12}+2D_{66})}{\sqrt{D_{11}D_{22}}} + \frac{3}{4}\lambda^2 \qquad 0 < \lambda < 1.662$$
$$K = \frac{m^4 + 8m^2 + 1}{\lambda^2(m^2+1)} + \frac{2(D_{12}+D_{66})}{\sqrt{D_{11}D_{22}}} + \frac{\lambda^2}{m^2+1} \qquad \lambda > 1.662$$

$$N_o = \frac{\pi^2}{b^2}\sqrt{D_{11}D_{22}}\,(K)$$
$$K = \frac{m^2}{\lambda^2} + \frac{2(D_{12}+2D_{66})}{\sqrt{D_{11}D_{22}}} + \frac{16\lambda^2}{3\,m^2}$$

$$N_o = \frac{\pi^2}{b^2}\sqrt{D_{11}D_{22}}\,(K)$$
$$K = \frac{4^2}{\lambda^2} + \frac{8(D_{12}+2D_{66})}{3\sqrt{D_{11}D_{22}}} + 4\lambda^2 \qquad 0 < \lambda < 1.094$$
$$K = \frac{m^4 + 8m^2 + 1}{\lambda^2(m^2+1)} + \frac{2(D_{12}+D_{66})}{\sqrt{D_{11}D_{22}}} + \frac{\lambda^2}{m^2+1} \qquad \lambda > 1.094$$

$$N_o = \frac{\pi^2}{b^2}\sqrt{D_{11}D_{22}}\,(K)$$
$$K = 12\,\frac{D_{66}}{\pi^2\sqrt{D_{11}D_{22}}} + \frac{1}{\lambda^2}$$

$$M_o = 0.047\,\pi^2 b^2 \sqrt{D_{11}D_{22}\,(K)}$$
$$K = \sqrt{\left(\frac{m^2}{\lambda^2} + \frac{2(D_{12}+2D_{66})}{\sqrt{D_{11}D_{22}}} + \frac{\lambda^2}{m^2}\right)\left(\frac{m^2}{\lambda^2} + \frac{8(D_{12}+2D_{66})}{\sqrt{D_{11}D_{22}}} + 16\frac{\lambda^2}{m^2}\right)}$$

$$\lambda = \frac{a}{b}\left(\frac{D_{22}}{D_{11}}\right)^{1/4}$$

$$N_o = \frac{\pi^2 \left[D_{11}m^4 + 2(D_{12}+2D_{66})m^2n^2(AR)^2 + D_{22}n^4(AR)^4 \right]}{a^2 \left(m^2 + kn^2(AR)^2\right)}$$

$$N_{xycrit} = \frac{4}{b^2}\left(D_{11}D_{22}^3\right)^{1/4}(K) \qquad \beta = \left(\frac{D_{11}}{D_{22}}\right)^{1/4}$$

$$K = 8.2 + 5\frac{(D_{12}+2D_{66})}{\sqrt{D_{11}D_{22}}} - \frac{1}{10\dfrac{(D_{12}+2D_{66})}{\sqrt{D_{11}D_{22}}}}\left(\frac{A}{\beta}+B\beta\right)$$

$$A = -0.27 + 0.185\frac{(D_{12}+2D_{66})}{\sqrt{D_{11}D_{22}}}$$

$$B = 0.82 + 0.46\frac{(D_{12}+2D_{66})}{\sqrt{D_{11}D_{22}}} - 0.2\left(\frac{(D_{12}+2D_{66})}{\sqrt{D_{11}D_{22}}}\right)^2$$

$$R_b^{1.76} + R_c = 1$$
$$R_b = \frac{M}{M_{crit}}, \quad R_c = \frac{N_x}{N_{xcrit}}$$

$$R_b^2 + R_s^2 = 1$$
$$R_b = \frac{M}{M_{crit}}, \quad R_s = \frac{N_{xy}}{N_{xycrit}}$$

$$R_c + R_s^2 = 1$$
$$R_c = \frac{N_x}{N_{xcrit}}, \quad R_s = \frac{N_{xy}}{N_{xycrit}}$$

$$\lambda = \frac{a}{b}\left(\frac{D_{22}}{D_{11}}\right)^{1/4}$$

$$a=b; \quad 0<\beta\le1$$

(Hint: start either from the energy expression or from the two equations obtained after differentiation and set $N_x = 0$).

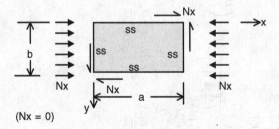

(Nx = 0)

Figure E6.1   Composite plate under shear

6.2 One of the spars of a wing is 101.6 cm deep. A bending moment $M = 11,290.3$ N m is acting on the spar web (Figure E6.2).

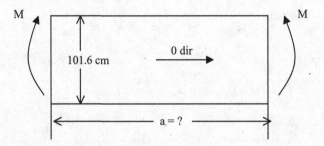

Figure E6.2   Spar web under in-plane bending moment M

The manufacturer of the spar has automated the process of laying up the following stacking sequence: [45/–45/0/90/0/–45/45] with the intent of simply stacking up multiples of this base laminate everywhere to keep the fabrication costs low. The basic material properties are:

$$E_{11} = 131 \text{ GPa}$$
$$E_{22} = 11.37 \text{ GPa}$$
$$G_{12} = 4.82 \text{ GPa}$$
$$\nu_{12} = 0.29$$
$$t_{ply} = 0.1524 \text{ mm}$$

To keep the number of basic laminates stacked together in the spar web low, the manufacturer/designer intends to use ribs to break up the spar. The rib spacing is a.

Create a graph that shows how the maximum allowable rib spacing a varies with the number n of basic laminates used for the spar not to buckle. What is the value of a when $n = 3$, that is when the web lay-up is: [45/–45/0/90/0/–45/45]$_3$?

6.3 Prove that for a simply supported square composite panel for which $D_{11} = D_{22}$, the number of half-waves m into which the panel buckles under compression is always 1. What should the condition be between D_{11} and D_{22} for the square panel to buckle in two half-waves? (Assume $D_{16} = D_{26} = 0$.)

6.4 A rectangular composite plate with simply supported sides all around is under compression and shear (Figure E6.3). A Gr/E unidirectional composite material is available with basic (single ply) properties:

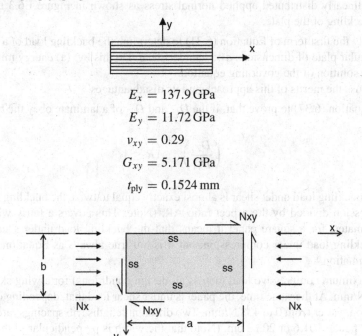

$$E_x = 137.9\,\text{GPa}$$
$$E_y = 11.72\,\text{GPa}$$
$$\nu_{xy} = 0.29$$
$$G_{xy} = 5.171\,\text{GPa}$$
$$t_{ply} = 0.1524\,\text{mm}$$

$a = 558.8\,\text{mm}; b = 304.8\,\text{mm}$

Figure E6.3 Composite plate under combined loading

The application is a wing skin (bending upwards as shown in Figure E6.4) with the following four loading conditions (note that here, $N_x > 0$ means compression):

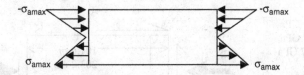

Figure E6.4 Composite plate with linearly distributed edge stress

Load case	1	2	3	4
N_x (N/mm)	4.99	31.54	23.84	34.34
N_{xy} (N/mm)	0.50	25.23	35.76	171.69

For ease of manufacture you want to only use lay-ups of the form $[45_n/-45_n/0_n/90_n]_s$.

(a)   If the D_{ij} terms of the D matrix for the basic lay-up $[45/-45/0/90]_s$ are known and denoted by D_{ijb}, determine the D_{ij} for any value of n as a function of D_{ijb}.

(b)   Use your result in (a) to determine the lowest value of n such that all load conditions are met without buckling of the plate. Do not use any knockdowns for environment, material scatter or damage.

(c)   For your final answer in (b), determine the maximum applied stress σ_{amax} for a linearly distributed applied normal stress as shown in Figure E6.3 that causes buckling of the plate.

6.5   Use only the first term of Equation (6.31) to determine the buckling load of a composite rectangular plate of dimensions $a \times b$ under shear. Do this by: (a) energy minimization and (b) solution of the governing equation.

Discuss the merits of this approach and its disadvantages.

6.6   Use Equation (6.37) to prove that, if the D_{11} and D_{22} of a laminate obey the relation:

$$\left(\frac{D_{22}}{D_{11}}\right)^{1/4} AR \leq 1$$

The buckling load under shear is almost exactly equal to twice the buckling load under compression divided by the aspect ratio AR. (Note: This covers a fairly wide variety of laminates; for a square panel it means that the buckling load under shear is twice the buckling load under compression; but it is not true always as Equation 6.37 is an approximation.)

6.7   The maximum compressive load (across all design conditions) for a wing skin panel is 131.4 N/mm. At the same time, the panel is under shear load that, depending on the load condition, varies from 0 to 438 N/mm. Two different lengths (rib spacings) are proposed for this panel: 101.6 or 203.2 cm. (Note that the width is perpendicular to the direction of the compression load). For each of the two lengths determine the maximum allowable width (i.e. stiffener spacing) for the panel not to buckle as a function of the applied shear load. Show your answer graphically as two curves of max width versus applied shear load (while the compression load is fixed). From other considerations, the skin lay-up has been fixed already to $[45/-45/0/90]_{2s}$ and the material used has the following basic ply properties:

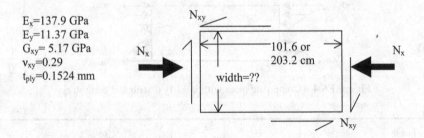

E_x=137.9 GPa
E_y=11.37 GPa
G_{xy}= 5.17 GPa
v_{xy}=0.29
t_{ply}=0.1524 mm

Assume the wing panel is simply supported all around.

6.8   Given the lay-up in Exercise 6.7, rearrange the stacking sequence so you can maximize the stiffener spacing for the same loading conditions. That is, keeping the same plies but reordering them, determine the symmetric lay-up that maximizes the plate width. This would be very helpful in minimizing the weight (fewer stiffeners for the same skin thickness) and the cost (fewer stiffeners mean less labour to make them and assemble them to the skin). Note: While it is possible to check all possible combinations, you should be able to argue what the best stacking sequence should be on the basis of physical arguments and equations given in this and previous chapters. For a plate length of 101.6 cm only, obtain the best stiffener spacing values and plot them on the same plot as in Exercise 6.7.

6.9   Using the two-term approach given in class, obtain an expression for the buckling load of a simply supported rectangular composite plate under tension and shear. Discuss what happens when the shear load is zero or low. Determine the critical value of k.

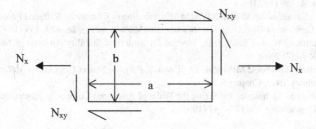

6.10 A rectangular composite panel of length a and width b is under compression. Two opposite sides are clamped and the other two are simply supported. The panel is made of fabric material with basic properties:

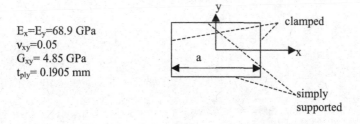

$E_x=E_y=68.9$ GPa
$v_{xy}=0.05$
$G_{xy}= 4.85$ GPa
$t_{ply}= 0.1905$ mm

One of the dimensions of the panel, a, is fixed at 50 cm. Determine the maximum value of the other dimension, b and the stacking sequence, such that:

(a)   The panel loaded in compression along a or along b has a buckling load which is less than or equal to the load corresponding to 4500 microstrain, that is buckling always happens first.

(b)   Given any loading direction, for a given applied load, the strain is the same as along the two laminate axes x,y.

(c)   The stacking sequence consists only of (±45) and (0/90) plies.

(d)   The weight is minimized.

References

[1] Whitney, J.M., *Structural Analysis of Laminated Anisotropic Plates*, Technomic Publishing, Lancaster, PA, 1987, Section 5.5.

[2] Tung, T.K. and Surdenas, J., Buckling of Rectangular Orthotropic Plates Under Biaxial Loading, *Journal of Composite Materials*, **21**, 124–128 (1987).

[3] Libove, C., Buckle Pattern of Biaxially Compressed Simply Supported Orthotropic Rectangular Plates, *Journal of Composite Materials*, **17**, 45–48 (1983).

[4] Barbero, E.J., *Introduction to Composite Material Design*, Taylor & Francis Inc, Philadelphia, PA, 1999, Chapter 9.

[5] Whitney, J.M., *Structural Analysis of Laminated Anisotropic Plates*, Technomic Publishing, Lancaster, PA, 1987, Section 5.7.

[6] Mura, T., and Koya, T., *Variational Methods in Mechanics*, Oxford University Press, New York, NY, 1992, Chapter 8.

[7] Thielemann, W., Contribution to the Problem of Buckling of Orthotropic Plates With Special Reference to Plywood, *NACA TM* 1263 (1950).

[8] Seydel, E., On the Buckling of Rectangular Isotropic or Orthogonal Isotropic Plates by Tangential Stresses, *Ingenieur Archiv*, **4**, 169 (1933).

[9] Kassapoglou, C., Simultaneous Cost and Weight Minimization of Composite Stiffened Panels Under Compression and Shear, *Composites Part A: Applied Science and Manufacturing*, **28**, 419–435 (1997).

[10] Sullins, R.T., Smith, G.W. and Spier, E.E., Manual for Structural Stability Analysis of Sandwich Plates and Shells, *NASA CR-1457*, 1969.

[11] Composite Materials Handbook, Mil-Hdbk-17-3F, vol. 3, Polymer Matrix Composites Materials Usage, Design and Analysis, January 1997, Chapter 5.7.

[12] Roberts, J.C., Analytic Techniques for Sizing the Walls of Advanced Composite Electronics Enclosure, *Composites Part B: Engineering*, **30**, 177–187 (1999).

7

Post-Buckling

Post-buckling is the load regime a structure enters after buckling (Figure 7.1). In several situations, dictated by robust design practices, externally imposed requirements, or even the degree of comfort a designer is willing to accept, especially for single load path critical structures, buckling may be taken to coincide with final failure. However, in general, there may be considerable load capacity beyond buckling before final failure occurs. This is true in particular for plates, which, in contrast to beams (Figure 7.2), may have significant load-carrying ability beyond buckling. This ability is often capitalized on to generate designs of lighter weight.

As is seen from Figure 7.2, after buckling ($P/P_{cr} >1$) the centre deflections of a beam increase rapidly compared with those of a plate. This means that, in a beam, high bending moments develop early in the post-buckling regime and will lead to failure. In a plate, on the other hand, the deflections increase more slowly with increasing load and the panel can withstand significant excursion in the post-buckling regime before the resulting bending moments become critical.

This ability of plates to withstand load in the post-buckling regime without failing makes such configurations very attractive for design. Thinner skins can be used in wings and fuselages, resulting in lighter structure. However, designing in the post-buckling regime requires knowledge and accurate quantification of failure modes that are not present below the buckling load. One such failure mode is the skin–stiffener separation in stiffened panels shown schematically in Figure 7.3.

The buckled pattern consists of a number of half-waves as shown in Figure 7.3. Depending on which way the skin deforms locally, there will be locations such as the one shown in Figure 7.3 where the skin tends to peel away from the stiffeners. Out-of-plane normal and shear stresses develop, which may exceed the material strength and lead to separation and final failure.

Even if the out-of-plane stresses that develop during post-buckling do not lead to skin–stiffener separation under static loading, they may lead to the creation of delaminations under repeated loading. Designing post-buckled panels that perform well under fatigue loading requires accurate knowledge of internal loads and the use of geometries and layups that delay the creation and growth of delaminations.

Design and Analysis of Composite Structures: With Applications to Aerospace Structures, Second Edition. Christos Kassapoglou.
© 2013 John Wiley & Sons, Ltd. Published 2013 by John Wiley & Sons, Ltd.

Figure 7.1 Post-buckled curved composite stiffened panel (see Plate 15 for the colour figure)

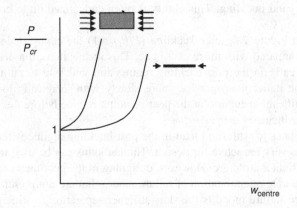

Figure 7.2 In-plane load versus centre deflection for plates and beams

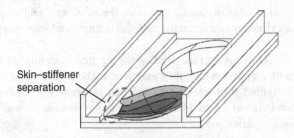

Figure 7.3 Post-buckled skin between stiffeners in a stiffened panel

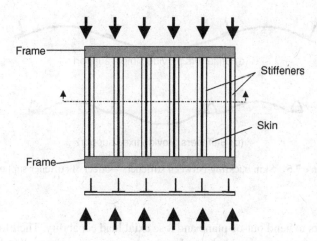

Figure 7.4   Stiffened panel under compressive load

The tendency for the stiffeners to separate from the skin or, more generally, for delaminations to form is higher for higher values of the ratio of the applied load to the buckling load. This ratio is referred to as the post-buckling ratio (PB) and, for post-buckled structures, is greater than 1. The PB ratio to be used in a design must be carefully selected, especially for load situations that include shear. Note that the post-buckling ratio should not be confused with the post-buckling factor defined later in Section 7.2.

At the conservative end of the design spectrum, PB ratio is not allowed to exceed 1.5. This means that the structure would buckle at limit load and fail at ultimate load. This protects against fatigue loading as the fatigue loads are lower than the limit load and, therefore, the structure does not buckle repeatedly during service. At the other end of the design spectrum, PB values greater than 5 are very challenging because the loads in the post-buckling regime, both static and fatigue, are significant and it is hard to design an efficient structure that will not fail after a relatively low number of cycles.

In a typical structure, such as a stiffened panel, a number of components or structural details may buckle. Selecting the sequence in which the various components of a structure will buckle is crucial for creating a lightweight design. For example, for the stiffened panel of Figure 7.4 the following buckling modes can be identified: panel buckles as a whole, the stiffeners serve to mainly increase the bending stiffness of the panel; skin between stiffeners buckles and stiffeners remain straight, stiffeners carry significant axial loads; stiffeners buckle as columns; and stiffener flanges buckle locally (crippling).

The panel in Figure 7.4 is loaded under compression, but the buckling modes mentioned are valid, with minor changes for any load situation that may induce buckling. For a panel under compression, it is usually more efficient to carry most of the compressive load by the stiffeners. The stiffener cross-sectional area required to carry compressive load is a smaller fraction of the total weight than the skin cross-section required to carry significant amounts of compressive load. This means that the stiffeners must remain straight (no column buckling and no crippling; see Sections 8.3 and 8.5) and the panel should not buckle as a whole which would

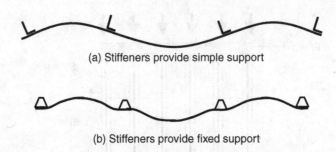

(a) Stiffeners provide simple support

(b) Stiffeners provide fixed support

Figure 7.5   Skin buckling between stiffeners—effect of stiffener support

force the stiffeners to bend out-of-plane and lose axial load capability. Therefore, usually, the buckling scenario for a post-buckled panel under compression requires that buckling of the skin between stiffeners happens first.

For a panel buckling under shear load, forcing the skin between stiffeners to buckle first is also desirable. For judiciously chosen stiffener spacing and with sufficient bending stiffness in the stiffeners, the buckling load of the skin between stiffeners can be increased or, more importantly, the amount of skin thickness required to have the skin between stiffeners buckle at a required PB ratio can be decreased. This decrease in skin thickness decreases the panel weight.

After the skin between stiffeners buckles, the load can be increased until the desired PB value is reached. At that point the next failure mode occurs which can be any of the buckling and failure modes mentioned above or material failure of any of the constituents. The preferred failure mode is skin material failure and/or crippling of the stiffeners. (Global) buckling of the panel as a whole or column buckling of the stiffeners is avoided because this would overload adjacent panels in the structure and might lead to catastrophic failure. Local skin or stiffener failures still leave some load-carrying ability in the panel and the load redistributed in adjacent panels is less. This results in a more damage-tolerant overall structure.

In addition to the failure modes and their sequence, the boundary conditions of the panel as a whole, and also of the skin between stiffeners, can be very important and, at least for the skin, are directly related to when skin buckling occurs. As was shown in Section 6.6, the boundary conditions can increase the buckling load by more than a factor of 2 (clamped versus simply supported conditions in Figure 6.15). This is directly related to the stiffener cross-section selected. A schematic of the two extreme behaviours is shown in Figure 7.5.

In both cases in Figure 7.5 it is assumed that the stiffeners have sufficient bending stiffness to stay straight and force the panel to buckle between them. This means that they act as panel breakers (see Section 9.2.1 for related discussion). In Figure 7.5a the torsional rigidity of the stiffeners is negligible (open cross-section stiffeners). As a result, they rotate with the skin locally and the corresponding boundary condition they impose is that of a simple support (zero deflection but nonzero rotation). In Figure 7.5b, the closed cross-section stiffeners have very high torsional rigidity and they locally force the skin to remain nearly horizontal. In such a case, the imposed boundary condition approaches that of a fixed support (zero deflection and slope).

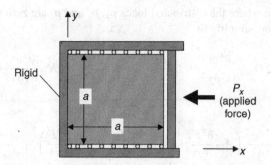

Figure 7.6 Square composite panel under compression

7.1 Post-Buckling Analysis of Composite Panels under Compression

The specific case of a square plate simply supported with three edges immovable and one loaded in compression (Figure 7.6) will be used as the example to bring out the most important characteristics of the behaviour.

The in-plane deflection u of the panel is zero at $x = 0$; also, the in-plane deflection v is zero at $y = 0$ and $y = a$ (immovable edges). This means that an in-plane transverse force P_y must develop at $y = 0$ and $y = a$ in order to keep the panel edges from moving. The out-of-plane deflection w of the panel is zero all around its boundary (simple support condition). The applied load P_x (units of force) at $x = a$ is a result of a uniformly applied deflection $-C$ at that location. The boundary conditions of the problem can then be written as

$$w = 0 \text{ at } x = y = 0 \text{ and } x = y = a$$
$$u = 0 \text{ at } x = 0$$
$$v = 0 \text{ at } y = 0 \text{ and } y = a \tag{7.1}$$
$$u = -C \text{ at } x = a$$

As the plate may undergo moderate to large deflections once it buckles, the governing equations are the two von Karman large-deflection equations (5.16) and (5.20) repeated here for convenience:

$$D_{11}\frac{\partial^4 w}{\partial x^4} + 2(D_{12} + 2D_{66})\frac{\partial^4 w}{\partial x^2 \partial y^2} + D_{22}\frac{\partial^4 w}{\partial y^4} = N_x\frac{\partial^2 w}{\partial x^2} + 2N_{xy}\frac{\partial^2 w}{\partial x \partial y} + N_y\frac{\partial^2 w}{\partial y^2}$$
$$-p_x\frac{\partial w}{\partial x} - p_y\frac{\partial w}{\partial y} + p_z \tag{5.16}$$

$$\frac{1}{A_{11}A_{22} - A_{12}^2}\left(A_{22}\frac{\partial^4 F}{\partial y^4} - 2A_{12}\frac{\partial^4 F}{\partial x^2 \partial y^2} + A_{11}\frac{\partial^4 F}{\partial x^4} + (A_{22} - A_{12})\frac{\partial^2 V}{\partial y^2}\right.$$
$$\left. + (A_{11} - A_{12})\frac{\partial^2 V}{\partial x^2}\right) + \frac{1}{A_{66}}\frac{\partial^4 F}{\partial x^2 \partial y^2} = \left(\frac{\partial^2 w}{\partial x \partial y}\right)^2 - \frac{\partial^2 w}{\partial x^2}\frac{\partial^2 w}{\partial y^2} \tag{5.20}$$

For the present case where the distributed loads p_x, p_y and p_z are zero (and the potential V is zero), these equations simplify to

$$D_{11}\frac{\partial^4 w}{\partial x^4} + 2(D_{12} + 2D_{66})\frac{\partial^4 w}{\partial x^2 \partial y^2} + D_{22}\frac{\partial^4 w}{\partial y^4} = N_x\frac{\partial^2 w}{\partial x^2} + 2N_{xy}\frac{\partial^2 w}{\partial x \partial y} + N_y\frac{\partial^2 w}{\partial y^2} \quad (5.16a)$$

and

$$\frac{1}{A_{11}A_{22} - A_{12}^2}\left(A_{22}\frac{\partial^4 F}{\partial y^4} - 2A_{12}\frac{\partial^4 F}{\partial x^2 \partial y^2} + A_{11}\frac{\partial^4 F}{\partial x^4}\right) - \frac{1}{A_{66}}\frac{\partial^4 F}{\partial x^2 \partial y^2}$$

$$= \left(\frac{\partial^2 w}{\partial x \partial y}\right)^2 - \frac{\partial^2 w}{\partial x^2}\frac{\partial^2 w}{\partial y^2} \quad (5.20a)$$

The solution to Equations (5.16a) and (5.20a) can be obtained using infinite series. Here, for simplicity, the series are truncated after the first few terms. The results will be of sufficient accuracy to show the basic trends.

The following expressions are assumed for w and F:

$$w = w_{11} \sin\frac{\pi x}{a}\sin\frac{\pi y}{a} \quad (7.2)$$

$$F = -\frac{P_x}{a}\frac{y^2}{2} - \frac{P_y}{a}\frac{x^2}{2} + K_{20}\cos\frac{2\pi x}{a} + K_{02}\cos\frac{2\pi y}{a} \quad (7.3)$$

with w_{11}, K_{20}, K_{02} and P_y unknowns.

It is readily seen that the expression for w satisfies the first of the boundary conditions (Equations 7.1). The expression for the Airy stress function F is constructed such that the average loads P_x, at any station x, and P_y, at any station y, are recovered. This can be seen by integrating the first two of Equations (5.17) with $V = 0$. The first is integrated with respect to y and the second with respect to x.

Using Equations (7.2) and (7.3) to substitute in Equation (5.20a) gives

$$\frac{A_{22}}{A_{11}A_{22} - A_{12}^2}K_{02}\frac{16\pi^4}{a^4}\cos\frac{2\pi y}{a} + \frac{A_{11}}{A_{11}A_{22} - A_{12}^2}K_{20}\frac{16\pi^4}{a^4}\cos\frac{2\pi x}{a}$$

$$= w_{11}^2\frac{\pi^4}{2a^4}\cos\frac{2\pi y}{a} + w_{11}^2\frac{\pi^4}{2a^4}\cos\frac{2\pi x}{a} \quad (7.4)$$

Matching coefficients of $\cos 2\pi x/a$ and $\cos 2\pi y/a$ gives

$$K_{02} = \frac{A_{11}A_{22} - A_{12}^2}{A_{22}}\frac{w_{11}^2}{32} \quad (7.5)$$

$$K_{20} = \frac{A_{11}A_{22} - A_{12}^2}{A_{11}}\frac{w_{11}^2}{32} \quad (7.6)$$

With these expressions for the coefficients K_{20} and K_{02}, the second von Karman equation (5.20a) is exactly satisfied.

Before proceeding to the first von Karman equation (5.16a), the transverse load P_y and the displacement $-C$ at $x = a$ corresponding to the applied load P_x are determined. The nonlinear strain displacement equation (5.13a) is rearranged:

$$\frac{\partial u}{\partial x} = \varepsilon_{xo} - \frac{1}{2}\left(\frac{\partial w}{\partial x}\right)^2$$

and the first of the inverted strain–stress equations (5.19) is used to substitute for the mid-plane strain ε_{xo}. This gives

$$\frac{\partial u}{\partial x} = \frac{A_{22}}{A_{11}A_{22} - A_{12}^2}N_x - \frac{A_{12}}{A_{11}A_{22} - A_{12}^2}N_y - \frac{1}{2}\left(\frac{\partial w}{\partial x}\right)^2$$

Now the first two of Equations (5.17) can be used to substitute for N_x and N_y in terms of F (with $V = 0$ as mentioned earlier):

$$\frac{\partial u}{\partial x} = \frac{A_{22}}{A_{11}A_{22} - A_{12}^2}\frac{\partial^2 F}{\partial y^2} - \frac{A_{12}}{A_{11}A_{22} - A_{12}^2}\frac{\partial^2 F}{\partial x^2} - \frac{1}{2}\left(\frac{\partial w}{\partial x}\right)^2 \tag{7.7}$$

Integrating over the entire plate,

$$\int_0^a\int_0^a \frac{\partial u}{\partial x}dxdy = \frac{A_{22}}{A_{11}A_{22} - A_{12}^2}\int_0^a\int_0^a \frac{\partial^2 F}{\partial y^2}dxdy - \frac{A_{12}}{A_{11}A_{22} - A_{12}^2}\int_0^a\int_0^a \frac{\partial^2 F}{\partial x^2}dxdy$$

$$-\frac{1}{2}\int_0^a\int_0^a \left(\frac{\partial w}{\partial x}\right)^2 dxdy$$

Equations (7.2) and (7.3) can be used to substitute for F and w. This leads to

$$a\left(u(a, y) - u(0, y)\right) = \frac{A_{22}a^2}{A_{11}A_{22} - A_{12}^2}\left(-\frac{P_x}{a}\right) - \frac{A_{12}a^2}{A_{11}A_{22} - A_{12}^2}\left(-\frac{P_y}{a}\right)$$
$$-\frac{1}{2}\iint \left(\left(w_{11}\frac{\pi}{a}\cos\frac{\pi x}{a}\sin\frac{\pi y}{a}\right)^2 dxdy \right. \tag{7.8}$$

But $u(a,y) = -C$ and $u(0,y) = 0$ from Equation (7.1). Substituting, performing the integration on the right-hand side, and rearranging,

$$C = \frac{aA_{22}}{A_{11}A_{22} - A_{12}^2}\frac{P_x}{a} - \frac{aA_{12}}{A_{11}A_{22} - A_{12}^2}\frac{P_y}{a} + w_{11}^2\frac{\pi^2}{8a} \tag{7.9}$$

In an exactly analogous fashion, but starting this time from Equation (5.13b),

$$\varepsilon_{yo} = \frac{\partial v}{\partial y} + \frac{1}{2}\left(\frac{\partial w}{\partial y}\right)^2$$

and using the third of Equations (7.1), the transverse load P_y is obtained as

$$P_y = P_x \frac{A_{12}}{A_{11}} - w_{11}^2 \frac{\pi^2}{8a} \frac{A_{11}A_{22} - A_{12}^2}{A_{11}} \tag{7.10}$$

It is interesting to note that for in-plane problems, where $w_{11} = 0$, Equation (7.10) gives

$$P_y = P_x \frac{A_{12}}{A_{11}}$$

Also, for isotropic plates, it can be shown that A_{12}/A_{11} equals the Poisson's ratio v and thus,

$$P_y = v P_x$$

as expected.

At this point, P_y is known from Equation (7.10) and one can substitute it in the first von Karman equation (5.16a). To do this, the following intermediate results are used:

$$\frac{\partial^4 w}{\partial x^4} = \frac{\partial^4 w}{\partial x^2 \partial y^2} = \frac{\partial^4 w}{\partial y^4} = w_{11} \left(\frac{\pi}{a}\right)^4 \sin\frac{\pi x}{a} \sin\frac{\pi y}{a}$$

$$\frac{\partial^2 w}{\partial x^2} = \frac{\partial^2 w}{\partial y^2} = -w_{11} \left(\frac{\pi}{a}\right)^2 \sin\frac{\pi x}{a} \sin\frac{\pi y}{a}$$

$$\frac{\partial^2 F}{\partial x^2} = -\frac{P_y}{a} - \left(\frac{2\pi}{a}\right)^2 \frac{A_{11}A_{22} - A_{12}^2}{A_{11}} \frac{w_{11}^2}{32} \cos\frac{2\pi x}{a}$$

$$\frac{\partial^2 F}{\partial y^2} = -\frac{P_x}{a} - \left(\frac{2\pi}{a}\right)^2 \frac{A_{11}A_{22} - A_{12}^2}{A_{22}} \frac{w_{11}^2}{32} \cos\frac{2\pi y}{a}$$

In addition, the following trigonometric identities are used:

$$\sin\frac{\pi x}{a} \cos\frac{2\pi x}{a} = \frac{1}{2}\left(\sin\frac{3\pi x}{a} - \sin\frac{\pi x}{a}\right)$$

$$\sin\frac{\pi y}{a} \cos\frac{2\pi y}{a} = \frac{1}{2}\left(\sin\frac{3\pi y}{a} - \sin\frac{\pi y}{a}\right)$$

Upon substituting in Equation (5.16a) there will be terms multiplying $\sin(\pi x/a)\sin(\pi y/a)$ and terms multiplying $\sin(3\pi x/a)\sin(\pi y/a)$ or $\sin(\pi x/a)\sin(3\pi y/a)$. The terms involving $3\pi x/a$ or $3\pi y/a$ are higher-order terms that would lead to additional equations if additional terms in the w expression (7.2) had been included. For the current expression for w with only one term, only coefficients of $\sin(\pi x/a)\sin(\pi y/a)$ are matched giving the following equation:

$$\frac{\pi^2}{a} \frac{(A_{11}A_{22} - A_{12}^2)A_{11} + 3A_{22}}{16A_{11}A_{22}} w_{11}^3$$

$$+ \left(\frac{\pi^2}{a}(D_{11} + 2(D_{12} + 2D_{66}) + D_{22}) - P_x\left(1 + \frac{A_{12}}{A_{11}}\right)\right) w_{11} = 0$$

which can be solved for w_{11} to give

$$w_{11} = \sqrt{\frac{16A_{11}A_{22}(D_{11} + 2(D_{12} + 2D_{66}) + D_{22})}{(A_{11}A_{22} - A_{12}^2)(A_{11} + 3A_{22})} \left[\frac{P_x}{\dfrac{\pi^2}{a} \dfrac{(D_{11} + 2(D_{12} + 2D_{66}) + D_{22})}{\left(1 + \dfrac{A_{12}}{A_{11}}\right)}} - 1 \right]}$$

(7.11)

With w_{11} known from Equation (7.11), P_y can be obtained from Equation (7.10) and K_{02} and K_{20} can be obtained from Equations (7.5) and (7.6). This completely determines the displacement w and the Airy stress function F from Equations (7.2) and (7.3), respectively.

Equation (7.11) has certain important implications. The denominator in the quantity in brackets under the square root is the buckling load (units of force) for a square plate with simply supported and immovable edges. This expression is exact for a square plate. By denoting this buckling load by P_{cr},

$$P_{cr} = \frac{\pi^2}{a} \frac{(D_{11} + 2(D_{12} + 2D_{66}) + D_{22})}{\left(1 + \dfrac{A_{12}}{A_{11}}\right)}$$

(7.12)

Equation (7.11) can be rewritten as

$$w_{11} = \sqrt{\frac{16A_{11}A_{22}(D_{11} + 2(D_{12} + 2D_{66}) + D_{22})}{(A_{11}A_{22} - A_{12}^2)(A_{11} + 3A_{22})} \left[\frac{P_x}{P_{cr}} - 1 \right]}$$

(7.11a)

It can be seen from Equation (7.11a) that the quantity under the square root is negative if the applied load P_x is less than the buckling load P_{cr}. In such a case, w_{11} does not exist. Out-of-plane deflections corresponding to a positive value of w_{11} are possible only after the plate has buckled and the applied load P_x is greater than the buckling load P_{cr}.

Additional implications are better understood through an example. Consider a square plate with layup $(\pm45)/(0/90)/(\pm45)$ made of plain weave fabric plies. Note that for such a symmetric layup D_{16} and D_{26} are always zero. The basic material properties are given below:

$$E_x = E_y = 68.94\,\text{GPa}$$

$$\nu_{xy} = 0.05$$

$$G_{xy} = 5.17\,\text{GPa}$$

$$t_{\text{ply}} = 0.19\,\text{mm}$$

With these properties the pertinent quantities in Equation (7.11) can be calculated:

D_{11}	659.7	N mm	A_{11}	28,912.44	N/mm
D_{12}	466.9	N mm	A_{12}	12,491.43	N/mm
D_{22}	659.7	N mm	A_{22}	28,912.44	N/mm
D_{66}	494.0	N mm	A_{66}	13,468.58	N/mm

and a plot of applied (normalized) load versus (normalized) centre deflection is given in Figure 7.7. The plate thickness is denoted by h.

As already discussed, the centre deflection w_{11} is zero for applied loads P_x lower than the buckling load P_{cr}. Once the applied load P_x exceeds the buckling load P_{cr}, the plate deflects out-of-plane and $w_{11} > 0$. As already suggested by the qualitative discussion of Figure 7.2, the load versus deflection curve is nonlinear and, for a plate, starts relatively flat and increases rapidly only after the centre deflection becomes significantly larger than the plate thickness ($w_{11}/h > 1$).

The distribution of the in-plane load N_x is also very interesting. N_x can be obtained from the first of Equations (5.17) with $V = 0$ after substituting into Equation (7.3) with P_y, K_{02} and K_{20} given by Equations (7.10), (7.5) and (7.6) respectively. A plot of N_x as a function of the transverse coordinate y is shown in Figure 7.8. N_x is normalized by the average N_x value which equals P_x/a, and the y coordinate is normalized by the plate dimension a. The peak value of N_x occurs at the panel edge, suggesting that failure of a post-buckled plate under compression will initiate there.

The variation of N_x is shown for different load ratios P_x/P_{cr}, starting with $P_x = P_{cr}$, which is the case when the plate just buckles under compression. In that case, the in-plane force N_x is constant across the plate, as indicated by the vertical line in Figure 7.8. As the applied load increases beyond the buckling load ($P_x/P_{cr} > 1$) the N_x distribution is no longer uniform. More load concentrates at the edges of the panel while the load at the centre is much lower. Already at $P_x/P_{cr} = 2$ the load at the panel edge is approximately twice the value at the centre as can be seen from Figure 7.8. At $P_x/P_{cr} = 5$, the load at the edge is, approximately, four times the load at the panel centre.

The reason for this nonuniform distribution is that once the plate buckles, its centre is softer than the edges where the supports are. So the load is diverted from the centre to the edge of

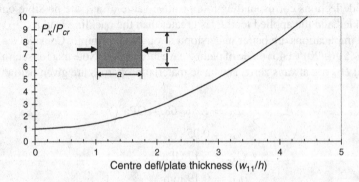

Figure 7.7   Load versus deflection for a square plate with layup (±45)/(0/90)/(±45)

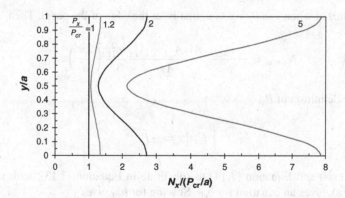

Figure 7.8   In-plane axial load N_x as a function of location and post-buckling ratio

the panel. This difference between the load at the centre and the edges of the panel becomes more and more pronounced as the load ratio P_x/P_{cr} increases.

This load redistribution can be used in design to generate simpler (conservative) design equations. The approach is based on approximating the actual N_x distribution by a step function that is zero at the panel centre and generates the same total applied force. This is shown schematically in Figure 7.9. At each of the loaded edges of the panel, the load N_x is localized at the two edges, is constant, and is acting over an effective width b_{eff}. The magnitude of N_x equals the maximum magnitude of N_x shown in Figure 7.8 for the respective P_x/P_{cr}.

The total force applied by this step-wise distribution must equal the applied load P_x. In terms of the force per unit width N_x this requirement can be expressed as

$$\int N_x \mathrm{d}y = 2(N_{x\,\mathrm{max}})b_{eff} \tag{7.13}$$

Now from Equation (5.17),

$$N_x = \frac{\partial^2 F}{\partial y^2} = -\frac{P_x}{a} - \frac{A_{11}A_{22} - A_{12}^2}{A_{22}} \frac{w_{11}^2}{32} \left(\frac{2\pi}{a}\right)^2 \cos\frac{2\pi y}{a}$$

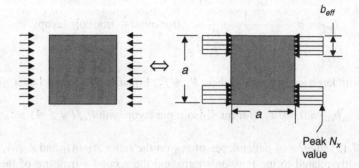

Figure 7.9. Equivalent in-plane compression in the post-buckling regime

which is maximized when $y = 0$ or $y = a$, that is, at the edge of the panel. Then,

$$N_{x\,\text{max}} = -\frac{P_x}{a} - \frac{A_{11}A_{22} - A_{12}^2}{A_{22}} \frac{w_{11}^2}{32} \left(\frac{2\pi}{a}\right)^2 \tag{7.14}$$

Also, by the definition of P_x,

$$\int N_x \mathrm{d}y = -P_x$$

Using this result and Equation (7.14) to substitute in Equation (7.13), with w_{11} given by Equation (7.11a), gives an equation for b_{eff}. Solving for b_{eff} gives

$$b_{\text{eff}} = a \frac{1}{2\left(1 + 2\left(1 + \dfrac{A_{12}}{A_{11}}\right)\left(1 - \dfrac{P_{\text{cr}}}{P_x}\right)\dfrac{A_{11}}{A_{11} + 3A_{22}}\right)} \tag{7.15}$$

This b_{eff} can be viewed as the effective portion of the skin over which applying the maximum N_x value given by Equation (7.14) gives a loading that is equivalent to the applied load P_x, but also conservative. It is conservative because a larger portion of the plate is exposed to the maximum value $N_{x\text{max}}$ than the exact N_x distribution suggests. As a result, designing a compressive panel in the post-buckling regime is equivalent to checking if the stress $N_{x\text{max}}/h$ (where h is the plate thickness) exceeds the allowable compression stress for the layup used and, if so, reinforcing the panel edges over a distance given by Equation (7.15) so that there is no failure.

It should be noted that for a quasi-isotropic layup,

$$\frac{A_{12}}{A_{11}} = \nu_{12}$$

$$\frac{A_{11}}{A_{11} + 3A_{22}} = \frac{1}{4}$$

$$\nu_{12} \approx 0.3$$

and substituting in Equation (7.15) gives

$$b_{\text{eff}} = a \frac{1}{2 + 1.3\left(1 - \dfrac{P_{\text{cr}}}{P_x}\right)} \quad \text{(for quasi} - \text{isotropic layup)} \tag{7.15a}$$

For cases with large loading ratios where $P_x > P_{\text{cr}}$ Equation (7.15a) becomes

$$b_{\text{eff}} = 0.303a \quad \text{(for quasi-isotropic layup with } P_x/P_{\text{cr}} \gg 1)$$

Equation (7.15) suggests a dependence of b_{eff} on the ratios A_{12}/A_{11} and A_{22}/A_{11}. The first ratio is indirectly related to the Poisson's ratio and the second a measure of the degree of

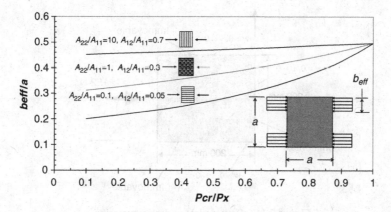

Figure 7.10   Variation of b_{eff} as a function of loading fraction and material properties

orthotropy. While these ratios are independent, for typical composite materials they lie within a range (e.g. a 0° ply of unidirectional material has a high degree of orthotropy, and A_{22}/A_{11} can be as low as 0.1, but the corresponding Poisson's ratio A_{12}/A_{11} is typically between 0.25 and 0.35). Based on typical composite material values, the three curves shown in Figure 7.10 can be constructed. The upper and lower curves correspond to extreme cases of high degree of orthotropy and the middle curve corresponds to a quasi-isotropic laminate. The two extreme curves give an idea of the range of variation of b_{eff} for typical composite materials. Note that as expected, all curves go through $b_{\text{eff}}/a = 0.5$ when $P_{\text{cr}}/P_x = 1$. This means that at buckling the entire skin is effective so the strip on each edge equals half the plate thickness.

It should be emphasized that the preceding discussion and derivation were based on single- or two-term expansions of the deflection w and Airy stress function F. The resulting post-buckled shape has a single half-wave across the entire plate. As such, while the basic conclusions of the present analysis are valid, the absolute numbers may not be sufficiently accurate for detailed analysis (but they are for preliminary design). This is particularly true for plates with aspect ratios different from 1, where the post-buckled shape involves more than the one half-wave assumed here. In such a case, more terms should be included in the analysis (see also Exercise 7.3).

Once w and F are known the internal forces (N_x, N_y and N_{xy}) and moments (M_x, M_y and M_{xy}) can be determined. Based on these, ply strains and stresses can be calculated and a failure criterion invoked. This can be a first-ply failure criterion (see Chapter 4) or a semi-empirically derived criterion based on test results.

7.1.1   Application: Post-Buckled Panel under Compression

Consider a square plate simply supported all around ($w = 0$) with three edges immovable (no displacement perpendicular to them in the plane of the plate) and one edge loaded by a force of 2152 N, as shown in Figure 7.11.

Two candidate layups are proposed using plain weave fabric material: layup A with stacking sequence $(\pm 45)/(0/90)_3/(\pm 45)$ and layup B with stacking sequence $(0/90)/(\pm 45)/(0/90)/(\pm 45)/(0/90)$. Note that the two layups have exactly the same thickness and plies used. Only the ordering of the plies is different.

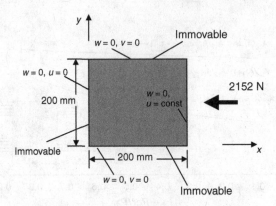

Figure 7.11   Square plate under compression

The basic material (ply) properties are given by

Property	Value
E_x	69 GPa
E_y	69 GPa
v_{xy}	0.05
G_{xy}	5.1 GPa
t_{ply}	0.19 mm

It is required to determine the location and magnitude of the highest N_x value and which of the two proposed layups is better for this application.

The boundary conditions and loading are the same as the post-buckling under compression situation analysed earlier in this section and the solution just derived applies. From classical laminated-plate theory the following properties are obtained for each of the two layups:

	layup A	layup B	
a	200	200	mm
b	200	200	mm
A_{11}	55,265	55,265	N/mm
A_{12}	13,821.5	13,821.5	N/mm
A_{22}	55,265	55,265	N/mm
A_{66}	15,452.5	15,452.5	N/mm
D_{11}	3412.967	4560.031	N mm
D_{12}	1809.787	662.723	N mm
D_{22}	3412.967	4560.031	N mm
D_{66}	1932.848	785.784	N mm
t	0.9525	0.9525	mm
E10	5.44E + 10	5.44E + 10	N/m^2
E20	5.44E + 10	5.44E + 10	N/m^2
E60	1.62E + 10	1.62E + 10	N/m^2

Applying Equation (7.12), the buckling load for layup A is found to be 718 N and for layup B 536 N, which is 25% smaller than layup A. This is due to the rearranging of the stacking sequence. It is interesting to note that placing (±45) plies on the outside as in layup A increases the buckling load. This can be seen from Equation (7.12) where the coefficient of D_{66}, which is 4, is higher than the coefficients of the remaining D_{ij} terms in the buckling load expression. Thus, increasing D_{66} increases the buckling load more than does the same percentage increase in other D_{ij} terms. And placing ±45° plies on the outside of a layup maximizes D_{66}.

Since the applied load is 2152 N, both layups have buckled and the post-buckling ratio for layup A is 2152/718 = 3.0 and for layup B is 2152/536 = 4.0. Using Equation (7.11a) the corresponding maximum centre deflections w_{11} for the two layups are found to be

$$w_{11A} = 1.67 \, \text{mm}$$

$$w_{11B} = 1.78 \, \text{mm}$$

These deflections are about twice the plate thickness and justify the use of large-deflection theory. Even though layup B has 25% lower buckling load, its centre deflection is only 6.5% higher than layup B, showing that increased bending stiffness has less of an effect in the post-buckling regime.

Using Equations (7.5) and (7.6), the constants K_{02} and K_{20} are found to be

	layup A	layup B
$K_{02} = K_{20}=$	4538.78	5112.312

The in-plane force N_x is now determined from Equations (5.17) and (7.3) as

$$N_x = -\left[\frac{P_x}{a} + \frac{4\pi^2}{a^2} K_{02} \cos \frac{2\pi y}{a} \right]$$

where P_x is the applied load, 2152 N, and a is the side of the plate, 200 mm.

The expression for N_x is independent of x, so it is the same for any x location along the plate. After substituting values, the plot of the magnitude of N_x as a function of y can be obtained and it is shown for both layups in Figure 7.12.

As can be seen from Figure 7.12 and as expected from the expression for N_x above and the earlier discussion, N_x reaches its maximum compressive values at the edges of the panel. It is also evident from Figure 7.12 that the maximum N_x values for the two layups differ only by 3.7%. Thus, a significant difference (25%) in the buckling load leads to a negligible difference in the maximum in-plane force in the plate. This suggests that the failure loads in the post-buckling regime for the two layups will be close to each other. Thus, significant differences in buckling performance do not translate to analogous differences in post-buckling performance.

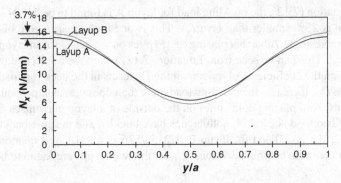

Figure 7.12   Axial force as a function of location

7.2   Post-Buckling Analysis of Composite Plates under Shear

A post-buckled stiffened composite plate under shear is shown in Figure 7.13. The buckling pattern consists of half-waves confined between the stiffeners. These half-waves make an angle α with the stiffener axis.

The situation of Figure 7.13 is idealized in Figure 7.14.

Assuming that the bending loads are taken by the two frames, the skin between stiffeners is under pure shear. The constant (applied) shear stress in each skin bay is then given by

$$\tau_a = \frac{V}{ht} \tag{7.16}$$

When the applied shear load V is low, the skin does not buckle and the shear stress τ_a can be resolved into a biaxial state of stress consisting of tension stress σ_t along a 45° line (see last bay in Figure 7.14) and a compression stress σ_c. It can be shown that the magnitudes of

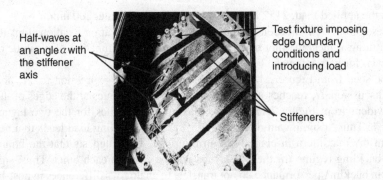

Figure 7.13   Stiffened composite panel in the post-buckling regime (see Plate 16 for the colour figure)

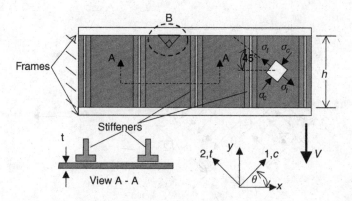

Figure 7.14   Stiffened skin under shear

σ_t and σ_c are equal. This can be derived from the standard stress transformation equations (3.35):

$$\begin{Bmatrix} -\sigma_c \\ \sigma_t \\ \tau_{12} \end{Bmatrix} = \begin{bmatrix} \cos^2\theta & \sin^2\theta & 2\sin\theta\cos\theta \\ \sin^2\theta & \cos^2\theta & -2\sin\theta\cos\theta \\ -\sin\theta\cos\theta & \sin\theta\cos\theta & \cos^2\theta - \sin^2\theta \end{bmatrix} \begin{Bmatrix} \sigma_x \\ \sigma_y \\ \tau_{xy} \end{Bmatrix} \quad (7.17)$$

In Equation (7.17), the original coordinate system x,y (see Figure 7.14) is rotated through the angle θ to the new 1,2 (or c,t) coordinate system. A minus sign appears in front of σ_c on the left-hand side to stay consistent with the orientation of σ_c in Figure 7.14 (the sign convention requires tensile normal stresses to be positive; σ_c is compressive). Given the sign convention in the x,y coordinate system, $\tau_{xy} = -\tau_a$.

Also, in the same coordinate system, $\sigma_x = \sigma_y = 0$. And for $\theta = 45°$, Equation (7.17) simplifies to

$$\sigma_c = \tau_a$$
$$\sigma_t = \tau_a \quad (7.18)$$
$$\tau_{12} = 0$$

The fact that the shear stress τ_{12} is zero in the 1,2 coordinate system implies that the 1–2 axes are principal axes. This is expected from the fact that the skin is under pure shear, which translates to pure biaxial loading (tension and compression) in a coordinate system rotated by 45° with respect to the original.

Equation (7.18) describes the situation until the skin buckles. Once the skin buckles, it is assumed that the compression direction of the skin cannot support any higher stress. So as the applied load is increased beyond the load that causes skin buckling, the compressive stress σ_c

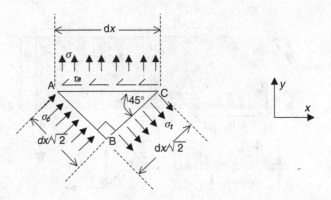

Figure 7.15   Free-body diagram of triangular skin element after buckling

stays constant and equal to its value at buckling. Letting τ_{cr} be the value of τ_a when the skin buckles, the compressive skin stress after the skin buckles is given by

$$\sigma_c = \tau_{cr} \text{ for } \tau_a > \tau_{cr} \text{ or } V > V_{cr} \tag{7.19}$$

where V_{cr} is the applied shear load at which the skin buckles.

The stresses in the skin after it buckles can be determined by considering the equilibrium of the triangular piece of skin with base length dx shown in detail B in Figure 7.14. The free-body diagram of that detail is shown in Figure 7.15.

With reference to Figure 7.15, if the length of segment AC is dx, then, by Pythagoras' theorem, the two segments AB and BC are

$$AB = BC = \frac{dx}{\sqrt{2}} \tag{7.20}$$

As already mentioned, sides AB and BC are under pure compression and pure tension, respectively (x and y axes are principal axes). On the other hand, both a shear stress τ_a and a normal stress σ are applied on side AC. Considering force equilibrium in the x direction,

$$\sigma_t \left(\frac{t dx}{\sqrt{2}} \right) \sin 45 + \sigma_c \left(\frac{t dx}{\sqrt{2}} \right) \sin 45 - \tau_a t dx = 0 \Rightarrow$$
$$\frac{\sigma_t + \sigma_c}{2} = \tau_a \tag{7.21}$$

Similarly, considering force equilibrium in the y direction,

$$-\sigma_t \left(\frac{t dx}{\sqrt{2}} \right) \cos 45 + \sigma_c \left(\frac{t dx}{\sqrt{2}} \right) \cos 45 + \sigma t dx = 0 \Rightarrow$$
$$\frac{\sigma_t - \sigma_c}{2} = \sigma \tag{7.22}$$

Now σ_c is constant and given by Equation (7.19) while τ_a is proportional to the applied load V and given by Equation (7.16). Therefore, Equations (7.21) and (7.22) form a system of two equations in the two unknowns σ_t and σ. Solving,

$$\sigma_t = 2\tau_a - \sigma_c$$

$$\sigma = \tau_a - \sigma_c$$

and using Equations (7.16) and (7.19),

$$\sigma_t = \frac{2V}{ht} - \tau_{cr} \tag{7.23}$$

$$\sigma = \frac{V}{ht} - \tau_{cr} \tag{7.24}$$

As mentioned earlier, τ_{cr} is the shear stress at which the skin buckles. This can be determined following the procedures of Sections 6.3–6.5 (e.g. Equation 6.28 for long plates or Equation 6.37 with appropriate adjustment to improve its accuracy).

Designing a composite skin under shear in the post-buckling regime would then require the determination of a layup that will not fail when stresses σ_t and σ_c are applied. Note that these are along the axes 1, 2 in Figure 7.14 and, therefore, the resulting layup would be in that coordinate system and would have to be rotated in order to express it in the original x,y coordinate system. In addition to the skin, the stiffeners and flanges would have to be designed taking into account the stresses σ and τ_a (for stiffener and flange design see Sections 8.3–8.7).

It is important to note that the preceding derivation only gives the average stresses in the skin and assumes that the angle of the principal axes remains constant and equal to 45° after buckling. In reality, the angle changes as the applied load increases. The solution given above is conservative and mainly underlines the fact that the skin is under increasing tension along diagonal lines (hence the term 'diagonal tension' for such situations) and constant compression perpendicular to these diagonal lines. For sufficiently high applied loads the effect of the compression stress σ_c is very small and can be neglected.

An improved analytical approach for post-buckled panels under shear was proposed by Wagner for isotropic materials [1, 2]. In this analysis the effect of stiffener spacing, flange geometry and skin dimensions is taken into account and an iterative set of equations is derived for the post-buckling angle α (Figure 7.16) and the stresses in the skin, stiffeners and flanges.

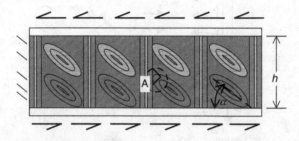

Figure 7.16   Post-buckled skin under shear showing post-buckling angle α

The analysis by Wagner was further improved by Kuhn *et al.* [3, 4] who accounted more accurately for the relative stiffnesses of the skin, stiffeners and flanges and used test results to derive some of their semi-empirical equations. In fact, the analysis by Kuhn *et al.* forms even today, with minor modifications, the basis for designing isotropic post-buckled panels under shear.

For the case of composite materials, the analysis by Kuhn *et al.* was modified by Deo *et al.* [5, 6]. The basic results from their work form the starting point for the discussion in Section 7.2.1. In what follows, the mathematical aspects of their method are improved upon in order to minimize or eliminate the need for iterations.

7.2.1 Post-Buckling of Stiffened Composite Panels under Shear

The situation is shown in Figure 7.16.

Before going into the equations that describe the state of stress in the panel of Figure 7.16, a qualitative discussion of how loads are shared between the different components might help visualize what happens. Consider that the panel of Figure 7.16 is a portion of a fuselage, with the two flanges (above and below) being frames and the vertical stiffeners being stringers.

It is simpler to visualize what happens if only a shear load is applied as in Figure 7.16. This could be the result of torsion in the fuselage. Combined load cases are briefly discussed in the next section. If the load is low enough and the skin does not buckle, the skin is under pure shear and there is no load in the stringers or the frames.

After the skin buckles in shear, it resists the applied load in diagonal tension along lines forming an angle α with the frames. A small amount of compression (see Equation 7.19) is also present. As the skin pulls away from the stiffeners it exerts both a tension load along each stiffener and a transverse load that would bend the stiffener in the plane of Figure 7.16. This is shown in Figure 7.17 where detail A from Figure 7.16 is put in equilibrium.

At the interface between the skin and stringer a normal stress σ_s and a shear stress τ_s must develop to put the skin in equilibrium. These, in turn, are exerted on the stringer. In order for the stringer to be in equilibrium axial loads P_{st1} and P_{st2} and bending moments must develop. In an analogous fashion, axial and bending loads develop in the frames. Therefore, determining the stresses or strains not only in the skins, but also in the stringers and frames becomes crucial for designing such structures in the post-buckling regime.

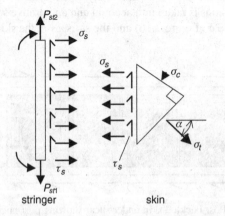

Figure 7.17 Equilibrium of detail A from Figure 7.16

The best analytical solution for composite panels in the post-buckling regime was developed by Deo, Agarwal and Madenci [5] and Deo, Kan and Bhatia [6]. In that work, the original methodology for metal panels developed by Kuhn, Peterson and Levin [3, 4] was modified to account for the anisotropy of composite panels and the additional failure mode of skin/stiffener separation, typically not present in metal structures.

This work relates the strains in the skin, stiffener and frame to the post-buckling angle α, which is the angle formed by the buckled shape of the skin and the stiffener axis (see Figures 7.16, 7.17). The equations are transcendental, and iterations are needed to eliminate α in order to obtain strains as a function of geometry and stiffness. An approach to simplify the algebra and solve the equations presented in [5, 6] is presented below.

The governing equations are as follows:

Post-Buckling Factor k

$$k = \tanh \left[\frac{1}{2} \ln \left(\frac{N_{xy}}{N_{xycr}} \right) \right] \tag{7.25}$$

The post-buckling factor (not to be confused with the post-buckling ratio (PB) introduced at the start of this chapter) ranges between 0 and 1 and gives a measure of how much of the applied shear load is taken by in-plane shear and how much by diagonal tension. A value of $k = 0$ denotes pure shear. A value of $k = 1$ denotes all the applied load is taken by diagonal tension.

Post-Buckling Angle α

$$\alpha = \tan^{-1} \sqrt{\frac{\varepsilon - \varepsilon_s}{\varepsilon - \varepsilon_f}} \tag{7.26}$$

Skin Strain ε in Diagonal Tension Direction

$$\varepsilon = \frac{Nxy}{E_{w\alpha}} \left[\frac{2k}{\sin 2\alpha} + \frac{E_{w\alpha}}{2G_{sk}} (1 - k) \sin 2\alpha \right] \tag{7.27}$$

Stiffener Strain ε_s

$$\varepsilon_s = \frac{-kN_{xy} \cot \alpha}{\left[\dfrac{\overline{EA}_s}{h_s t_{sk}} + \dfrac{1}{2}(1 - k)E_{ws} \right]} \tag{7.28}$$

Frame Strain ε_f

$$\varepsilon_f = \frac{-kN_{xy} \tan \alpha}{\left[\dfrac{\overline{EA}_f}{h_f t_{sk}} + \dfrac{1}{2}(1 - k)E_{wf} \right]} \tag{7.29}$$

where
N_{xy} is the applied shear load (force/length);
N_{xycr} the buckling load under shear of skin between adjacent stiffeners and frames;
ε the skin strain in diagonal tension direction;

ε_s the strain in the stringer averaged over its length;

ε_f the strain in the frame averaged over its length;

t_{sk} the skin thickness;

h_s the stiffener spacing;

h_f the frame spacing;

$E_{w\alpha}$ the skin (or web) modulus in the diagonal tension direction;

E_{ws} the skin (or web) modulus along the stiffener direction;

E_{wf} the skin (or web) modulus along the frame direction;

G_{sk} the skin shear modulus;

$\overline{EA}_s$ the axial stiffness of the stiffener $(= EA_s(EI_s)/\overline{EI}_s)$;

$\overline{EA}_f$ the axial stiffness of frame $(= EA_f(EI_f)/\overline{EI}_f)$;

EI_s, EI_f the bending stiffnesses about stiffener and frame neutral axis; and

$\overline{EI}_s$, $\overline{EI}_f$ the corresponding bending stiffnesses about the skin midsurface.

It is important to note that the axial stiffnesses $\overline{EA}$ (= Young's modulus E multiplied by area A) are corrected by the bending stiffnesses EI (= Young's modulus E multiplied by moment of inertia I). This is done to account for the fact that, in general, for a composite beam, the membrane stiffness and the bending stiffnesses are different (see Sections 3.3 and 8.2).

It can be seen from Equations (7.26) to (7.29) that trigonometric functions of α and the strains ε, ε_s and ε_f all appear in the governing equations. Traditionally, the approach to solving them is to assume a value of α (about 40° is a good starting value) and substitute in Equations (7.27), (7.28) and (7.29) to get the strains ε, ε_s and ε_f. These strains are then substituted in Equation (7.26) to obtain an updated value for α. The procedure is repeated until two successive values of α are equal to within some preset tolerance value.

This approach is not very efficient because it involves iterations. These iterations would be repeated for each candidate design during an optimization run and would slow the process tremendously. It would be advantageous if these iterations were minimized and another way to solve Equations (7.26) to (7.29) were found.

It turns out that if $E_{w\alpha}$, the skin modulus in the direction of the diagonal tension angle α is assumed constant, Equations (7.26) to (7.29) can be solved exactly without iterations. First, the trigonometric expressions involving α are expressed in terms of $\tan \alpha$:

$$\sin^2\alpha = \frac{\tan^2\alpha}{1 + \tan^2\alpha}$$

$$\cos^2\alpha = \frac{1}{1 + \tan^2\alpha}$$

$$\sin 2\alpha = 2\frac{\tan\alpha}{1 + \tan^2\alpha}$$

Then, using these expressions, the three strains from Equations (7.27) to (7.29) are written as

$$\varepsilon = A\frac{1 + \tan^2\alpha}{2\tan\alpha} + B\frac{2\tan\alpha}{1 + \tan^2\alpha} \tag{7.27a}$$

$$\varepsilon_s = \frac{C}{\tan\alpha} \tag{7.28a}$$

$$\varepsilon_f = D\tan\alpha \tag{7.29a}$$

where

$$A = \frac{N_{xy}2k}{t_{sk}E_{w\alpha}} \tag{7.30}$$

$$B = \frac{N_{xy}(1-k)}{t_{sk}G_{sk}} \tag{7.31}$$

$$C = \frac{-kN_{xy}}{t_{sk}\left(\dfrac{EAs}{h_s t_{sk}} + \dfrac{1}{2}(1-k)E_{ws}\right)} \tag{7.32}$$

$$D = \frac{-kN_{xy}}{t_{sk}\left(\dfrac{EAf}{h_r t_{sk}} + \dfrac{1}{2}(1-k)E_{wf}\right)} \tag{7.33}$$

If Equation (7.26) is now used to solve for $\tan\alpha$ and substitute in Equations (7.27a) to (7.29a), the angle α is eliminated and three equations in the three unknowns ε, ε_r and ε_f are obtained. After some manipulation, ε and ε_s can be eliminated and a single equation in

$$z = \left(\frac{\varepsilon_f}{D}\right)^2 \tag{7.34}$$

is obtained as follows:

$$z^3 + z^2\frac{2D - A - 4B}{2D - A} + z\frac{A - 2C + 4B}{2D - A} + \frac{A - 2C}{2D - A} = 0 \tag{7.35}$$

From the theory of cubic equations (see [7]), it can be shown that since A and B have the same sign as N_{xy} and C and D have sign opposite to that of N_{xy}, Equation (7.35) has three real and unequal solutions. The solutions are given by

$$z_1 = 2\sqrt{-Q}\cos\frac{\theta}{3} - \frac{a_2}{3}$$

$$z_2 = 2\sqrt{-Q}\cos\left(\frac{\theta + 2\pi}{3}\right) - \frac{a_2}{3} \tag{7.36}$$

$$z_3 = 2\sqrt{-Q}\cos\left(\frac{\theta + 4\pi}{3}\right) - \frac{a_2}{3}$$

with

$$\theta = \cos^{-1}\left(\frac{R}{\sqrt{-Q^3}}\right)$$

$$R = \frac{9a_1a_2 - 27a_o - 2a_2^3}{54} \tag{7.37}$$

$$Q = \frac{3a_1 - a_2^2}{9}$$

and

$$a_o = \frac{A - 2C}{2D - A}$$

$$a_1 = \frac{A - 2C + 4B}{2D - A}$$

$$a_2 = \frac{2D - A - 4B}{2D - A}$$

(7.38)

Of the three solutions in Equation (7.36), only the positive ones are acceptable because, otherwise, the right-hand side of Equation (7.34), which is positive, would be a negative number. If there is more than one positive solution, the lowest one should be selected.

As was mentioned earlier, the solution of Equations (7.36) to (7.38) assumes that the stiffness of the skin in the direction of α, $E_{w\alpha}$, is constant. However, without knowing α a priori E_{wa} is not known exactly. A small number of iterations (typically significantly fewer than those required with the traditional approach mentioned earlier) is required after all. To determine the skin stiffness along any direction α, it is assumed that a tension load is applied in that direction and the stress–strain equations are solved for. This parallels the derivation of Equation (5.19) in chapter 5. For a symmetric and balanced skin, the normal stresses σ_{11} and σ_{22} in a coordinate system with the 1 axis aligned with α are given by the following expressions (see Equation 3.32):

$$\sigma_{11} = Q_{11}\varepsilon_{11} + Q_{12}\varepsilon_{22}$$

$$\sigma_{22} = Q_{12}\varepsilon_{11} + Q_{22}\varepsilon_{22}$$

But $\sigma_{22} = 0$ since only a tension load is applied in the 1 (or α) direction. Therefore, solving for ε_{22},

$$\varepsilon_{22} = -\frac{Q_{12}}{Q_{22}}$$

and substituting in the equation for σ_{11},

$$\sigma_{11} = \left(\underbrace{Q_{11} - \frac{Q_{12}^2}{Q_{22}}}_{E_{wa}} \right)\varepsilon_{11}$$

from which

$$E_w = Q_{11} - \frac{Q_{12}^2}{Q_{22}}$$

(7.39)

with the standard transformation giving the stiffnesses Q_{11}, Q_{12} and Q_{22} (see Equations 3.33):

$$Q_{11} = Q_{xx} \cos^4 \alpha + Q_{yy} \sin^4 \alpha + 2 \sin^2 \alpha \cos^2 \alpha (Q_{xy} + 2Q_{ss})$$

$$Q_{22} = Q_{xx} \sin^4 \alpha + Q_{yy} \cos^4 \alpha + 2 \sin^2 \alpha \cos^2 \alpha (Q_{xy} + 2Q_{ss}) \qquad (7.40)$$

$$Q_{12} = \sin^2 \alpha \cos^2 \alpha (Q_{xx} + Q_{yy} - 4Q_{ss}) + (\sin^4 \alpha + \cos^4 \alpha)Q_{xy}$$

and from Equations (3.27) to (3.29):

$$Q_{xx} = \frac{E_{xx}}{1 - \nu_{xy}\nu_{yx}}$$

$$Q_{yy} = \frac{E_{yy}}{1 - \nu_{xy}\nu_{yx}} \qquad (7.41)$$

$$Q_{xy} = \frac{\nu_{yx} E_{xx}}{1 - \nu_{xy}\nu_{yx}}$$

$$Q_{ss} = G_{sk}$$

where E_{xx}, E_{yy}, G_{sk}, ν_{xy} and ν_{yx} are engineering constants for the entire skin laminate with x coinciding with the stiffener direction and y coinciding with the frame direction.

The solution procedure is then as follows:

1. Select a value of $E_{w\alpha}$. Typically, since $\alpha \approx 45°$ select $E_{w\alpha}$ corresponding to the 45° direction.
2. Calculate the coefficients in Equation (7.35) using Equations (7.37) and (7.38).
3. Calculate z_1, z_2, z_3 from Equation (7.36).
4. Pick the positive z value from step 3. If there is more than one positive value, use the lowest one.
5. Calculate a new value of $E_{w\alpha}$ using Equations (7.39) to (7.41). If it is equal to the previous value of $E_{w\alpha}$ within a preset tolerance, the diagonal tension analysis is complete. If not, go to step 2 above and repeat the process.

7.2.1.1 Application: Post-Buckled Stiffened Fuselage Skin under Shear

Consider the portion of fuselage skin enclosed by two adjacent stiffeners and frames as shown in Figure 7.18. The skin is under pure shear N_{xy}.

A notional sketch of the post-buckling shape with the half-waves inclined at an angle α to the stiffener axis is also included in Figure 7.18. The length of the stiffeners and frames is set at typical values of 508×152.4 mm, respectively. Two different skin layups of the same thickness are used: (a) $(\pm 45)_5$ and (b) $(\pm 45)/(0/90)_3/(\pm 45)$. The material used for the skin is plain weave fabric with the following properties:

$$E_x = E_y = 68.9 \, \text{GPa}$$

$$G_{xy} = 4.82 \, \text{GPa}$$

$$\nu_{xy} = 0.05$$

$$t_{ply} = 0.1905 \, \text{mm}$$

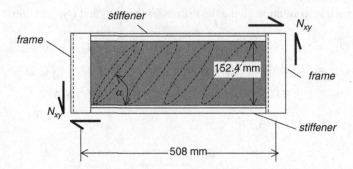

Figure 7.18   Stiffened skin under shear load

The details of the stiffener and frame layups are of no interest at this point other than the fact that, for both skin layups, EA for the stiffener is 6953 kN and for the frame is 75,828 kN. The buckling load for the skin of case (a) is determined to be 10.82 N/mm and for the skin of case (b) is 10.88 N/mm. Note that the buckling loads are essentially the same for the two cases.

The results for the post-buckling behaviour for both cases are shown in Figures 7.19–7.21. The post-buckling angle α (see Figure 7.18) for the two different skin layups is given in Figure 7.19 as a function of the applied load normalized by the buckling load.

For both cases, the post-buckling angle starts at 45° when the applied load N_{xy} equals the buckling load N_{xycrit} and decreases towards an asymptote around 25° for high values of N_{xy}/N_{xycrit}. The post-buckling angle for the $(\pm45)/(0/90)_3/(\pm45)$ skin is slightly higher.

The strains in the skin, stiffener and frame for the $(\pm45)_5$ skin layup are shown as a function of the applied load in Figure 7.20. It is seen that the stiffener and frame are always in compression. While the skin strains are relatively linear, the stiffener and frame strains are nonlinear and they increase more rapidly than the skin strains. If now the cut-off strain of 4500 μs calculated in Section 5.1.6 is used for the stiffener and frame, which are in compression, and a corresponding value of 6000 μs is used for the skin, which is in tension (tension allowable is higher than compression allowable for most layups used in practice), it can be seen from Figure 7.20 that the stiffener will fail first at a value of N_{xy}/N_{xycrit}, slightly higher than 11. This is interesting because at lower loads the skin has higher strains than the stiffener, but they are increasing more slowly as the load increases.

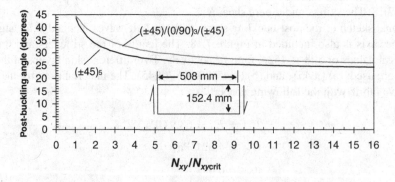

Figure 7.19   Post-buckling angle as a function of applied shear load

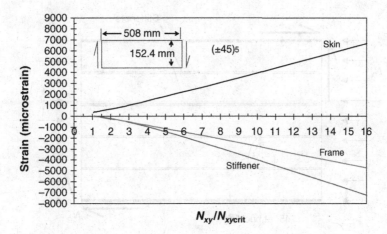

Figure 7.20   Strains in the post-buckling regime for $(\pm45)_5$ skin

It is important to note that the use of cut-off strains does not explicitly account for the layup. This is a conservative approach for generating or evaluating a preliminary design. A more detailed analysis would require knowledge of the specific layups and geometries for the skin and stiffener. Also, additional failure modes such as crippling of the stiffener or frame (see Section 8.5) and skin–stiffener separation (see Section 9.2.2) would have to be included in the evaluation. The present discussion gives a good starting point for generating a viable design.

The corresponding strains for the second case with skin layup $(\pm45)/(0/90)_3/(\pm45)$ are shown in Figure 7.21. Again, the stiffener and frame are in compression while the skin is in tension. Unlike the case of the $(\pm45)_5$ skin where the stiffener strains rapidly exceeded in magnitude the skin strains, here the skin strains are always higher in magnitude. This is due to the fact that the skin in this case is significantly stiffer and absorbs more load, thus unloading the stiffeners to some extent. However, the stiffeners are still critical as the cut-off value of 4500 μs for the stiffeners is reached before the cut-off value of 6000 μs is reached for the skin.

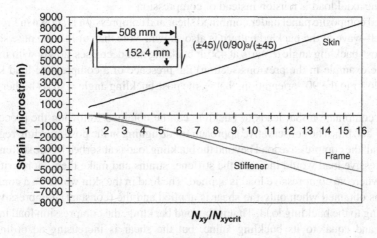

Figure 7.21   Strains in the post-buckling regime for $(\pm45)/(0/90)_3/(\pm45)$ skin

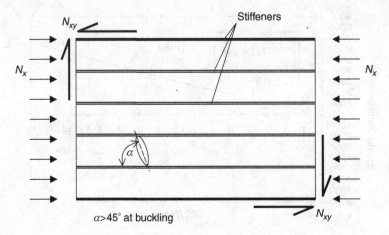

Figure 7.22   Post-buckled panel under compression and shear

Comparing the results in Figures 7.20 and 7.21 it is seen that, even though the dimensions, the stiffeners and the frames are identical and the skin buckling loads are the same, the post-buckling behaviour in the second case is significantly different. For example, at low applied loads the skin strains are twice as high for the second case. This difference decreases as the applied load increases, but is still at least 25% at high applied loads.

7.2.2   Post-Buckling of Stiffened Composite Panels under Combined Uniaxial and Shear Loading

When both shear and an axial load (tension or compression) act on a stiffened panel, the state of stress developing in the post-buckling skin is quite complicated and very hard to obtain without a good computational model, usually based on finite elements. Also, the situation changes if the axial load is tension instead of compression.

A stiffened composite panel under combined shear and compression is shown in Figure 7.22. A typical half-wave of the buckled pattern is also shown. Note that, unlike the pure shear case where the post-buckling angle α starts at 45° at buckling and decreases slowly with increasing load (see the example in the previous section) the presence of a compressive load keeps the half-wave closer to the 90° orientation, that is, the post-buckling angle α starts higher than 45° at buckling.

The load combination that leads to buckling can be obtained following the procedures of Sections 6.5 and 6.6. The conservative approach for designing in the post-buckling regime is to assume that all the compressive load beyond the buckling load is absorbed by the stiffeners. This extra compressive load would increase the stiffener strains and make them more critical than in the case where no compressive load is applied. The load in the skin would be a combination of the strains obtained when only the shear is applied and the (constant) compressive strains corresponding to the buckling load. That is, beyond buckling, the compression load in the skin is constant and equal to its buckling value, but the shear is increasing according to the post-buckling analysis given in the previous section. The fact that the skin is loaded by the compressive strains that were exerted at the buckling load in addition to the diagonal tension strains resulting from the shear load is more critical than in the case where only shear load was applied.

In contrast to the combined compression and shear case, a tension and shear case is, usually, less critical. First, the magnitude of the buckling load under tension and shear is higher (depending on the direction of the shear load and on which edges the tension load is applied). Then, in the post-buckling regime, the tension strains caused by the applied tensile load are split between the skin and stiffeners according to, roughly, their respective EA ratios. This means that the compression strains in the stiffeners caused by the shear load are relieved and the stiffeners are less critical. In the skin, the diagonal compression strains are relieved while the diagonal tension strains are increased. However, since in most designs the stiffeners are much stiffer than the skin (have much higher EA) the amount of tension left in the skin is small and most of it is taken by the stiffeners.

The procedure for preliminary design and analysis of a composite stiffened panel under combined uniaxial load and shear is summarized in Figure 7.23. Note that in this figure $N_x > 0$ corresponds to tension. This procedure should be viewed as approximate because it requires combining skin strains caused by shear load applied alone and a portion of strains from compression applied alone. This implies some kind of superposition is being used. However, linear superposition is not valid in the post-buckling regime because the deflections are large and the problem is nonlinear. Only for applied loads that do not exceed the buckling loads significantly is superposition (approximately) valid. Therefore, the results of this process are approximate. They can be very useful in determining a good starting design for further more detailed analysis.

Exercises

7.1 Refer to the application discussed in Section 7.1. A fabric material is made available that has the basic properties given in the following table:

Property	Value
E_x	69 GPa
E_y	69 GPa
ν_{xy}	0.05
G_{xy}	5.1 GPa
t_{ply}	0.19 mm

The material is to be used in a skin application. The skin panel of interest is square of dimensions 200 × 200 mm and the applied ultimate load is 2152 N. The following two skin layups are proposed for this application: (a) $(\pm 45)/(0/90)_3/(\pm 45)$ and (b) $(0/90)/(\pm 45)/(0/90)/(\pm 45)/(0/90)$. Given the highest value of N_x determined in Section 7.1 and its location, comment on or discuss the following. Based on the highest N_x value, which of the two layups is stronger and why? What is better: placing plies with fibres aligned with the load away from the laminate mid-plane or placing plies with fibres at 45° to the load away from the laminate mid-plane?

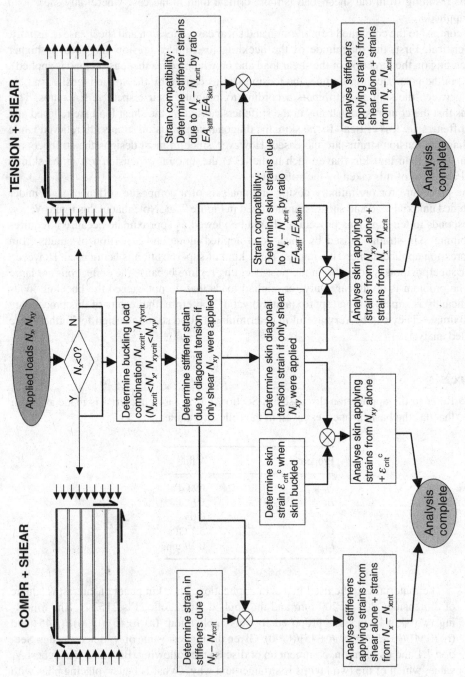

Figure 7.23   Design/analysis procedure for stiffened panels under combined uniaxial load and shear

7.2 The layup for the skin in a certain application is dictated to be $[(45/-45)_2/0/\overline{90}]_s$ (total of 11 plies). The basic material properties are given below:

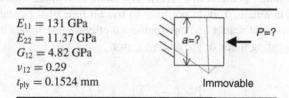

$E_{11} = 131$ GPa
$E_{22} = 11.37$ GPa
$G_{12} = 4.82$ GPa
$\nu_{12} = 0.29$
$t_{ply} = 0.1524$ mm

$a=?$ $P=?$

Immovable

This skin is square and under compression and is allowed to post-buckle. Three of the edges of the skin are immovable and the fourth one is under load P (N). If the maximum deflection of the skin is not allowed to exceed 6.35 mm (to avoid interference with adjacent structure) determine (a) the largest dimension a the skin can have and (b) the corresponding maximum allowable load P if the axial strain in the plate is not to exceed 5000 microstrain (this includes scatter, environmental effects and damage). In order to correlate load to strain, use the effective width concept and assume that $\sigma = E\varepsilon$ is the constitutive relation (Hooke's law) with E the engineering membrane stiffness of the laminate at hand.

7.3 Using the same assumed expressions for w and F (Equations 7.2 and 7.3) re-derive the post-buckling solution for a non-square plate, that is, determine w_{11} and b_{eff} for a rectangular plate of dimensions $a \times b$. Once you derive the expression for the centre deflection, verify that it coincides with the expression (7.11) for the case $a = b$. Determine the range of aspect ratios a/b for which the buckling mode has only one half-wave ($m = 1$). Use this result to suggest over what range your post-buckling solution in this problem is accurate.

7.4 A square composite plate of side 254 mm simply supported all around is loaded on one side by a force F. The remaining three edges are fixed so they do not move in-plane. Determine the location and magnitude of the maximum strains ε_{xo} and ε_{yo}.

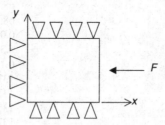

7.5 For the case of Exercise 7.4, a composite material with properties

$$E_x = 137.9\,\text{GPa}$$

$$E_y = 11.72\,\text{GPa}$$

$$\nu_{xy} = 0.29$$

$$G_{xy} = 5.171\,\text{GPa}$$

$$t_{ply} = 0.1524\,\text{mm}$$

is made available. If the applied load is $F = 31.1$ kN determine the 10-ply symmetric and balanced laminate consisting of only $45°$, $-45°$, $0°$ and $90°$ plies that has the lowest value of maximum ε_{xo} found in Exercise 7.4. Do not use more than two $0°$ or two $90°$ plies per half-laminate. You do not have to use all four ply orientations. Accounting for damage, material scatter and environmental effects, determine either the margin of safety or the loading index or the reserve factor.

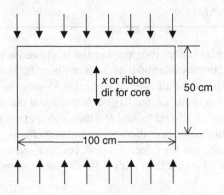

7.6 (May be done in conjunction with Exercise 10.5). You are to design a composite panel under compressive load using a skin–stiffener configuration. The panel dimensions are 100×50 cm and the applied load is 1750 N/mm acting parallel to the 50 cm dimension. Two composite materials are available, with properties as follows:

Unidirectional tape graphite/epoxy	Plain weave fabric graphite/epoxy
$E_x = 131$ GPa	68.9 GPa
$E_y = 11.4$ GPa	68.9 GPa
$\nu_{xy} = 0.31$	0.05
$G_{xy} = 5.17$ GPa	5.31 GPa
$t_{ply} = 0.1524$ mm	0.1905 mm
$X_t = 2068$ MPa	1378.8 MPa
$X_c = 1723$ MPa	1378.8 MPa
$Y_t = 68.9$ MPa	1378.8 MPa
$Y_c = 303.3$ MPa	1378.8 MPa
$S = 124.1$ MPa	119.0 MPa
$P = 1611$ kg/m^3	1611 kg/m^3

Once you determine any strength values needed for any of the layups selected, you are to assume the same knockdowns mentioned in Section 5.1.6 for environment, material scatter and damage.

Of the seven shapes below select one for the stiffeners.

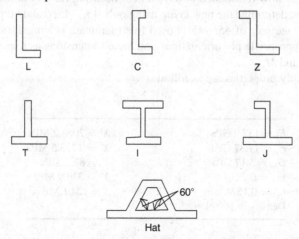

You are to select the layup for the skin and each member of the stiffener cross-section for a post-buckling factor of 2.5 and a post-buckling factor of 5. You may use one of the two composite materials or a combination of both. You will need to decide on a stiffener spacing and use the solution to Exercise 7.3. It is up to you to decide if you want to reinforce the skin at the edges (near the stiffeners) over the effective width in order to get a lighter design or simply use the same layup for the skin everywhere. Note that the stiffener height cannot exceed 10 cm and no horizontal flange of the stiffener can exceed 5.5 cm. The skin or any member of the stiffener cannot be thinner than 0.57 mm. Make sure that you account for all failure modes that apply in this case. Assume that the stiffeners are co-cured with the skin.

Determine the layup of each member of each stiffener and its dimensions observing the following design rules: (a) laminates are symmetric and balanced; (b) at least 10% of the fibres are in each of the four principal directions $0°$, $45°$, $-45°$ and $90°$; (c) no more than four unidirectional plies of the same orientation may be next to each other; (d) use only $0°$, $45°$, $-45°$ and $90°$ plies. Provide a simple sketch of the cross-section of stiffeners that shows the plies, layup, dimensions, etc. Calculate the corresponding weights for skin/stiffened panel with PB = 2.5, skin/stiffened panel with PB = 5.0 and, if available, compare with the results from Exercise 10.5.

7.7 A square fuselage panel of layup $[45/-45/0_2/90]s$ is under compression and is buckling critical. The side is 50.8 cm long. Big stiff stringers and a frame are attached on the three edges of the panel as shown below:

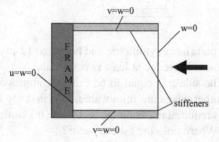

The panel is now redesigned to have a post-buckling ratio of at least 2, 4 or 7. For each PB ratio, determine the new layup using only 45, −45, 0 and 90 plies and making sure at least one pair of 45/−45 is used (also laminate is symmetric and balanced). Determine appropriate ply orientations and local dimensions as required. Neglect N_y, N_{xy}, M_x, M_y and M_{xy}.

The basic ply properties are as follows:

$E_x = 137.9$ GPa	$X^t = 2068.2$ MPa
$E_y = 11.37$ GPa	$X^c = 1723.5$ MPa
$G_{xy} = 5.17$ GPa	$Y^t = 82.7$ MPa
$\nu_{xy} = 0.29$	$Y^c = 330.9$ MPa
$t_{ply} = 0.1524$ mm	$S = 124.1$ MPa
Density $= 1609$ kg/m^3	

7.8 Redo problem 7.7 using Al design for the same applied load and PB ratios. For Al use $E = 68.94$ GPa, $\nu = 0.3$, and compression yield stress $\sigma_{cy} = 440.5$ MPa. The density of Al is 2774 kg/m^3. To do this, you should derive an expression for the required plate thickness. Then use that to generate the corresponding weights for the Al plates. Compare the weight with your result in problem 1 in a plot of weight versus PB values with PB = 1, 2, 4 and 7.

7.9 In view of the composite and Al curves in your answer in Problem 7.8 how would you change the composite layup to further reduce the composite weight? Do not do any calculations, just provide a brief argument.

7.10 It is reasonable to assume that in a post-buckled panel, the lower the out-of-plane deflections, for a given load, the lower the possibility of failure. Bending moments are lower and in-plane strains due to membrane effects are lower. For a square panel with three sides immovable and a compressive load applied on the fourth, discuss in words which parameters or quantities pertaining to the laminate stacking sequence one should change and in which direction to get as low out-of-plane deflections as possible.

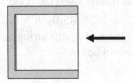

7.11 A square composite plate has a symmetric and balanced 12-ply layup consisting of only 45, −45, 0 and 90 plies. There are at least two +45 and two −45 plies and at least one 0 and one 90 ply. The side a is equal to 65 cm. A compressive load is applied along one edge while all other edges are immovable. Use first-ply failure and rearrange the stacking sequence as required to obtain the layup that maximizes the failure load. What is that failure load? Where does the failure occur?

The material properties are

$$E_x = 20\text{E6 psi (lb/in.}^2)$$
$$E_y = 1.7\text{E6 psi}$$
$$\nu_{xy} = 0.29$$
$$G_{xy} = 0.7\text{E6 psi}$$
$$t_{ply} = 0.012 \text{ in.}$$
$$X_t = 300 \text{ ksi (1 ksi} = 1000 \text{ psi)}$$
$$X_c = 200 \text{ ksi}$$
$$Y_t = 14 \text{ ksi}$$
$$Y_c = 48 \text{ ksi}$$
$$S = 18 \text{ ksi}$$

References

[1] Wagner, H., Structures of Thin Sheet Metal, Their Design and Construction, *NACA Memo 490*.

[2] Wagner, H., Flat Sheet Girder with Thin Metal Web, Part I (*NACA TM 604*), Part II (*NACA TM 605*), and Part III (*NACA TM 606*).

[3] Kuhn, P., Peterson, J.P. and Levin, L.R., A Summary of Diagonal Tension Part 1 – Methods of Analysis, *NACA TN 2661* (1952).

[4] Kuhn, P., Peterson, J.P. and Levin, L.R., A Summary of Diagonal Tension Part 2 – Experimental Evidence, *NACA TN 2662* (1952).

[5] Deo R.B., Agarwal, B.L. and Madenci, E., Design Methodology and Life Analysis of Postbuckled Metal and Composite Panels, *AFWAL-TR-85-3096*, Vol. I (1985).

[6] Deo, R.B., Kan, H.P. and Bhatia, N.M., Design Development and Durability Validation of Postbuckled Composite and Metal Panels, Volume III – Analysis and Test Results, *Northrop Corp., WRDC-TR-89-3030*, Vol. III (1989).

[7] http://mathworld.wolfram.com/CubicEquation.html. Last accessed 26 November, 2012.

(Plate 1) Figure 1.1 Akaflieg Phönix FS-24 (Courtesy of Deutsches Segelflugzeugmuseum)

(Plate 2) Figure 1.2 Aerospatiale SA 341G Gazelle (Copyright Jenny Coffey, printed with permission)

(Plate 3) Figure 1.3 Long EZ and Vari-Eze. (Vari-Eze photo; courtesy of Stephen Kearney; Long EZ photo: courtesy of Ray McCrea)

(Plate 4) Figure 1.4 LearAvia LearFan 2100 (Copyright Thierry Deutsch)

Design and Analysis of Composite Structures: With Applications to Aerospace Structures, Second Edition. Christos Kassapoglou.
© 2013 John Wiley & Sons, Ltd. Published 2013 by John Wiley & Sons, Ltd.

(Plate 5) Figure 1.5 Beech (Raytheon Aircraft) Starship I (Courtesy of Brian Bartlett)

(Plate 6) Figure 1.6 Airbus A-320 (Courtesy of Brian Bartlett)

(Plate 7) Figure 1.7 Boeing 777 (Courtesy of Brian Bartlett)

(Plate 8) Figure 1.8 Airbus A-380 (Courtesy of Bjoern Schmitt, World of Aviation.de)

(Plate 9) Figure 1.9 Boeing 787 Dreamliner (Courtesy of Agnes Blom)

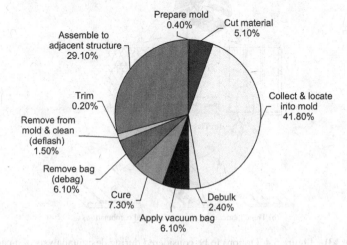

(Plate 10) Figure 2.1 Process steps for hand layup and their cost as fractions of total recurring cost

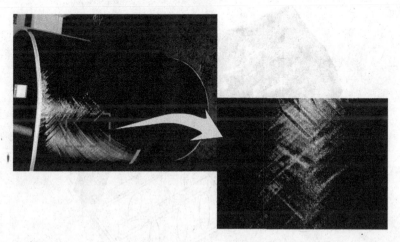

(Plate 11) Figure 2.5 Composite cylinder with steered fibres fabricated by automated fibre placement (made in a collaborative effort by TUDelft and NLR)

(Plate 12) Figure 2.9 Co-cure of large complex parts (Courtesy of Aurora Flight Sciences)

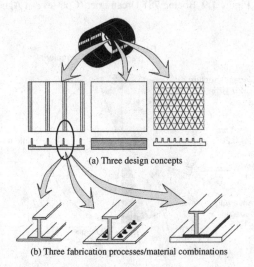

(a) Three design concepts

(b) Three fabrication processes/material combinations

(Plate 13) Figure 5.4 Options to be considered during design/analysis of a part

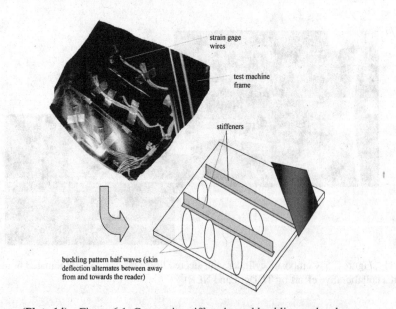

strain gage wires

test machine frame

stiffeners

buckling pattern half waves (skin deflection alternates between away from and towards the reader)

(Plate 14) Figure 6.1 Composite stiffened panel buckling under shear

(Plate 15) Figure 7.1 Post-buckled curved composite stiffened panel

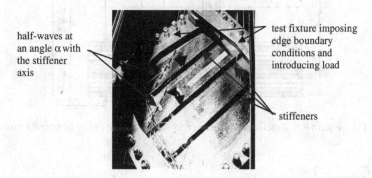

half-waves at an angle α with the stiffener axis

test fixture imposing edge boundary conditions and introducing load

stiffeners

(Plate 16) Figure 7.13 Stiffened composite panel in the post-buckling regime

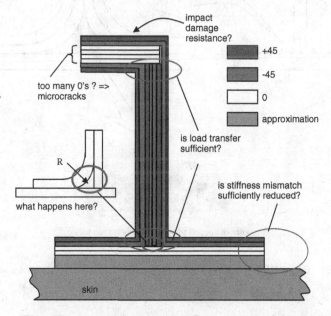

impact damage resistance?

+45

-45

0

approximation

too many 0's ? => microcracks

is load transfer sufficient?

R

is stiffness mismatch sufficiently reduced?

what happens here?

skin

(Plate 17) Figure 8.4 Improved stiffener cross-section design

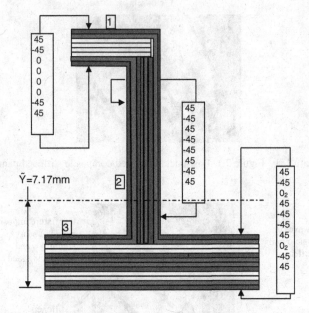

(Plate 18)   Figure 8.6 Baseline J stiffener cross-section made out of composite materials

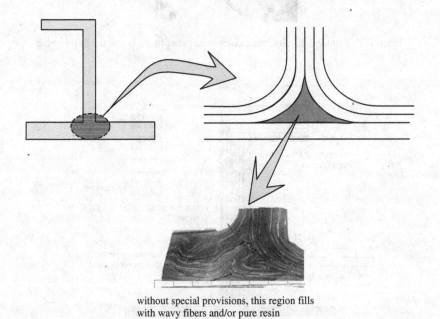

without special provisions, this region fills
with wavy fibers and/or pure resin

(Plate 19)   Figure 8.27 Resin pocket formed at web/flange intersection of a stiffener

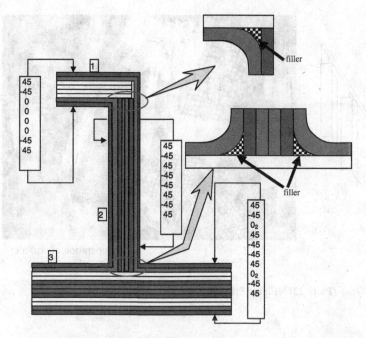

(Plate 20)   Figure 8.31 J stiffener cross-section with filler material

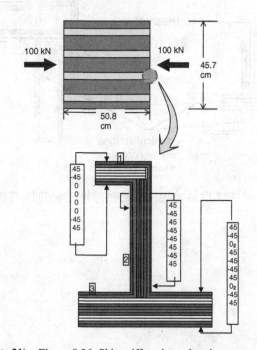

(Plate 21)   Figure 8.36 Skin-stiffened panel under compression

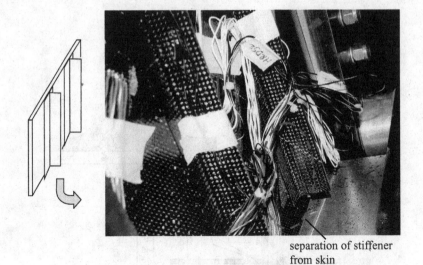

separation of stiffener
from skin

(**Plate 22**)   Figure 9.13 Skin–stiffener separation failure mode

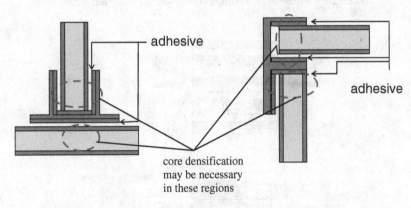

adhesive

adhesive

core densification
may be necessary
in these regions

(**Plate 23**)   Figure 10.21 Alternate means of joining sandwich structures

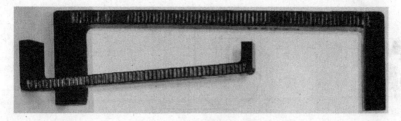

(Plate 24) Figure 10.22 Core transitioning to monolithic laminate without ramp-down (Courtesy of Aurora Flight Sciences)

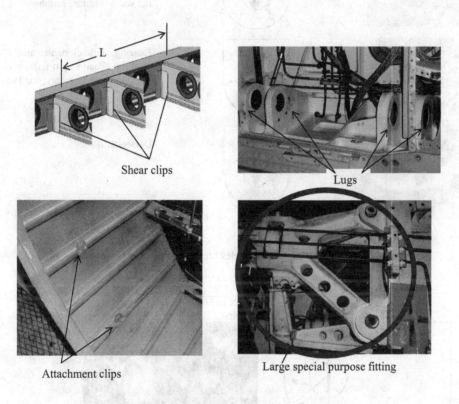

L

Shear clips

Lugs

Attachment clips

Large special purpose fitting

(Plate 25) Figure 11.1 Various types of fittings

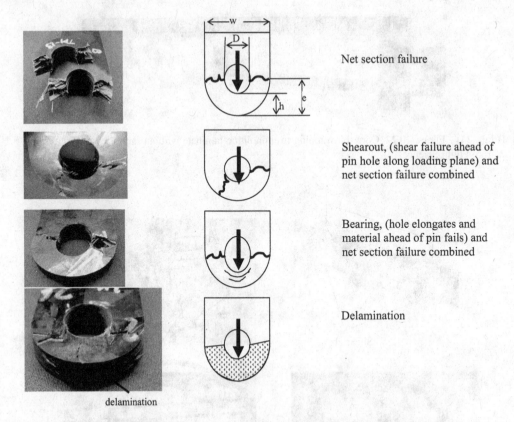

w

D

Net section failure

e

h

Shearout, (shear failure ahead of
pin hole along loading plane) and
net section failure combined

Bearing, (hole elongates and
material ahead of pin fails) and
net section failure combined

Delamination

delamination

(Plate 26)   Figure 11.15 Failure modes for a composite lug under axial load

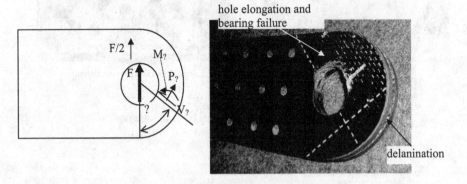

hole elongation and
bearing failure

F/2

M?

F

P?

?

V?

delamination

(Plate 27)   Figure 11.20 Lug under transverse load

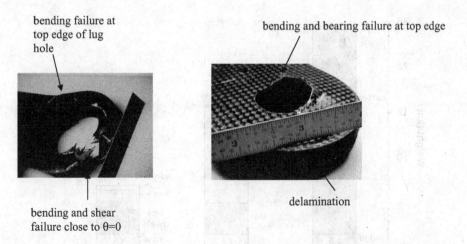

bending failure at top edge of lug hole

bending and bearing failure at top edge

bending and shear failure close to $\theta=0$

delamination

(Plate 28) Figure 11.24 Failure modes of lugs tested under transverse loading

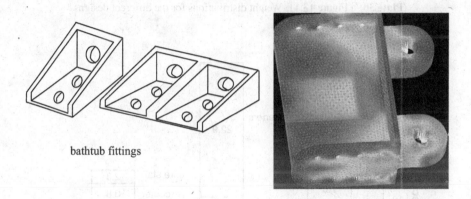

bathtub fittings

root fitting with stress contours from finite element analysis

(Plate 29) Figure 11.27 Other common fittings

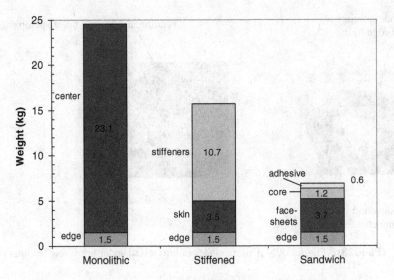

(Plate 30)   Figure 13.11 Weight distributions for the different designs

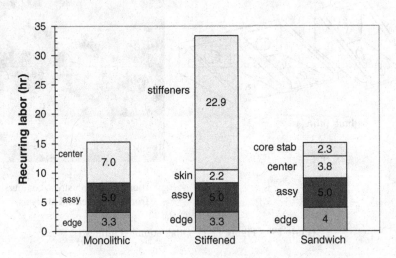

(Plate 31)   Figure 13.12 Cost distributions for the different designs

8

Design and Analysis of Composite Beams

The term beams is used here as a generic term referring to all one-dimensional parts that may be used in a structure. These include stiffeners, stringers, panel breakers, etc. There are many cross-sectional shapes that are used in practice. Of those, the ones used most frequently are (Figure 8.1) L or angle, C or channel, Z, T or blade, I, J and Hat or omega.

8.1 Cross-Section Definition Based on Design Guidelines

For a composite beam such as the one shown in Figure 8.2, each member may have a different layup. This would result in different stiffnesses and strengths for each of the flanges and web and would allow more efficient usage of composite materials through tailoring. Typically, the letter b with appropriate subscript is used to denote the longer dimension of each member and the letter t the shortest (the thickness).

As beams tend to be used in stability-critical situations, cross-sections with high moments of inertia are preferred. Besides the obvious implications for the beam geometry (high b_2 value, e.g. in Figure 8.2), there are certain guidelines that relate to the layup, which, when implemented, also contribute to robust performance.

With reference to Figure 8.2, stiff material must be located as far from the neutral axis as possible. Defining the 0 direction to be aligned with the beam axis (perpendicular to the plane of Figure 8.2), this stiffness requirement would result in the two flanges, the one next to the skin and the one away from the skin, being made up of mostly $0°$ plies.

Another clue can be deduced from the theory of joints (see, e.g. [1, 2]). It has been demonstrated that as the thickness of the adherends decreases, the strength of the joint increases because the peak stresses at the end of each adherend, where the load transfer to the adhesive is completed, are lower. This implies that the stiffness mismatch caused by the adherend termination is less and the associated stress concentration is reduced. A similar situation occurs in testing coupons (in tension) using bevelled tabs. The bevel in the tabs reduces the local stresses and helps eliminate the possibility of specimen failure at tab termination. With

Design and Analysis of Composite Structures: With Applications to Aerospace Structures, Second Edition. Christos Kassapoglou.
© 2013 John Wiley & Sons, Ltd. Published 2013 by John Wiley & Sons, Ltd.

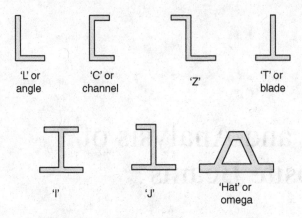

Figure 8.1   Typical beam cross-sections

this background, it is easy to deduce that by decreasing the stiffness mismatch between the flange next to the skin and the skin itself, the possibility of skin failure or flange/skin separation is reduced. This means that

(a) the flange should be made as thin as possible, still meeting other load requirements; and
(b) the stiffness of the flange should be as close as possible to that of the skin.

Finally, to improve performance against shear loads (parallel to the web axis) the web must have high strength and stiffness under shear loads, which means it should consist mostly of 45° plies. The importance of 45° plies is discussed in Section 8.5.1 where the pertinent stiffness term Q_{66} is shown to be maximum for 45° plies.

Combining these into a design would result in the preliminary configuration shown in Figure 8.3. The flange away from the skin consists of only 0° plies (for increased stiffness

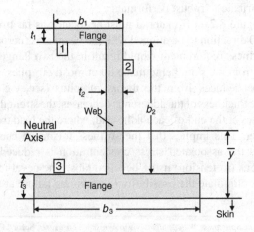

Figure 8.2   J stiffener cross-section

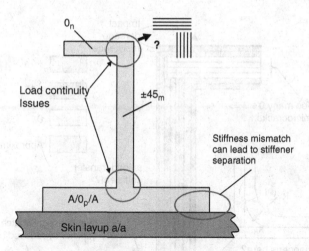

Figure 8.3   Preliminary design of stiffener cross-section based on first (crude) guidelines

away from the neutral axis). The flange next to the skin consists of a combination of some 0°
plies, for increased stiffness, sandwiched between the two halves of the skin layup in order to
minimize the stiffness mismatch between skin and flange. The web consists of 45° and −45°
plies for increased shear stiffness and strength.

As it stands, the design in Figure 8.3 is inadequate. There are several issues readily apparent:

(a) at the two corners where the web meets the flanges, dissimilar plies meet and load transfer
would rely on the matrix, which is grossly inefficient;
(b) the layup of the flange next to the skin may be too thick if the number p of 0° plies is too
high and/or it may still have very different stiffness from the skin;
(c) using only one ply orientation in the flange away from the skin and the web to satisfy
the respective requirements results in less than adequate performance of the beam when
(secondary) conditions or requirements other than the ones used here are considered.

To improve on this design, additional guidelines, developed on the basis of experience
and good engineering practice, are implemented. In order to get better load continuity, some
plies in the web must continue into the flanges. Also, to protect against secondary load cases,
ply orientations that cover the basic possible load directions, 0, 45, −45 and 90, are used.
Usually a very small number of load conditions (quite often a single one) are the critical
conditions that size a structure. This does not mean that there are no other load conditions,
only that their corresponding loads are lower (see, e.g. Section 5.1.1 for the implications of
multiple design load cases). If the layup selected were optimized for a single load case (or
few load cases), it might not have enough fibres in other load directions where the loads may
be significantly lower, but could still lead to premature failure if not enough fibres are present
in those directions. Finally, robust performance under impact requires +45°/−45° plies to be
placed on the outside of laminates susceptible to impact. (This is more a consequence of the
observation that layups with 45° fabric material on the outside tend to more effectively contain
impact damage and minimize ply splitting.)

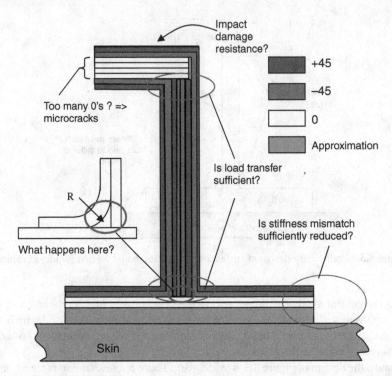

Figure 8.4   Improved stiffener cross-section design (see Plate 17 for the colour figure)

Introducing these requirements to the design in Figure 8.3 results in the improved configuration of Figure 8.4.

In this case, a 45/−45 pair is on the outside to improve impact resistance. It is not, however, clear whether this is sufficient especially in view of the complete layup of, say, the top flange where four 0° plies are stacked next to the two outer 45° plies. Also, it is known that stacking too many plies of the same orientation next to each other leads to the creation of sizeable microcracks during cure or during loading perpendicular to the fibres. The reason is that a matrix crack forming between fibres in a ply can progress easily to the next ply since there are no fibres perpendicular to it in the next ply to stop or slow down its growth. So the improved design in Figure 8.4 still has issues associated with the layup of the flanges.

Furthermore, the continuity of plies around the corners may or may not be sufficient and it will depend on the applied loads. At the top and bottom of the web several plies terminate causing stress concentrations. In particular, adjacent to the terminated plies are plies turning away from the web and into the flange. Typically, it is very hard to force the turning plies to conform perfectly to the 90° turn, and a small radius as shown in the enlarged detail of Figure 8.4 will be present. The resulting gap between terminated and turned plies is usually filled with resin, creating a weak spot for the entire cross-section.

Finally, reaching a compromise layup for the flange next to the skin so that the stiffness mismatch at the flange termination is minimized is difficult and more information about applied loading and skin layup is needed for further improvement. The design configuration of Figure 8.4 will be periodically revisited and improved upon in future chapters as a better understanding of designing to specific requirements is developed.

8.2 Cross-Sectional Properties

The axial (EA) and bending (EI) stiffnesses of a beam are very often used in design and analysis of such structures and, therefore, accurate determination of their values for a composite cross-section is very important. There are some significant differences from metal cross-sections stemming from the fact that, for a cross-section made using composite materials, different members may have different layups and thus stiffnesses.

These differences become evident in the calculation of the location of the neutral axis for a cross-section made up of composite materials. With reference to Figure 8.2, the neutral axis is located at

$$\bar{y} = \frac{\sum (EAy)_j}{\sum (EA)_j} \tag{8.1}$$

Note that E here is either the membrane or bending modulus of each member (see Section 3.3 and Equations 3.65). The two moduli are, in general, different. This means that one should differentiate between axial and bending problems and use the appropriate moduli. In what follows, the axial stiffness EA (= modulus × cross-sectional area), which is a quantity needed in uniaxial loading situations, is calculated first and the bending stiffness EI (= modulus × moment of inertia), which is used in bending problems, is calculated afterwards.

In order to determine the axial stiffness (EA) of a cross-section, assume, for simplicity, that the layup of each member is symmetric and balanced. Then, denoting the beam axis as the x-axis, a uniaxial loading situation for member i is represented by

$$(N_x)_i = (A_{11})_i (\varepsilon_x)_i + (A_{12})_i (\varepsilon_y)_i$$
$$(N_y)_i = (A_{12})_i (\varepsilon_x)_i + (A_{22})_i (\varepsilon_y)_i \tag{8.2}$$

where the subscript i denotes the ith member. This is the same as Equation (3.49) adjusted to uniaxial loading ($N_{xy} = 0$) and symmetric and balanced layups (B matrix $= A_{16} = A_{26} = 0$).

If now only load N_x is applied, $N_y = 0$ and substituting in Equation (8.2) and solving for $(\varepsilon_y)_i$ give

$$(\varepsilon_y)_i = -\left(\frac{A_{12}}{A_{22}}\right)_i (\varepsilon_x)_i \tag{8.3}$$

This result can now be substituted into the first of Equations (8.2) to obtain

$$(N_x)_i = \left(A_{11} - \frac{A_{12}^2}{A_{22}}\right)_i (\varepsilon_x)_i \tag{8.4}$$

If both sides of Equation (8.4) are divided by the thickness of the member t_i the left-hand side becomes the applied stress:

$$(\sigma_x)_i = \underbrace{\frac{1}{t_i}\left(A_{11} - \frac{A_{12}^2}{A_{22}}\right)_i}_{E_i} (\varepsilon_x)_i \tag{8.5}$$

It can be seen by inspection of Equation (8.5) that the quantity multiplying the strain in the right-hand side is the equivalent axial modulus of the member, which was also given in Section 3.3, Equation (3.59), as a slightly different (but equivalent) expression:

$$E_i = \frac{1}{t_i}\left(A_{11} - \frac{A_{12}^2}{A_{22}}\right) = \frac{1}{(a_{11})_i t_i} \tag{8.6}$$

with a_{11} the 11 entry of the inverse of the A matrix.

The axial or membrane stiffness EA of member i can now be written as

$$(EA)_i = E_i b_i t_i \tag{8.7}$$

with b_i and t_i the width and thickness of the member, respectively.

Consider now that an axial force F_{TOT} is applied to the entire cross-section. Because of the different EA values for each member, the corresponding forces acting on each member will be different. For the three-member cross-section of Figure 8.2, the total force equals the sum of the forces acting on the individual members:

$$F_{TOT} = F_1 + F_2 + F_3 \tag{8.8}$$

but the force F acting on each member is related to the corresponding force per unit width N_x via

$$(N_x)_i = \frac{F_i}{b_i} \tag{8.9}$$

Now for a uniaxial loading case with the load applied at the neutral axis, uniform extension or compression results, which means the strains in all members of the cross-section are equal:

$$(\varepsilon_x)_1 = (\varepsilon_x)_2 = (\varepsilon_x)_3 = \varepsilon_a \tag{8.10}$$

Combining Equations (8.4), (8.6), (8.7), (8.9) and (8.10) it can be shown that

$$\frac{F_1}{(EA)_1} = \frac{F_2}{(EA)_2} = \frac{F_3}{(EA)_3} = \frac{F_{TOT}}{(EA)_{eq}} \tag{8.11}$$

with $(EA)_{eq}$ the equivalent membrane stiffness for the entire cross-section.

Combining Equations (8.11) with Equation (8.8) and solving for the individual forces on each member it can be shown that the force on member i is given by

$$F_i = \frac{(EA)_i}{\sum_{j=1}^{3}(EA)_j} F_{\text{TOT}} = \frac{E_i b_i t_i}{\sum_{j=1}^{3} E_j b_j t_j} F_{\text{TOT}} \tag{8.12}$$

with $(EA)_j$ given by Equation (8.7).

Equations (8.11) and (8.12) can be combined in order to determine the equivalent axial stiffness for the entire cross-section. Eliminating the forces gives

$$(EA)_{\text{eq}} = \sum_{j}(EA)_j \tag{8.13}$$

The situation for pure bending is shown in Figure 8.5. Each member contributes to the EI calculation for the entire cross-section according to

$$(EI)_i = E_{bi}\left[\frac{(\text{width})_i(\text{height})_i^3}{12} + A_i d_i^2\right] \tag{8.14}$$

where A_i is the area of the ith member $(= b_i t_i)$ and d_i is the distance of the neutral axis of the ith member from the neutral axis of the entire cross-section determined by Equation (8.1) and the bending modulus is given by Equation (3.63):

$$E_{bi} = \frac{12}{t_i^3(d_{11})_i} \tag{8.15}$$

If a bending moment M_{TOT} is applied to the beam, the individual bending moments and overall bending stiffness of the cross-section are calculated in a manner analogous to the case

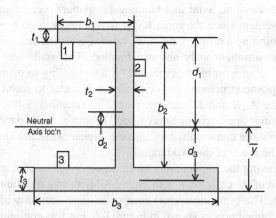

Figure 8.5 Definition of pertinent quantities for a beam cross-section in bending

of uniaxial loading. However, instead of the strain compatibility condition that required that the strains in all members be equal, here the requirement is that the radii of curvature R_{ci} for all members are all equal to that of the neutral axis of the entire cross-section. Therefore,

$$R_{c1} = R_{c2} = R_{c3} = R_{ca} \tag{8.16}$$

In addition,

$$M_{TOT} = M_1 + M_2 + M_3 \tag{8.17}$$

Also, the local radius of curvature is given by the well-known moment–curvature relation of simple beam theory,

$$R_{ci} = \frac{(EI)_i}{M_i} \tag{8.18}$$

Combining Equations (8.16) to (8.18) and solving for the moments acting on each individual member give

$$M_i = \frac{(EI)_i}{\sum\limits_{j=1}^{3} (EI)_j} M_{TOT} \tag{8.19}$$

with $(EI)_i$ given by Equation (8.14). This relation is the analogous relation to Equation (8.12) for the axial forces on the members of the cross-section.

With the individual moments given by Equation (8.19), the overall bending stiffness of the cross-section can be obtained from Equations (8.16), (8.18) and (8.19):

$$(EI)_{eq} = \sum\limits_{j} (EI)_j \tag{8.20}$$

In the preceding discussion, axial and bending behaviours were completely uncoupled, which was very convenient since the same layup would undergo each type of deformation (axial or bending) exhibiting a different modulus value. Care must be exercised if both modes of deformation occur simultaneously and are coupled. This issue was first mentioned in Section 7.2.1 and points to a problem associated with attempting to oversimplify the design and analysis of composite structures. In such cases it is better to resort to the constitutive relations involving the A, B and D matrices (see, e.g. Equation 3.49). As a simpler, less accurate, but conservative approach, one may select the one of the two moduli (membrane or bending) that leads to more conservative results (e.g. higher post-buckling deflections which are caused by using the lower of the two moduli).

As an example showing the implications of the equations presented so far in this section consider a comparison of the cross-section of Figure 8.4 with an aluminium cross-section with the same dimensions. The (graphite/epoxy) composite cross-section with the flange next to the skin completely defined now is shown in Figure 8.6. For this configuration, the pertinent quantities are shown in Table 8.1.

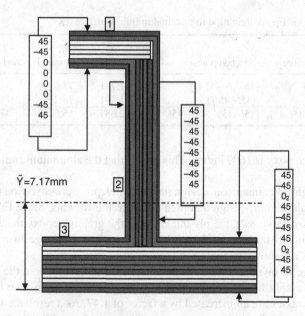

Figure 8.6 Baseline J stiffener cross-section made out of composite materials (see Plate 18 for the colour figure)

Using Equation (8.1) the neutral axis is found to be located 7.17 mm away from the outer edge of the bottom flange (Figure 8.6). Using Table 8.1 and Equations (8.13) and (8.20) the membrane (EA) and bending (EI) stiffnesses of the cross-section of Figure 8.6 are found and compared with the case of aluminium with the exact same geometry in Table 8.2.

Table 8.2 shows that the same geometry made with aluminium has significantly higher stiffness both in-plane (47% higher) and bending (122% higher). Since the geometries are identical between aluminium and composite, it is easy to compare the respective weights. The weight ratio will equal the density ratio. The density of aluminium is 2774 kg/m³ while the

Table 8.1 Properties for baseline composite configuration of Figure 8.6

Member	b (mm)	t (mm)	E_m (GPa)	E_b (GPa)
1	12.7	1.2192	75.6	32.4
2	31.75	1.2192	18.2	17.9
3	38.1	1.8288	56.5	47.9

Table 8.2 Composite versus same geometry aluminium

	Al	Composite	Δ (%)
EA (kN)	8525	5803	46.9
EI (Nm²)	1401	631	121.8

Table 8.3 Revised Gr/Ep configuration to match aluminium stiffnesses

Member	layup before	layup now	b (mm)	t (mm)	E_m (stays same)	E_b before (GPa)	E_b now (GPa)
1	$[45/-45/0_2]_s$	$[45/0_2/-45]_s$	12.7	1.791	75.6	32.4	62.7
2	$[(45/-45)_2]_s$	$[(45/-45)_2]_s$	31.75	1.791	18.2	17.9	17.9
3	$[45/-45/0_2/45/-45]_s$	$[45/-45/0_2/45/-45]_s$	38.1	2.687	56.5	47.9	47.9

density of graphite/epoxy is 1609 kg/m^3. This means that the aluminium configuration is 72% heavier.

For a more insightful comparison, design the graphite/epoxy cross-section to have the same stiffnesses as the aluminium design and then compare the weights. Since EA for aluminium is 46.9% higher (see Table 8.2), the ply thickness of the graphite/epoxy material is increased by 46.9% and a minor reshuffling of the stacking sequence of the flange away from the skin is done. The changes are shown in Table 8.3.

The only change in layup is in the flange away from the skin where the starting 45/−45 combination is now split by moving the pair of 0 plies between them. The b values stay the same but the thicknesses are all increased by a factor of 1.47. As a result the bending stiffness of the flange away from the skin is now increased from 32.4 to 62.7 GPa. It should be noted that simply scaling the ply thickness as was suggested here is not usually possible. The raw material is available only in limited ply thicknesses, so increasing to a specified value would require rounding up to the next integral multiple of ply thickness. So the results of this example are only approximate.

With the changes of Table 8.3, the graphite/epoxy cross-section now matches (or is very close to) the stiffnesses of the aluminium cross-section as is shown in Table 8.4.

Now the weight comparison includes both the density and thickness difference. The weight ratio of the two configurations is given by

$$\frac{W_{\text{Gr/Ep}}}{W_{\text{Al}}} = 0.58 \frac{1.469}{1} = 0.852$$

$$\uparrow$$

density ratio for carbon/epoxy

or, the composite design is, approximately, 15% lighter.

The previous example points to the important fact that once the design and best-practices rules and other constraints are imposed on the composite design, the weight savings are drastically reduced compared with the nominal savings one would obtain based on the tension or compression strength of a unidirectional ply. The savings of 15% found here is typical of the

Table 8.4 Comparison of aluminium and G/E stiffnesses for revised configuration

	Al	Composite	Δ (%)
EA (kN)	8525	8525	0.0
EI (Nm^2)	1401	1441	2.8

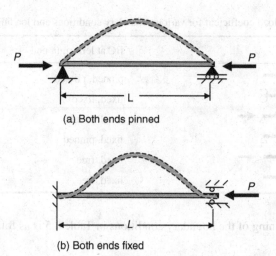

Figure 8.7   Buckling of simply supported (pinned) and clamped (fixed) beams

performance of modern composite materials when used on airframe structures. A somewhat higher value of 38% savings was found in Section 6.2 for a buckling application. In general, the weight savings rarely exceeds 30% and not without detailed evaluation of all possible failure modes and use of a good, robust optimization scheme.

8.3   Column Buckling

In column buckling a beam under compression suddenly deflects perpendicular to its axis. With EI defined in the previous section, the standard buckling expressions can be used for the two cases shown in Figure 8.7. It should be noted that, for buckling load calculations, the membrane modulus E_m given by Equation (8.6) should be used.

The solutions for the buckling loads are well known and can be found elsewhere in the literature, for example [3, 4]. For convenience, they are provided here without derivation:

$$P_{cr} = \frac{\pi^2 EI}{L^2} \text{(pinned ends)} \tag{8.21}$$

$$P_{cr} = \frac{4\pi^2 EI}{L^2} \text{(fixed ends)} \tag{8.22}$$

As is seen from the coefficient in the right-hand side of Equations (8.21) and (8.22), the boundary conditions at the beam ends play a big role in the value of the buckling load. A brief compilation of buckling load values for different boundary conditions and loadings is given in Table 8.5. In all cases the buckling load is given by

$$P_{cr} = \frac{c\pi^2 EI}{L^2} \tag{8.23}$$

and the coefficient c is given in Table 8.5.

Table 8.5   Buckling load coefficient for various boundary conditions and loadings

Configuration	BC at left, right end	c
	pinned, pinned	1.88
	fixed, fixed	7.56
	fixed, pinned	2.05
	fixed, pinned	5.32
	fixed, free	0.25
	fixed, free	0.80

The physical meaning of the boundary conditions in Table 8.5 is as follows:

- free: free rotation and free translation
- pinned: free rotation and fixed translation
- fixed: fixed rotation and fixed translation

8.4   Beam on an Elastic Foundation under Compression

The situation is shown in Figure 8.8. A beam rests on an elastic foundation which has a spring constant k. In general, the beam ends have linear (K_1 and K_2) and torsional (G_1 and G_2) springs restraining them. Depending on the spring stiffnesses, the end boundary conditions can range from free to clamped and can achieve any intermediate value.

To gain insight into the problem, the case of a simply supported beam ($G_1 = G_2 = 0$, $K_1 = K_2 = \infty$) is solved first in detail. This is done using energy methods.

Referring to the discussion in Section 5.4 the one-dimensional counterpart of the energy expression is

$$\Pi_c = \frac{1}{2} \int_0^L EI \left(\frac{d^2 w}{dx^2} \right)^2 dx + \frac{1}{2} \int_0^L (-P) \left(\frac{dw}{dx} \right)^2 dx + \frac{1}{2} \int_0^L k w^2 dx \qquad (8.24)$$

where w is the out-of-plane displacement of the beam.

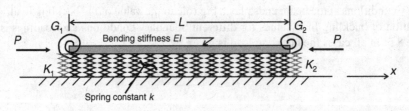

Figure 8.8   Beam on an elastic foundation

The first term in the right-hand side of Equation (8.24) is the potential energy stored in bending the beam. The second term is the work done by the external force P and the third term is the energy stored in the spring foundation. The units of k are force/area.

An expression for w is assumed such that the boundary conditions that $w = 0$ at the two ends of the beam are satisfied:

$$w = \sum A_m \sin \frac{m\pi x}{L} \tag{8.25}$$

where L is the length of the beam and A_m are unknown coefficients.

Substituting in the energy expression (8.24) and performing the integrations results in

$$\Pi_c = \sum \left[\frac{(\mathrm{EI})m^4\pi^4}{4L^3} - \frac{Pm^2\pi^2}{4L} + \frac{kL}{4} \right] A_m^2 \tag{8.26}$$

The energy must be minimized with respect to the unknowns A_m which leads to

$$\frac{\partial \Pi_c}{\partial A_m} = 0 \tag{8.27}$$

Carrying out the differentiation and setting the result equal to zero yield the following equation:

$$2\left[\frac{(\mathrm{EI})m^4\pi^4}{4L^3} - \frac{Pm^2\pi^2}{4L} + \frac{kL}{4} \right] A_m = 0 \tag{8.28}$$

This is a matrix equation with a diagonal matrix multiplying the vector A_m:

$$\begin{bmatrix} \frac{\mathrm{EI}\pi^4}{4L^3} + \frac{kL}{4} - \frac{P\pi^2}{4L} & 0 & 0 & \cdots \\ 0 & \frac{\mathrm{EI}(16)\pi^4}{4L^3} + \frac{kL}{4} - \frac{P(4)\pi^2}{4L} & 0 & \cdots \\ 0 & 0 & \frac{\mathrm{EI}(81)\pi^4}{4L^3} + \frac{kL}{4} - \frac{P(9)\pi^2}{4L} & \cdots \\ & \cdots & \cdots & \cdots \end{bmatrix} \begin{Bmatrix} A_1 \\ A_2 \\ A_3 \\ A_4 \\ \vdots \end{Bmatrix} = 0$$

The obvious possibility $A_m = 0$ corresponds to the uniform compression pre-buckling case. Therefore, for out-of-plane deflections to be possible (A_m must be different from zero) the determinant of this matrix must equal zero. Defining

$$K_{mm} = \frac{\pi^2 \mathrm{EI}}{L^2} \left(m^2 + \frac{kL^4}{\pi^4(\mathrm{EI})m^2} \right) \tag{8.29}$$

the matrix equation can be rewritten in the form

$$
\begin{bmatrix}
K_{11} - P & 0 & 0 & 0 & 0 \\
0 & K_{22} - P & 0 & 0 & 0 \\
0 & 0 & K_{33} - P & 0 & 0 \\
0 & 0 & 0 & \cdots & \\
0 & 0 & \cdots & \cdots & \cdots
\end{bmatrix}
\begin{Bmatrix}
A_1 \\
A_2 \\
\cdots \\
\cdots \\
\cdots
\end{Bmatrix} = 0
\qquad (8.30)
$$

Since the matrix in the left-hand side is diagonal, setting its determinant equal to zero is equivalent to setting the product of the diagonal terms equal to zero:

$$
(K_{11} - P)(K_{22} - P)(K_{33} - P) \cdots = 0 \qquad (8.31)
$$

There are as many solutions to Equation (8.31) as there are terms. Of them, the solution that results in the lowest buckling load $P = P_{cr}$ is selected:

$$
P_{cr} = \min(K_{ii}) \qquad (8.32)
$$

From Equations (8.31) and (8.29), the buckling load has the form

$$
\frac{P_{cr}}{\dfrac{\pi^2 EI}{L^2}} = m^2 + \frac{kL^4}{\pi^4 EI} \frac{1}{m^2} \qquad (8.33)
$$

where K_{mm} from Equation (8.29) was rearranged to bring the term $\pi^2 EI/L^2$ to the left-hand side.

As a special case of Equation (8.33) consider the situation in which $k = 0$. This would be the case of a pinned beam under compression. Then, the critical buckling load would be given by

$$
P_{cr} = \frac{\pi^2 EI}{L^2} m^2 \qquad (8.34)
$$

which is minimized for $m = 1$. If $m = 1$, Equation (8.34) is identical to Equation (8.21) and the exact solution for this case is recovered.

In the general case when $k \neq 0$, the value of m that minimizes the right-hand side of Equation (8.32) or (8.33) depends on the value of k itself. This can be seen more easily graphically where the normalized buckling load (left-hand side of Equation 8.33) is plotted as a function of the parameter $kL^4/(\pi^4 EI)$ which appears in the right-hand side of Equation (8.33). This plot is shown in Figure 8.9.

For each value of m, the right-hand side of Equation (8.33) is a straight line. The straight lines corresponding to different values of m are shown in Figure 8.9. The bold black line giving the envelope of the lowest values of the buckling load defines the critical buckling load

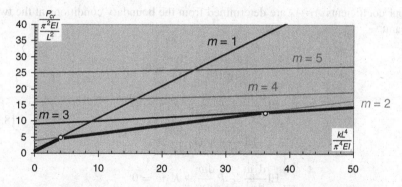

Figure 8.9 Buckling load of a beam on elastic foundation

for a given value of the parameter $kL^4/(\pi^4EI)$. It is seen that for low values of this parameter, $m = 1$ (one half-wave over the entire length of the beam) gives the lowest buckling load. As the value of this parameter increases, the buckling mode progressively switches to $m = 2$ (two half-waves along the beam length), $m = 3$, etc. Unlike the case of no elastic foundation where the beam always buckles in one half-wave ($m = 1$), the presence of an elastic foundation changes the buckling mode. The higher the value of $kL^4/(\pi^4EI)$, the higher the number of half-waves (value of m) into which the beam buckles.

For boundary conditions other than pinned ends, the governing equation

$$EI\frac{d^4w}{dx^4} + P\frac{d^2w}{dx^2} + kw = 0 \tag{8.35}$$

can be solved. The solution has the form

$$w = Ae^{px} \tag{8.36}$$

with the exponent p given by

$$p = \pm\sqrt{\frac{-\dfrac{P}{EI} \pm \sqrt{\left(\dfrac{P}{EI}\right)^2 - \dfrac{4k}{EI}}}{2}} \tag{8.37}$$

Thus, there are four solutions of the form (8.36):

$$w = A_1e^{p_1x} + A_2e^{p_2x} + A_3e^{p_3x} + A_4e^{p_4x} \tag{8.38}$$

The four coefficients A_1–A_4 are determined from the boundary conditions at the two ends of the beam:

$$-\text{EI}\frac{d^2w}{dx^2} + G_1\frac{dw}{dx} = 0$$

$$\text{EI}\frac{d^3w}{dx^3} + P\frac{dw}{dx} + K_1w = 0$$

$$-\text{EI}\frac{d^2w}{dx^2} + G_2\frac{dw}{dx} = 0 \qquad \text{(8.39a–d)}$$

$$\text{EI}\frac{d^3w}{dx^3} + P\frac{dw}{dx} + K_2w = 0$$

Equations (8.39a) and (8.39c) are statements of moment equilibrium at the two beam ends respectively, that is, the moment caused by the torsional spring $G(dw/dx)$ equals the beam bending moment at that end. Equations (8.39b) and (8.39b) express shear force equilibrium at the same locations.

Detailed results for various values of the spring constants G_1, G_2, K_1 and K_2 can be found in [5]. Following the approach in that reference, the following parameters are defined (subscript $i = 1, 2$ denotes end $x = 0$ or end $x = L$):

$$R_i = \frac{G_iL}{\text{EI}}$$

$$\rho_i = \frac{1}{1 + \dfrac{3}{R_i}} \qquad \text{(8.40)}$$

Then, $\rho_i = 0$ implies no torsional stiffness at end i or the beam is free to attain any slope locally (pinned end). Also, $\rho_i = 1$ implies infinite torsional stiffness at end i or the beam has zero slope at that end (fixed end). As an example, the case where the beam is pinned at the left end and has variable stiffness at the other end is shown in Figures 8.10 and 8.11.

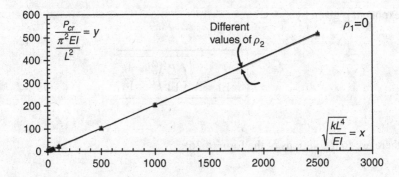

Figure 8.10 Buckling load of beam on elastic foundation pinned at one end and with variable rotational restraint at the other

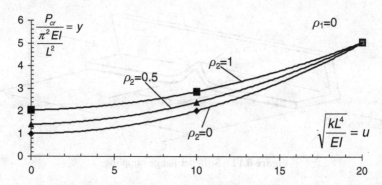

Figure 8.11 Buckling load of beam on elastic foundation pinned at one end and with variable rotation restraint at the other (detail of Figure 8.10 for low values of x)

In addition to Figures 8.10 and 8.11, approximate equations for a range of values of the parameters were obtained by best-fitting the results in [5]. These approximate equations are given in Table 8.6.

8.5 Crippling

Crippling is a stability failure where a flange of a stiffener locally buckles and then collapses. This is shown graphically in Figure 8.12. Under compressive load one (or more) of the flanges buckles locally with a half-wavelength ℓ, which is much smaller than the length L of the stiffener. Once the flange buckles it can support very little load in the post-buckling regime

Table 8.6 Buckling load y as a function of elastic foundation stiffness x and boundary rigidity $\rho_1 = 0$

ρ_2	x	y	r^2 (goodness of fit)
0	$0 \leq x \leq 20$	$0.0099x^2 + 0.0041x + 1$	1.0000
0	$20 \leq x \leq 100$	$0.0003x^2 + 0.1579x + 1.6161$	0.9987
0.5	$0 \leq x \leq 20$	$0.0083x^2 + 0.0169x + 1.4069$	1.0000
0.5	$20 \leq x \leq 100$	$0.0002x^2 + 0.1714x + 1.4518$	0.9988
1	$0 \leq x \leq 20$	$0.0069x^2 + 0.01134x + 2.046$	1.0000
1	$20 \leq x \leq 100$	$7 \times 10^{-5}x^2 + 0.1924x + 1.2722$	0.9998

$\rho_1 = \rho_2$	x	y	r^2 (goodness of fit)
0.2	$0 \leq x \leq 20$	$0.099x^2 + 0.0039x + 1.28$	1.0000
0.2	$20 < x \leq 100$	$0.0004x^2 + 0.1517x + 1.9512$	0.9987
0.5	$0 \leq x \leq 20$	$0.0099x^2 + 0.0019x + 1.916$	1.0000
0.5	$20 < x \leq 100$	$0.0003x^2 + 0.1539x + 2.5361$	0.999
1	$0 \leq x \leq 20$	$0.0051x^2 + 0.0265x + 4$	1.0000
1	$20 < x \leq 100$	$-0.0003x^2 + 0.2385x + 2.3368$	0.9943

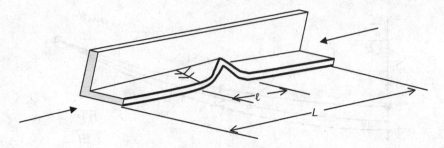

Figure 8.12   Stiffener flange crippling

and fails (collapses). Its load is then shared by other members of the cross-section (if they have not failed) until the entire cross-section collapses.

Crippling is one of the most common failure modes in a composite airframe. It may occur on stiffeners, stringers, panel breakers, beams, ribs, frame caps and all other members that are stability critical and do not fail by global buckling. An approximate distribution of failure modes for fuselage and wing of an aircraft is shown in Figure 8.13.

It can be seen from Figure 8.13 that crippling designs as much as one quarter of the parts in a composite airframe. The distribution in Figure 8.13 should not be viewed as exact and it will vary significantly from one application to the next. The main message however, about the importance of certain failure modes such as crippling, carries over to most applications.

In a one-dimensional structure under compression such as a stiffener, crippling competes with (at least) two other failure modes: material failure and column buckling. Typically, for a robust design, material failure (not preceded by some stability failure) is not the driver because it leads to heavy designs. Between crippling and column buckling failure, crippling is preferred

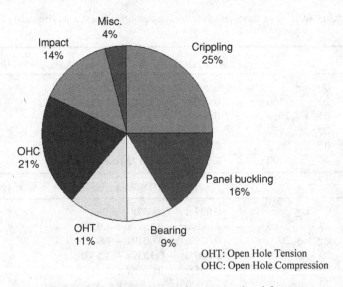

OHT: Open Hole Tension
OHC: Open Hole Compression

Figure 8.13   Failure modes in a composite airframe

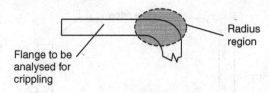

Figure 8.14 Radius region at the root of a flange

as the primary failure mode. The reason is that it typically occurs on one member (or portion of a member in bending situations as will be discussed in Section 8.5.3) and when the crippled flange collapses there is a good chance that the remaining members of the cross-section may be able to absorb some or all of the load originally in the failed flange, thus precluding or delaying complete failure of the stiffener. In column buckling the whole stiffener fails. As a result, crippling has a better chance of leading to a robust design and is preferred as the designing failure mode. This does not mean that there are no cases where column buckling is the driver, especially in long beams. In such cases, if a weight-competitive design can be generated, the approach is to either increase the bending stiffness of the entire cross-section or shorten the unsupported length of the beam so that column buckling happens at a higher load than flange crippling.

To analyse crippling in detail one would first have to obtain the portion of the total load that acts on each flange and then determine the corresponding buckling load. The flange in that case would be modelled as a long plate with three edges simply supported and one edge free. Then, a post-buckling solution similar to that in Chapter 7 would have to be carried out to determine deflection strains and stresses and some failure criterion applied to determine the load for final collapse. Such an approach is cumbersome and relies on many simplifying assumptions to make the solution tractable. As a result the solution is not accurate enough. In addition, there are issues with modelling the boundary condition at the root of the flange. Typically there is a radius region (Figure 8.14) and there is some finite stiffness, meaning the boundary condition is somewhere between simply supported and clamped. The exact type of boundary condition depends on the radius, thickness, layup and in- and out-of-plane stiffnesses in a complex way further complicating the possibility of generating accurate analytical predictions for this failure mode.

For isotropic configurations, attempts have been made [6] to account for the local specifics of the boundary conditions, in particular for beams that are nominally cantilevered. But more work is needed in this area with an extension to composites before more accurate analytical models for crippling analysis can be developed.

A semi-empirical approach has been favoured instead. Two cases are distinguished as shown in Figure 8.15: (a) one-edge-free (OEF) and (b) no-edge-free (NEF). In the OEF case, one end of the flange is constrained, for example by being attached to a web or other member, and the other end is free. In the NEF case, both flange ends are constrained from moving.

8.5.1 One-Edge-Free (OEF) Crippling

Consider the situation shown in Figure 8.16. The cross-section has three OEF flanges, one at the top and two at the bottom on either side of the vertical web. The buckling load for each

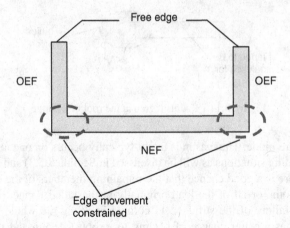

Figure 8.15 One-edge-free and no-edge-free flanges in a cross-section

flange corresponds to that of a long plate with three sides simply supported and one side free. This case was addressed in Chapter 6 for finite and infinite flange length (see Equations 6.12a and 6.13a). For convenience, the buckling load (infinitely long flange) is repeated here:

$$N_{x\text{crit}} = \frac{12D_{66}}{b^2} \tag{6.13a}$$

Looking at Equation (6.13a), for a given flange width b, to maximize the buckling load one should maximize the twisting stiffness D_{66}. One way to see how D_{66} can be maximized for a given laminate thickness is to consider how each ply contributes to the D_{66} term for the entire laminate.

Consider the situation shown in Figure 8.17 for a symmetric and balanced laminate.

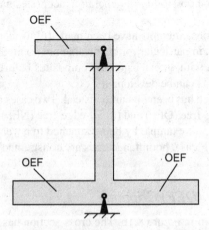

Figure 8.16 OEF flanges in a 'J' stiffener

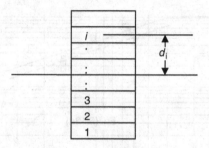

Figure 8.17   Section cut of a symmetric and balanced laminate

The equation that determines the contribution of the *i*th ply to the D_{66} term for that laminate (Equation 3.47) can be recast into the form

$$(D_{66})_c = 2D_{66}^{(i)} + 2A_{66}^{(i)}d_i^2 \tag{8.41}$$

where the superscript (i) denotes quantities for the *i*th ply with respect to its own mid-plane. Equation (8.41) is essentially the same as Equation (8.14) per unit width with the first term of Equation (8.14) being replaced by the corresponding D term of the *i*th ply and the second term replaced by the EA term of the *i*th ply. The factors of 2 in the right-hand side of Equation (8.41) account for the contribution of two plies symmetrically located with respect to the mid-plane of the laminate.

Based on Equation (8.41), the biggest contribution to the D_{66} term for the entire laminate comes from the second term in the right-hand side of Equation (8.41) because of the presence of the distance d_i. Terms away from the mid-plane contribute more to the bending stiffness. Therefore, one should place material with the high A_{66} term as far away from the mid-plane as possible. Now maximizing A_{66} for a single ply amounts to maximizing the corresponding expression for Q_{66} (see Equation 3.33d repeated below):

$$Q_{66} = \left(Q_{xx} + Q_{yy} - 2\left(Q_{xy} + Q_{ss}\right)\right)\sin^2\theta\cos^2\theta + Q_{66}\left(\sin^4\theta + \cos^4\theta\right)) \tag{3.33d}$$

By differentiating the right-hand side with respect to θ and setting the result equal to zero, it can be shown that Q_{66}, and therefore A_{66}, is maximized when $\theta = 45°$. Therefore, on the basis of Equation (6.13a), to maximize crippling performance one should select a flange layup that has the 45° plies as far from the mid-plane as possible.

This conclusion contradicts another of the design guidelines, stated in Section 8.1, that the 0° plies should be placed as far from the mid-plane as possible. This is not an inconsistency as the 0° ply guideline is for increasing the bending stiffness D_{11} and applies to column buckling while the 45° ply guideline is for increasing the bending stiffness D_{66} and applies to crippling. This situation occurs frequently in practice where different design requirements point to different directions and a compromise between them must be reached. In addition, as will be discussed below, the requirement of as many 45° plies away from the mid-plane as possible is not sufficient to guarantee optimum crippling performance and other ply orientations are also necessary.

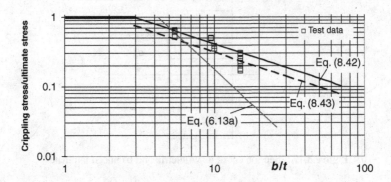

Figure 8.18   OEF crippling test results compared to buckling predictions

Equation (6.13a) is compared with test results for various layups and stiffener geometries in Figure 8.18. It is customary and insightful to plot crippling stress normalized by the compressive strength of the respective layup as a function of the ratio b/t (width divided by thickness) for the respective flange. For low b/t values the crippling strength is essentially the same as the compressive strength of the flange. This would be the case of a thick flange where buckling is delayed because of the high bending stiffness and material strength is the operative failure mode. For high b/t values the crippling strength drops rapidly as b/t increases, showing the sensitivity to reduced bending stiffnesses of the flange.

As is seen from Figure 8.18, the theoretical prediction of Equation (6.13a) is higher than the test results for low b/t values ($b/t < 6$) but close to them. However, it becomes very conservative for high b/t values. The reasons for this, as already mentioned above, are related to the boundary conditions at the edge of the flange that are not captured by Equation (6.13a), which assumes a simply supported edge, and to the post-buckling capability of the flange, which becomes more and more pronounced for larger b values. This would be the case where the flange stops behaving as a one-dimensional structure and behaves more like a plate, which, as was discussed in relation to Figure 7.2 in the previous chapter, would result in improved post-buckling load-carrying ability.

In view of these differences between analysis and experimental results and the difficulties associated with improving the analysis without resorting to expensive computational approaches, a semi-empirical approach has been adopted where the crippling strength is correlated with the b/t ratio. Over a large set of test results with different materials and layups, it has been found [7] that the following expression fits the data well:

$$\frac{\sigma_{\text{crip}}}{\sigma_c^u} = \frac{2.151}{\left(\dfrac{b}{t}\right)^{0.717}} \tag{8.42}$$

valid for $b \geq 2.91t$; for $b < 2.91t$, $\sigma_{\text{crip}} = \sigma_c^u$.

The two constants in Equation (8.42) are determined by best-fitting the data. For design, Equation (8.42) is modified to guarantee that at least 90% of the tests are higher than the prediction (see B-Basis definition in Section 5.1.3 and Figure 5.8). The design equation is then:

$$\frac{\sigma_{\text{crip}}}{\sigma_c^u} = \frac{1.63}{\left(\dfrac{b}{t}\right)^{0.717}} \tag{8.43}$$

valid for $b \geq 1.98t$; for $b < 1.98t$, $\sigma_{\text{crip}} = \sigma_c^u$.

In Equations (8.42) and (8.43) σ_c^u is the ultimate compressive strength of the flange which can be determined, for example, as the first-ply failure of the flange under compression (see Chapter 4 for first-ply-failure criteria). The predictions of these two equations are also shown in Figure 8.18. It can be seen that Equation (8.42) fits the present data well while Equation (8.43) is below most of the data and thus could be used as a design equation. It is important to note that the test results shown in Figure 8.18 were not used in generating the semi-empirical curves of Equation (8.42) or (8.43) so the agreement seems to reinforce the usefulness and applicability of these two equations.

An important note on applicability: As the test data on which Equations (8.42) and (8.43) are based come from laminates with at least 25% 0° plies and 25% 45° plies, these equations should be used only with layups that fall in this category. Extending to other layups with less 0° and/or 45° plies is not recommended. In any case, most flange designs do obey this requirement of at least 25% 0° and at least 25% 45° plies as a compromise between the two design requirements, already presented, of 0° plies for high D_{11} and 45° plies for high D_{66} and respective high EA $\times$ d^2 contribution.

A final point relating to the presence of 0° and 45° plies is in order. As was mentioned above, 45° plies away from the laminate mid-plane maximize D_{66} and thus the buckling load of the flange as given by Equation (6.13a). However, especially for large b/t values where the flange behaves as a plate and has significant post-buckling capability, using mostly 45° plies in the flange is not recommended. 0° plies are also required to increase the post-buckling strength. This is captured in Equation (8.43) in σ_c^u, the compression strength of the flange. As a result, in practice flange layups with at least 25% 0° and 25% 45° plies are used. They have been demonstrated by test to have better crippling performance.

8.5.2 No-Edge-Free (NEF) Crippling

The situation is shown in Figure 8.19. The vertical web in this figure is supported at the two ends by the flanges and is treated as no-edge-free (NEF) web.

The web in this case can be modelled as a long plate that is simply supported all around its boundary. This case was examined in detail in Section 6.2. Starting from the buckling load given by Equation (6.7),

$$N_o = \frac{\pi^2}{a^2}\left[D_{11}m^2 + 2(D_{12} + 2D_{66})(AR)^2 + D_{22}\frac{(AR)^4}{m^2}\right] \tag{6.7}$$

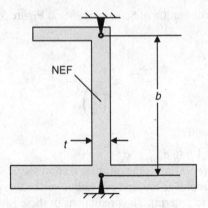

Figure 8.19   NEF web in a J stiffener

the corresponding expression for a very long plate ($a \to \infty$) can be determined as follows: The term a^2 is brought inside the brackets and a factor b^2 is factored out using the fact that $AR = a/b$. Also, the square root of the product $D_{11}D_{22}$ is factored out, giving

$$N_{x\text{crit}} = \frac{\pi^2}{b^2}\sqrt{D_{11}D_{22}} \left[\frac{m^2 b^2}{a^2\sqrt{\dfrac{D_{22}}{D_{11}}}} + \frac{2(D_{12}+2D_{66})}{\sqrt{D_{11}D_{22}}} + \sqrt{\frac{D_{22}}{D_{11}}}\frac{a^2}{b^2 m^2} \right] \tag{8.44}$$

To determine the number of half-waves m that minimizes Equation (8.46), the right-hand side is differentiated with respect to m and the result set equal to zero. This results in the equation

$$\frac{m b^2}{a^2\sqrt{\dfrac{D_{22}}{D_{11}}}} = \frac{a^2}{b^2 m^3}\sqrt{\frac{D_{22}}{D_{11}}}$$

and solving for m,

$$m = \frac{a}{b}\left(\frac{D_{22}}{D_{11}}\right)^{1/4} \tag{8.45}$$

Note that, since m is an integer, the right-hand side of Equation (8.45) must be rounded up or down to the next (or previous) integer, whichever minimizes the right-hand side of Equation (8.44). For a long plate (dimension a is large), m is large and using Equation (8.45) instead of the nearest integer that minimizes Equation (8.44) is justified. Then, using Equation (8.45) to substitute in Equation (8.44) gives

$$N_{x\text{crit}} = \frac{2\pi^2}{b^2}\left[\sqrt{D_{11}D_{22}} + (D_{12}+2D_{66})\right] \tag{8.46}$$

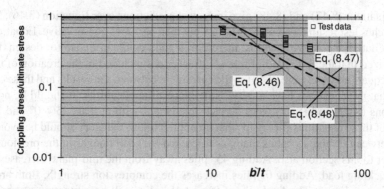

Figure 8.20 NEF crippling test results compared to buckling predictions

Equation (8.46) gives the buckling load of a long plate under compression and can be used to correlate with NEF crippling test results. It is the counterpart of Equation (6.13a), which was used to predict buckling of OEF flanges in the previous section. It is interesting to note that, in terms of laminate stiffnesses, the right-hand side of Equation (8.46) is most sensitive to D_{66} because of the factor of 2 multiplying that term. A fractional change in any other of the terms, D_{11}, D_{22} or D_{12}, will result in smaller increase of the buckling load for the same fractional change in D_{66}. Thus, similar to the OEF case, to maximize the crippling load one should maximize the D_{66} term which, as was shown in the previous section, is equivalent to maximizing the number of $45°/-45°$ plies and locating them as far from the mid-plane as possible.

Equation (8.46) is compared with test results for NEF crippling in Figure 8.20. Just as for the OEF case, the test results are slightly lower for low b/t values ($b/t < 15$) and significantly higher at high b/t values. The same arguments presented in the previous section for OEF flanges are also valid here. The post-buckling ability of the web or flange (not accounted for by Equation 8.46) and the specifics of the boundary condition at the roots of the web or flange are two of the main reasons for the discrepancy between the prediction and test results in Figure 8.20.

Fitting a curve to the test data [8] results in the expression

$$\frac{\sigma_{\text{crip}}}{\sigma_c^u} = \frac{14.92}{\left(\dfrac{b}{t}\right)^{1.124}} \tag{8.47}$$

where σ_c^u is the compression strength of the flange. Equation (8.47) is valid for $b \geq 11.07t$. For $b < 11.07t$, $\sigma_{\text{crip}} = \sigma_c^u$.

For design, a curve that is lower than 90% of the test results (B-Basis value) is used and is given by [8]

$$\frac{\sigma_{\text{crip}}}{\sigma_c^u} = \frac{11.0}{\left(\dfrac{b}{t}\right)^{1.124}} \tag{8.48}$$

for $b \geq 8.443t$ and $\sigma_{\text{crip}} = \sigma_c^u$ for $b < 8.443t$.

Both Equations (8.47) and (8.48) give improved predictions over Equation (8.46). Equation (8.47) matches test results well up to b/t of 25 but then becomes conservative. Equation (8.48) has all test data lying above it and is, therefore, a good equation to use for design. It should be emphasized that the test data in Figure 8.20 were not included in the creation of the semi-empirical equations (8.47) and (8.48) so the agreement between test results and these equations suggests that the equations have a fairly wide range of applicability. It should be noted that, as Equations (8.47) and (8.48) were derived for flanges with at least 25% 0° and 25% 45° plies, use of these equations for layups that do not fall in this category should be avoided. The tradeoff between 0° and 45° plies that was discussed for OEF flanges in the previous section carries over to this section also. Adding 45° plies away from the mid-plane increases D_{66} and thus the buckling load. Adding 0° plies increases the compression strength. Both are needed for an optimum design. The final mix of 0° and 45° plies will be a function of applied load and geometry.

Finally, by comparing the test results between Figures 8.18 and 8.20, it can be seen that a NEF flange always has greater crippling strength (as a fraction of the compression strength) than an OEF flange with the same b/t ratio.

8.5.3 Crippling under Bending Loads

If bending loads are applied to a stiffener (Figure 8.21), then some of the flanges or portions of flanges may still be under compression and can still be crippling-critical.

The recommended approach is to determine the portion of the flange that is under compressive loads and use that portion as the b value in the crippling analysis. Also, as applied load, the average compressive load exerted on that portion is used. This is shown in Figure 8.22 for a case of combined compression and bending loading. For this case, the portion that is under compression is a fraction of the entire flange and is denoted by b in Figure 8.22. Also, the

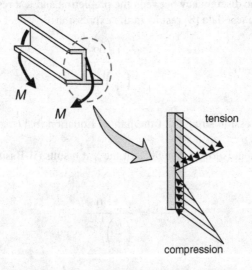

Figure 8.21 Stiffener under bending loads

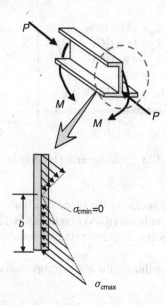

Figure 8.22 Stiffener cross-section under combined compression and bending

minimum compressive stress $\sigma_{c\min}$ is zero. Then, the average compressive stress acting over b is given by

$$\sigma_c = \frac{\sigma_{c\max} + \sigma_{c\min}}{2} = \frac{\sigma_{c\max}}{2} \tag{8.49}$$

The analysis then would consist of determining the crippling stress σ_{crip} for a NEF or an OEF flange (depending on the case; it is NEF for the example of Figure 8.22) and comparing it with the applied stress σ_c given by Equation (8.49). If the crippling stress exceeds σ_c then there is no failure.

Note that the example of Figure 8.22 assumes that the bending moment M is large enough to create high tension on the upper end of the flange which exceeds the compression stress due to the applied load P. If M were not high enough the entire flange would be under compression. In that case $\sigma_{c\min}$ is not zero, but equal to

$$\sigma_{c\min} = \frac{P}{A} - \frac{Mc}{I}$$

with A and I the area and moment of inertia of the entire cross-section and c the distance from the neutral axis. Also, in this case b is equal to the entire width of the web and not a portion of it.

8.5.3.1 Application: Stiffener Design under Bending Loads

Consider the L stiffener under bending moment $M = 22.6$ Nm shown in Figure 8.23. The stiffener layup is the same for both members: $[45/-45/0_2/\overline{90}]_s$ with the $0°$ fibres aligned with

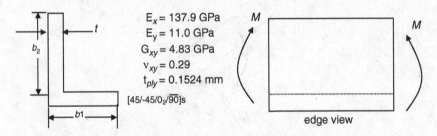

Figure 8.23 Stiffener under bending moment

the axis of the stiffener (perpendicular to the plane of Figure 8.23). The ultimate strain for this layup at room temperature ambient (RTA) conditions is 12,000 µs. The width b_1 of the horizontal flange is fixed at 17.78 mm. Determine the maximum value of b_2 so that the stiffener does not fail in crippling.

Using classical laminated-plate theory, the elastic properties of the stiffener flange and web are as follows:

A_{11} (N/mm)	113,015	D_{11} (N mm)	12,893.18	E_1memb	75.2	GPa
A_{12} (N/mm)	23,327.5	D_{12} (N mm)	6219.661	E_1bend	38.1	GPa
A_{22} (N/mm)	54,670	D_{22} (N mm)	8265.409	t	1.3716	mm
A_{66} (N/mm)	25,532.5	D_{66} (N mm)	6564.006			

where axis 1 is aligned with the stiffener axis and axis 2 is in the plane of the web or flange accordingly.

Using Figure 8.24, the stiffener cross-sectional properties are determined as follows:

$$\bar{y} = \frac{b_2 t \left(t + \dfrac{b_2}{2}\right) + b_1 t \dfrac{t}{2}}{b_1 t + b_2 t} = \frac{t \left(b_2 + \dfrac{b_1}{2}\right) + \dfrac{b_2^2}{2}}{b_1 + b_2}$$

$$I = \frac{t b_2^3}{12} + b_2 t \left(t + \frac{b_2}{2} - \bar{y}\right)^2 + \frac{b_1 t^3}{12} + b_1 t \left(\bar{y} - \frac{t}{2}\right)^2$$

with t the thickness of the laminate used ($=1.3716$ mm).

Assuming engineering bending theory is valid, the maximum compressive stress and strain in the stiffener can be shown to be

$$\sigma_{\text{comp}} = \frac{M (b_2 + t - \bar{y})}{I}$$

$$\varepsilon_{\text{comp}} = \frac{\sigma_{\text{comp}}}{E_{1\text{bend}}} = \frac{M (b_2 + t - \bar{y})}{E_{1\text{bend}} I}$$

Note that for small t values, using $E_{1\text{memb}}$ is more representative than $E_{1\text{bend}}$.

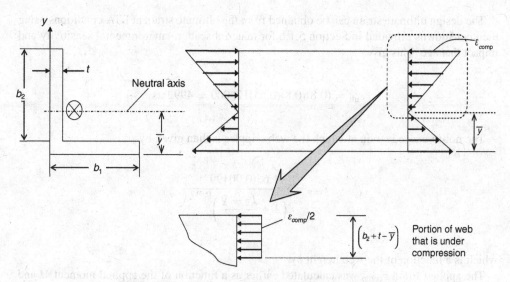

Figure 8.24 Stiffener geometry and neutral axis location

The upper portion of the web in Figure 8.24 is under compression and the lower portion, along with the flange, is under tension. Therefore, only the portion above the neutral axis in Figure 8.24 can fail in crippling. This means that the length of the web that may fail in crippling is (see insert of Figure 8.24) $b_2 + t - \bar{y}$. The linear strain distribution shown in Figure 8.24 is approximated as a constant compressive strain equal to the average compressive strain over the portion of the stiffener web that is under compression (see insert of Figure 8.24).

The portion of the stiffener web that is under compression is stabilized at the neutral axis and free at the top so it is OEF. Using Equation (8.43),

$$\frac{\sigma_{\text{crip}}}{\sigma_c^u} = \frac{1.63}{\left(\dfrac{b}{t}\right)^{0.717}}$$

By multiplying numerator and denominator by the axial stiffness E, the crippling equation in terms of strains can be obtained:

$$\frac{\sigma_{\text{crip}}}{\sigma_{\text{ult}}} = \frac{E\varepsilon_{\text{crip}}}{E\varepsilon_{\text{ult}}} = \frac{\varepsilon_{\text{crip}}}{\varepsilon_{\text{ult}}} = \frac{1.63}{\left(1 + \dfrac{b_2 - \bar{y}}{t}\right)^{0.717}}$$

where $b = b_2 + t - \bar{y}$ was substituted in the equation.

The design ultimate strain can be obtained from the ultimate strain at RTA conditions using the knockdowns provided in Section 5.1.6 for material scatter, environmental sensitivity and impact damage. This gives

$$\varepsilon_{\text{ult}} = (0.8)(0.8)(0.65)12,000 = 4992\,\mu s$$

The not-to-exceed strain at which the web cripples is then given by

$$\varepsilon_{\text{crip}} = \frac{1.63(0.004992)}{\left(1 + \dfrac{b_2 - \bar{y}}{t}\right)^{0.717}}$$

which is a function of the web height b_2.

The applied strain $\varepsilon_{\text{comp}}$ was calculated earlier as a function of the applied moment M and stiffener geometry. That strain must be below the crippling strain $\varepsilon_{\text{crip}}$ to avoid failure. A plot of the applied and crippling strains is shown in Figure 8.25 as a function of the web height b_2.

As can be seen from Figure 8.25, the crippling strain is lower than the applied strain for small b_2 values. In that range, the applied strain exceeds the crippling strain and failure occurs. This is because the moment of inertia I of the stiffener is low and, as a result, the bending strains are high. Now as b_2 increases, the crippling strain is reduced, which would be expected from the form of the crippling strain equation with b_2 in the denominator. Therefore, the situation would get 'worse' from a crippling perspective. However, the rate at which the crippling strain decreases is much lower than the rate at which the applied strain decreases. This is because a change in b_2 increases the moment of inertia (and thus decreases the applied strain) to a larger extent than the same change decreases the crippling strain. As a result, a value of b_2 can be found, 43.2 mm beyond which the applied strain is lower than the crippling strain and no failure occurs. Therefore, the minimum allowable value of b_2 is 43.2 mm.

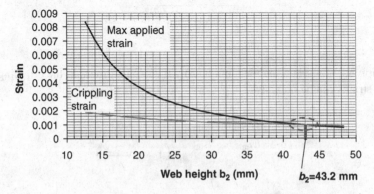

Figure 8.25   Comparison of maximum applied strain to the crippling strain of a stiffener under bending moment

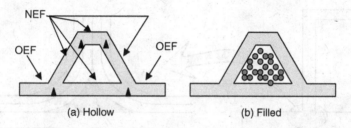

(a) Hollow   (b) Filled

Figure 8.26 Closed cross-section stiffeners (crippling considerations)

8.5.4 Crippling of Closed-Section Beams

For closed-section beams such as the hat stiffener shown in Figure 8.26, two cases are distinguished:

(a) the beam is hollow, and
(b) the beam is filled by foam or other material.

In the first case, the crippling analysis proceeds as in the previous sections by analysing each flange of the closed section as NEF. In the second case, an analysis of a beam on an elastic foundation can be carried out (provided a reliable post-buckling analysis is available) or, which is preferred, each flange is treated as a facesheet of a sandwich failing in wrinkling (see Section 10.3).

8.6 Importance of Radius Regions at Flange Intersections

It was briefly mentioned in Section 8.1 (see also Figure 8.4) that turning plies around 90° corners at flange/web intersections is very difficult without the creation of a 'pocket'. This happens irrespective of the fibre orientation, but is most pronounced in the case of 90° plies and least in the case of 0° plies (where 0° is the direction perpendicular to the page of Figure 8.27). This situation is shown (exaggerated and not to scale) in Figure 8.27.

Wavy fibres in the radius region compromise the strength of the cross-section. Resin-rich areas in the radius region suggest that there are resin-starved areas in adjacent plies, again leading to reduced strength and stiffness, especially under compression or shear. The size of the 'pocket' is a function of the layup (plies with fibres aligned with the turn are harder to turn 90° corners following a tight radius), tooling (concave tooling into which the material is placed results in larger pockets, as opposed to convex tooling over which material is draped), cure pressure (higher pressures tend to decrease the size of the pocket), resin flow and bleeding during cure, etc.

Since the existence of the pocket is unavoidable in such configurations, efforts are usually made to reinforce it by incorporating a piece of unidirectional tape or roving material. With reference to Figure 8.28, the area of the 'pocket' is found to be

$$A_f = 2 \left[R_i + \frac{t}{2} \right]^2 \left(1 - \frac{\pi}{4} \right) \tag{8.50}$$

where R_i and $t/2$ are the inner radius and thickness of the turning flange, respectively.

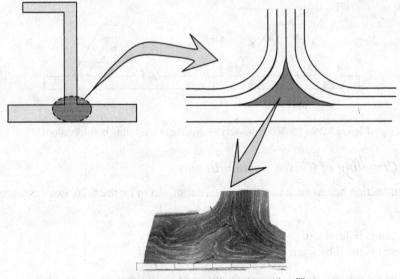

Without special provisions, this region fills
with wavy fibers and/or pure resin

Figure 8.27   Resin pocket formed at web/flange intersection of a stiffener (see Plate 19 for the colour figure)

For the case of uniaxial tension or (pre-buckling) compression, strain compatibility requires that the strain in the pocket be the same as the strain in every other member of the cross-section. Then, using Equation (8.12) the force in the pocket can be found to be

$$F_f = \frac{E_f A_f}{\sum E_j A_j} F_{\text{TOT}}$$

$$(8.51)$$

The significance of the force absorbed by the filler material can best be seen through an example. Consider the stiffener cross-section shown in Figure 8.29. It is assumed that the material used is typical graphite/epoxy. Then, for typical layups, the axial stiffness of the web and flanges is given by $E_1 = 89.6$ GPa, $E_2 = 31.0$ GPa, $E_3 = 48.3$ GPa, with the subscripts

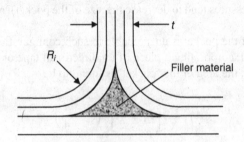

Figure 8.28   Pocket geometry

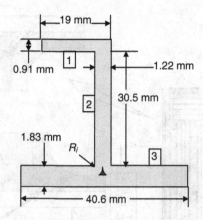

Figure 8.29 Stiffener cross-section with filler material at the interface of the web and bottom flange

referring to the three members in Figure 8.29. The filler material can be anything from pure resin (no filler) whose stiffness is $E_f = 3$ GPa, to completely filled by unidirectional material, in which case the stiffness would be $E_f = 138$ GPa. Finally, the inside radius R_i of the turning flange (see Figure 8.28) is assumed to vary in a typical range of 2.5–6.35 mm.

Using Equation (8.51) the force acting in the filler region can be determined as a fraction of the total applied force. The results are shown in Figure 8.30. It can be seen that the force on the filler can be a significant fraction of the total applied force, especially when it is filled with unidirectional material. In general, even when unidirectional or roving material is used, the force on the filler is neglected during the design phase. This increases the load on the other members of the cross-section, making the design more conservative. For a detailed analysis and for comparison with test results the force on the filler must be taken into account.

In view of the importance of the filler material in load sharing and alleviating some of the load in the stiffener web and flanges, the design of the J cross-section from Figure 8.6 is now revisited in Figure 8.31. Besides a filler material, the conclusions of the discussion on crippling have been applied to the flange and an attempt has been made to combine, in the

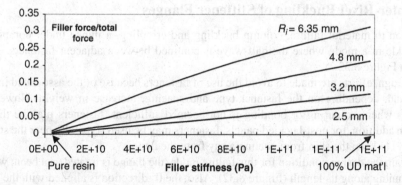

Figure 8.30 Force on filler region as a function of filler stiffness and flange inside radius

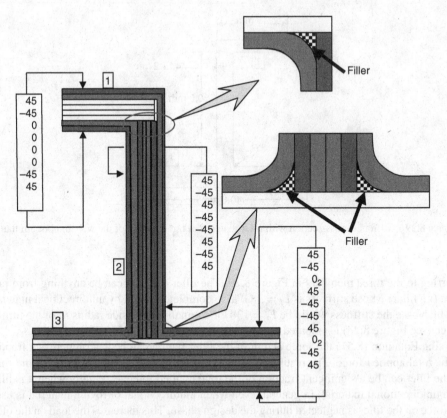

Figure 8.31 J stiffener cross-section with filler material (see Plate 20 for the colour figure)

bottom flange, 45° plies (for increased D_{66}) with 0° plies (for increased moment of inertia and compressive strength, which in turn increases the crippling strength).

8.7   Inter-Rivet Buckling of Stiffener Flanges

In addition to material failure, column buckling and crippling, a flange under compression may buckle in a mode where the half-wave is confined between adjacent fasteners. This is shown in Figure 8.32.

In a design, efforts are made to avoid the use of fasteners because of the associated increase in cost and, depending on the fastener type and spacing, increase in weight. However, in situations where co-curing or bonding is not deemed sufficient, fasteners may be the only option. In addition, for post-buckled panels, fasteners may be used (typically near the stiffener ends only) to keep the skin from peeling away from the flange.

To obtain the design condition for this failure mode, the flange is treated as a beam with the x-axis running along its length (Figure 8.32). Also, the 0° direction is aligned with the x-axis. Assuming the flange to be symmetric and have $D_{16} = D_{26} = 0$, the governing equation for

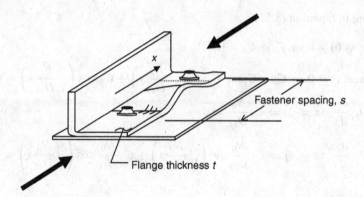

Figure 8.32 Flange inter-rivet buckling

the out-of-plane displacements w is given by Equation (5.16) applied to a one-dimensional problem $(\partial/\partial y = 0)$ with no distributed loads $(p_x = p_y = p_z = 0)$:

$$D_{11}\frac{\partial^4 w}{\partial x^4} + 2(D_{12}+2D_{66})\frac{\partial^4 w}{\partial x^2 \partial y^2} + D_{22}\frac{\partial^4 w}{\partial y^4} = N_x\frac{\partial^2 w}{\partial x^2} + 2N_{xy}\frac{\partial^2 w}{\partial x \partial y}$$

$$+N_y\frac{\partial^2 w}{\partial y^2} - p_x\frac{\partial w}{\partial x} - p_y\frac{\partial w}{\partial y} + p_z$$

(5.16)

which for a compressive load $N_o = -N_x$ simplifies to

$$D_{11}\frac{\partial^4 w}{\partial x^4} + N_o\frac{\partial^2 w}{\partial x^2} = 0$$

(8.52)

which has the general solution

$$w = C_o + C_1 x + C_2 \sin\left(\sqrt{\frac{N_o}{D_{11}}}x\right) + C_3 \cos\left(\sqrt{\frac{N_o}{D_{11}}}x\right)$$

(8.53)

Note that the partial derivatives of Equation (8.52) are, in fact, total derivatives in this case because there is no dependence on y.

If the fasteners are assumed to provide simple support to the flange at $x = 0$ and $x = s$, the boundary conditions are

$$w(x = 0) = w(x = s) = 0$$

$$-D_{11}\frac{d^2 w}{dx^2} = M = 0 \quad \text{at} \quad x = 0, \, x = s$$

(8.54)

Substituting in Equation (8.54)

$$w(x = 0) = 0 \Rightarrow C_o + C_3 = 0$$

$$w(x = s) = 0 \Rightarrow C_o + C_1 s + C_2 \sin\left(\sqrt{\frac{N_o}{D_{11}}} s\right) + C_3 \cos\left(\sqrt{\frac{N_o}{D_{11}}} s\right) = 0$$

$$-D_{11}\frac{d^2 w}{dx^2}(x = 0) \Rightarrow C_3 \frac{N_0}{D_{11}} = 0 \qquad\qquad (8.55\text{a–d})$$

$$-D_{11}\frac{d^2 w}{dx^2}(x = s) \Rightarrow C_2 \frac{N_0}{D_{11}} \sin\left(\sqrt{\frac{N_0}{D_{11}}} s\right) + C_3 \frac{N_0}{D_{11}} \cos\left(\sqrt{\frac{N_0}{D_{11}}} s\right) = 0$$

From Equation (8.55c),

$$C_3 = 0$$

which substituted in Equation (8.55a) gives

$$C_o = 0$$

Then, Equation (8.55b) can be used to obtain a relation between C_1 and C_2:

$$C_1 = -C_2 \frac{1}{s} \sin\left(\sqrt{\frac{N_o}{D_{11}}} s\right)$$

Finally, Equation (8.55d) gives

$$\sin\left(\sqrt{\frac{N_o}{D_{11}}} s\right) = 0 \Rightarrow \sqrt{\frac{N_o}{D_{11}}} s = n\pi \Rightarrow N_o = \frac{n^2 \pi^2}{s^2} D_{11}$$

which gives the buckling load N_o for the flange. The lowest buckling load ($n = 1$) is the one of interest. Therefore, the inter-rivet (buckling) stress is

$$\sigma_{ir} = \frac{N_o}{t} = \frac{\pi^2 D_{11}}{t s^2}$$

where t is the flange thickness.

This equation corresponds to simply supported ends at the fasteners. However, depending on the type of fastener, the support provided at the ends can be more restricting than a simple support with the slope locally constrained to a degree. The edge condition can range from simply supported (countersunk fasteners) to nearly fixed (protruding fasteners). This range of boundary conditions is represented by a coefficient of fixity c and the inter-rivet stress expression is generalized to

$$\sigma_{ir} = \frac{c \pi^2 D_{11}}{t s^2} \qquad\qquad (8.56)$$

with $c = 1$ for countersunk fasteners and $c = 3$ for protruding-head fasteners.

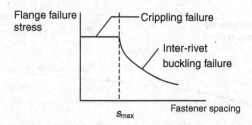

Figure 8.33   Flange failure modes as a function of fastener spacing

The fastener spacing and flange stiffness determine the failure mode of the flange. Contrasting inter-rivet buckling with crippling, for example, it can be seen that for relatively wide fastener spacings and soft flanges (s large and D_{11} small) the flange will fail by inter-rivet buckling. For narrow spacings and large bending stiffness the flange will fail in crippling. This means that, given a flange layup, there is a threshold fastener spacing value that if exceeded will result in the failure mode switching from crippling to inter-rivet buckling. This is shown schematically in Figure 8.33.

Typically, the flange attached to the skin that may fail by inter-rivet buckling is of the one-edge-free (OEF) type. Then, Equations (8.43) and (8.56) can be combined to determine the critical fastener spacing. Equating the inter-rivet buckling stress to the crippling stress and solving for s give

$$s_{max} = \sqrt{\frac{c\pi^2 D_{11}}{1.63 t\sigma_c^u}\left(\frac{b}{t}\right)^{0.717}} \qquad (8.57)$$

which gives the maximum fastener spacing for crippling to occur.

The implications of Equation (8.57) can be seen more clearly through an example. Consider the two flange layups given in Table 8.7. The first is a stiff flange and the second a very soft flange.

The maximum fastener spacing determined from Equation (8.57) is plotted in Figure 8.34 for the first layup, $[45/0_2/-45/0_4]_s$. As the width to thickness ratio b/t for the flange increases the value of s_{max} increases.

The maximum fastener spacing for the soft layup, $[(\pm 45)/(0/90)/[(\pm 45)]$, is shown in Figure 8.35. The trends are the same as in Figure 8.34, but an important problem not present in the stiff flange of Figure 8.34 is now evident: The maximum allowed fastener spacing is too small for typical b/t ratios. Fastener spacing values less than 20 mm are avoided in practice

Table 8.7   Properties of two potential flange layups

layup	$[45/0_2/-45/0_4]_s$	$[(\pm 45)/(0/90)/[(\pm 45)]$
σ_c^u (MPa)	762	529
D_{11} (Nm)	67.5	0.66
t (mm)	2.032	0.572

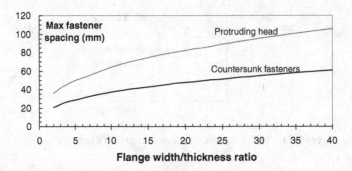

Figure 8.34   Fastener spacing to cause inter-rivet buckling for $[45/0_2/-45/0_4]_s$ flange

because the interaction between adjacent fasteners leads to increased bearing loads. For a (protruding-head) fastener spacing of 20 mm, Figure 8.34 suggests a b/t value of about 25 which corresponds to very low crippling failure loads (see Figure 8.18). The situation is even worse for countersunk fasteners.

This discussion brought to the surface some of the issues associated with fasteners and fastener spacing. While the interaction between inter-rivet buckling and crippling suggests that relatively large fastener spacings may be necessary, bolted joint analysis of multi-fastener joints shows that, for composites, lower fastener spacings should be preferred because they tend to maximize the net section strength [9]. As is often the case when multiple failure modes and constraints come into play, design guidelines can be generated that, at times, conflict with one another. Care must be exercised in such situations to generate designs that yield the best compromise. While an exhaustive discussion of bolted joints is beyond the scope of this book, some basic guidelines derived from the above discussion and more elaborate analyses of fastener joints [9] are summarized below:

1. If no other requirements dominate, use a fastener spacing of 4–5D (where D is the diameter of the fastener)
2. Minimum fastener spacing should be no less than 20 mm
3. Use skin thickness/diameter ratio < 1/3 to minimize fastener bending

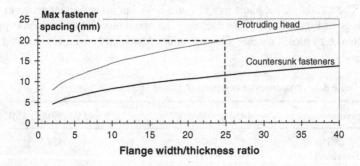

Figure 8.35   Fastener spacing to cause inter-rivet buckling for $[(\pm45)/(0/90)/[(\pm45)]$ flange

4. Use skin thickness/countersunk depth > 3/2 to avoid pulling countersunk fastener through the skin when loads perpendicular to the skin are applied (see section 11.2.1.1)
5. Use at least 40% 45°/–45° plies around fasteners for better load transfer

8.8 Application: Analysis of Stiffeners in a Stiffened Panel under Compression

A highly loaded stiffened panel is shown in Figure 8.36. The stiffener cross-section is the same, in terms of layup, as in Figure 8.31. The effect of the filler material at the radius regions is neglected. Skin and stiffener properties are summarized in Table 8.8. The axial stiffness of the skin is found from Equation (8.6) to be $E_{skin} = 41.15$ GPa. The portion of skin under compression, b_{eff} if the post-buckling load distribution is represented by a piecewise constant distribution, is given by Equation (7.15), resulting in $b_{eff} = 0.292a = 4.45$ cm.

Using Equation (6.7), the buckling load of the skin between the stiffeners is found to be N_o = 182 N/mm. This corresponds to a total force of $182 \times 457 = 83\ 174$ N. So, under the applied load of 100 kN, the skin buckles and the PB ratio is 100/83.17 = 1.20. It is now assumed that once the skin buckles all the excess load (between 100 and 83 kN) is taken by the stiffeners

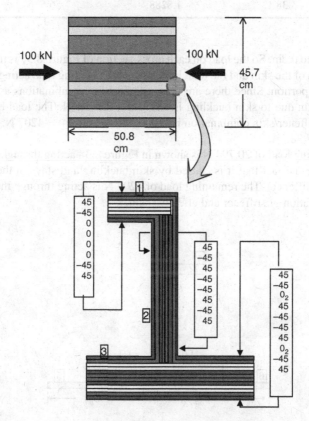

Figure 8.36 Skin-stiffened panel under compression (see Plate 21 for the colour figure)

Table 8.8 Skin and stiffener properties

	Skin	
D_{11}	65970	N mm
D_{12}	46690	N mm
D_{22}	65970	N mm
D_{66}	49400	N mm
	Skin thickness = 0.57 mm	
A_{11}	28,912.44	N/mm
A_{12}	12,491.43	N/mm
A_{22}	28,912.44	N/mm
A_{66}	13,468.58	N/mm
	Stiffener	

Member	b (mm)	t (mm)	E_m (GPa)	E_b (GPa)
1	12.7	1.2192	75.6	32.4
2	31.75	1.2192	18.2	17.9
3	38.1	1.8288	56.5	47.9

and b_{eff} skin next to them. So the load on each cross-section of Figure 8.37 is the buckling load on the b_{eff} portion of the skin and the total load minus the buckling load acting on the stiffener plus the b_{eff} skin portion. Since there are four stiffener/skin combinations as in Figure 8.37, the load in the skin due to skin buckling is 83174/4 = 20794 N. The load beyond buckling acting on each stiffener/skin combination is (100000 − 83174)/4 = 4207 N. The situation is shown in Figure 8.38.

The skin buckling load of 20,794 N is shown in Figure 8.38 acting through the skin neutral axis to emphasize the fact that it is caused by skin buckling and stays in the skin (does not transfer into the stiffener). The remaining load of 4207 N is acting through the neutral axis of the entire combination of stiffener and effective skin.

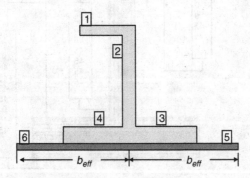

Figure 8.37 Cross-section carrying load in the post-buckling regime

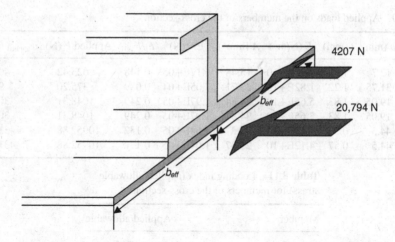

Figure 8.38   Loads on representative cross-section after the skin buckles

Using the geometry of Table 8.8 and the crippling equations (8.43) and (8.48) the crippling stresses in each member of the cross-section can be determined as a fraction of the compression strength of each member. Then, a first-ply failure criterion (Tsai–Wu failure theory, see Section 4.4) is used to determine the compression strength of each member and, from that, the crippling failure stresses in each member. The results are shown in Table 8.9.

Equation (8.12) can be used to determine the applied load on each of the members of the cross-section. This equation is applied to the 4207 N load (see Figure 8.38) while the 20,794 N load is added to the skin members 5 and 6 in addition to the contribution coming from the 4207 N load. The resulting loads are given in Table 8.10.

The last column of Table 8.9 gives the stress at which the corresponding member will fail (allowable stress). The last column of Table 8.10 gives the applied stress. The ratio of the two stresses is shown in Table 8.11. If the ratio of applied to allowable (also termed the loading index) is greater than 1, the corresponding member fails. As can be seen from Table 8.11, the last two members, marked in bold, that is, the effective skin portion, will fail in crippling.

Table 8.9   Crippling analysis of members of the cross-section

Member	b (mm)	t (mm)	OEF/NEF	b/t	$\sigma_{crip}/\sigma_{cu}$	σ_{cu} (N/mm^2)	σ_{fail} (N/mm^2)
1	12.7	1.22	OEF	10.42	0.304	494.64	150.2344
2	31.75	1.22	NEF	26.04	0.282	283.88	80.04247
3	19.05	1.83	OEF	10.42	0.304	351.75	106.8331
4	19.05	1.83	OEF	10.42	0.304	351.75	106.8331
5	44.5	0.57	NEF	78.07	0.082	529.14	43.43236
6	44.5	0.57	NEF	78.07	0.082	529.14	43.43236

Table 8.10 Applied loads on the members of the cross-section

Member	b (mm)	t (mm)	E (N/m²)	A (mm²)	EA (N)	F_i/F_{TOT}	Applied F (N)	$\sigma_{applied}$ (N/mm²)
1	12.7	1.22	7.56E+10	15.48	1.17E+05	0.148	623.42	40.26
2	31.75	1.22	1.82E+10	38.71	7.05E+04	0.089	375.20	9.69
3	19.05	1.83	5.65E+10	34.84	1.97E+05	0.249	1048.31	30.09
4	19.05	1.83	5.65E+10	34.84	1.97E+05	0.249	1048.31	30.09
5	44.5	0.57	4.12E+10	25.37	1.04E+05	0.132	10952.88	431.81
6	44.5	0.57	4.12E+10	25.37	1.04E+05	0.132	10952.88	431.81

Table 8.11 Loading index (applied/allowable stress) for members of the cross-section

Member	Applied/allowable
1	0.268
2	0.121
3	0.282
4	0.282
5	**9.942**
6	**9.942**

Exercises

8.1 A T stiffener is used in a compression application. It has the following configuration:

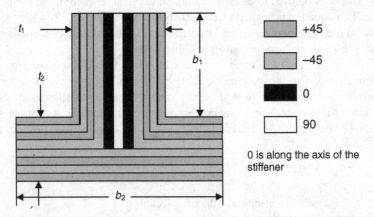

0 is along the axis of the stiffener

The material used has the properties:

$$E_x = 137.9\,\text{GPa}$$

$$E_y = 11.0\,\text{GPa}$$

$$G_{xy} = 4.82\,\text{GPa}$$

$$\nu_{xy} = 0.29$$

$$t_{ply} = 0.1524\,\text{mm}$$

The length L of the stiffener is 304.8 mm. The stiffener is pinned at the two ends and rests on an elastic foundation of spring constant k. Manufacturing considerations do not permit b_2 to be smaller than 19.05 mm or b_1 to be smaller than 12.7 mm. (a) If k is allowed to vary between 1,378,800 and 5,515,200 N/m^2, create a plot that shows how b_1 and b_2 vary with k so that the weight is minimized and the stiffener does not buckle below 31.115 kN. (b) What is the optimum value of k to use in this application (taking 'optimum' to mean the value that minimizes the stiffener weight)?

8.2 A stiffener terminates in the middle of nowhere on a skin. The stiffener is loaded on one end by a compressive load of 22.2 kN. The outer mould line of the stiffener cross-section (i.e. the outer shape) is fixed because a tool to make it is already available. But the stiffener thickness is variable. The situation is shown in the figure below.

The basic material properties are the same as in Exercise 8.1.

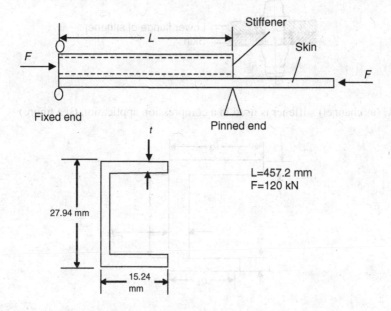

Use only 45°, −45°, 0° and 90° plies to create a symmetric and balanced laminate, which includes at least one of each of these principal four orientations, to determine the lowest thickness laminate that does not buckle under the applied load.

8.3 You are now given the basic strength values for the material of Exercise 8.2:

$$X^t = 2068\,\text{MPa (tension strength parallel to the fibres)}$$

$$X^c = 1378\,\text{MPa (compression strength parallel to the fibres)}$$

$$Y^t = 103.4\,\text{MPa (tension strength perpendicular to the fibres)}$$

$$Y^c = 310.2\,\text{MPa (compression strength perpendicular to the fibres)}$$

$$S = 124.1\,\text{MPa (shear strength)}$$

Check your solution of Exercise 8.3 for crippling of the upper flange (flange away from the skin) and the vertical web. If your solution fails in crippling, discuss how you would go about changing different parameters of the problem (layups, lengths, widths) to avoid failure with the lowest possible weight increase. If your solution does not fail in crippling, discuss what parameters of the problem (layups, length, widths, etc.) you should change to reduce the weight of the structure as much as possible without failing in crippling or buckling. Do not run any numbers, simply discuss what you should change and why you think it is the most effective.

8.4 The stiffener of Exercise 8.3 is riveted to the skin. Using the layup you obtained in Exercise 8.3, determine the maximum allowed rivet spacing for the rivets shown below. For this problem, disregard the crippling failure.

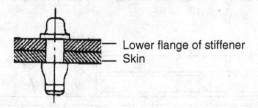

8.5 A C (or channel) stiffener is used in a compression application (see figure)

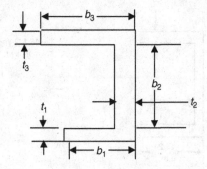

Member 1 is next to the skin and its layup is fixed to [45/–45/0/90/0/–45/45] with 0 running along the axis of the stiffener. The dimension b_1 is also fixed at 19.05 mm. The basic material properties are as follows:

E_x	1.31E+11 Pa	X^t	1.7235E+09 Pa
E_y	1.14E+10 Pa	X^c	1.379E+09 Pa
G_{xy}	4.83E+09 Pa	Y^t	8.273E+07 Pa
ν_{uxy}	2.90E-01	Y^c	3.033E+08 Pa
t_{ply}	1.52E-01 mm	S	8.809E+07 Pa

Dimensions b_2 and b_3 are allowed to vary between 12.7 and 48.26 mm. Use the layup for member 1 above and change it (if needed) according to the following rules:

1. no fewer plies than the base layup are allowed;
2. keep +45/−45 on the outside;
3. layup is symmetric;
4. at least one 0° and one 90° ply are present in the layup;
5. only 45°, −45°, 0° and 90° plies are used;
6. no layup has more than a total of 13 plies

to create candidate layups for members 2 and 3.

(a) (you will need access to do first-ply failure analysis). If the total compressive load applied to the cross-section is 26.67 kN, determine the optimum layup(s) and dimensions for members 2 and 3 so that the cross-section does not fail in crippling. Note that member 1 being next to the skin is reinforced by the effective skin and is assumed not to fail in crippling (so you do not need to do any failure analysis for member 1). For crippling equations, assume that the general equations given in this chapter are valid even if, for some of your layups, the requirement of at least 25% 0°, 25% 45° plies is not satisfied. Optimum here means that for each set of layups that do not fail in crippling, determine the one with the lowest cross-sectional area. Do not reduce the strength value calculated to account for environmental effects, impact and material scatter.

(b) Among the optimum layups determined in part (a) determine, as the best layup(s) to use, the one(s) that make most sense from a robust design and manufacturing perspective.

8.6 You are to design the cross-sectional shape and layup for a composite stiffener for an application under compressive load.

Of the seven shapes below select three (if the hat stiffener is not included in your selection) or two (if the hat stiffener is one of them). Also note that you are not allowed to include both the C and Z stiffeners in your selection. If you like both the C and the Z you must only include one of them in your analysis.

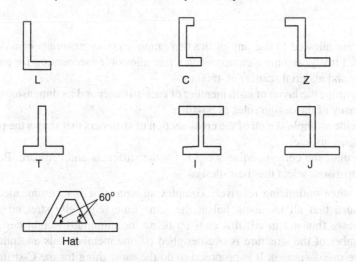

The stiffeners must fit within a rectangle of height 80 mm and width 50 mm. These are the maximum dimensions, but they can be smaller than that.

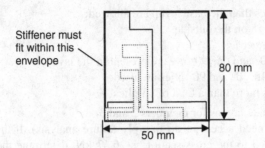

Stiffener must fit within this envelope

80 mm

50 mm

The applied load is 35,000 N (assume it is acting at the centre of gravity of the selected cross-section). The length ℓ of the stiffener is 550 mm.

Two composite materials are available, with properties as follows:

Unidirectional tape Gr/epoxy	Plain weave fabric Gr/epoxy
$E_x = 131$ GPa	68.9 GPa
$E_y = 11.4$ GPa	68.9 GPa
$\nu_{xy} = 0.31$	0.05
$G_{xy} = 5.17$ GPa	5.31 GPa
$t_{ply} = 0.1524$ mm	0.1905 mm
$\rho = 1611$ kg/m^3	1611 kg/m^3

You are allowed to use any of the two graphite/epoxy materials or a combination thereof. Finally, assume a compression strain allowable (accounting for environment, damage and material scatter) of 4500 μs.

1. Determine the layup of each member of each stiffener and its dimensions, observing as many of the design rules as possible.

2. Provide a simple sketch of the cross-section of stiffeners that shows the plies, layup, dimensions, etc.

3. Calculate the corresponding weights for the stiffeners and compare. Based on the comparison, select the 'best' design.

8.7   Often, when optimizing relatively complex structures, it is recommended to design them such that all members fail at the same time (e.g. fully stressed trusses). In some cases (but not in all) this ends up being the minimum weight design because no member of the structure is overdesigned (if one member fails and others do not, they are overdesigned). It is proposed to do the same thing for the C-stiffener shown below.

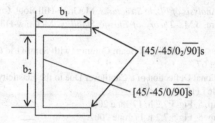

$[45/-45/0_2/90]$s

$[45/-45/0/90]$s

Derive a condition such that when a compressive load is applied to this cross-section, all members fail by crippling at the same time. This condition should give the length b_1 as a function of the length b_2 and all other pertinent quantities (i.e. derive a relation in the form $b_1 = \ldots$). For the specific layups shown in the figure above, generate a plot of b_1 as a function of b_2 (b_2 is on the x-axis). Due to manufacturing constraints, none of the members can be less than 12.7 mm. The material properties are

$$E_x = 20\text{E6 psi(lb/in}^2)$$

$$E_y = 1.7\text{E6 psi}$$

$$\nu_{xy} = 0.29$$

$$G_{xy} = 0.7\text{E6 psi}$$

$$t_{\text{ply}} = 0.012 \text{ in}$$

$$X^t = 300 \text{ ksi (1 ksi} = 1000 \text{ psi)}$$

$$X^c = 200 \text{ ksi}$$

$$Y^t = 14 \text{ ksi}$$

$$Y^c = 48 \text{ ksi}$$

$$S = 18 \text{ ksi}$$

Now take the case of $b_2 = 40.64$ mm and calculate the corresponding total force applied to the stiffener. For the SAME total force, examine by trial and error (or other means) what happens if you change b_2 to a value 10% larger or 10% smaller. Can you come up with a different b_1 value such that the overall weight is less? Describe what happens as you increase or decrease b_2.

8.8 For the case in Problem 8.7 where $b_2 = 38.1$ mm, and the stiffener under compression is fixed on one end and free on the other, determine the length L of the stiffener so that it buckles at the same time it cripples.

References

[1] Hart-Smith, L.J., The Key to Designing Durable Adhesively Bonded Joints, *Composites*, **25**, 895–898 (1994).
[2] Kairouz, K.C. and Matthews, F.L., Strength and Failure Modes of Bonded Single Lap Joints Between Cross-Ply Adherends, *Composites*, **24**, 475–484 (1993).

[3] Rivello, R.M., *Theory and Analysis of Flight Structures*, McGraw-Hill Book Co., New York, 1969, Chapter 14.

[4] Timoshenko, S.M. and Gere, J.M., *Theory of Elastic Stability*, McGraw-Hill Book Co., New York, 1961, Section 2.1.

[5] Aristizabal-Ochoa, J.D., Classical Stability of Beam Columns with Semi-Grid Connections on Elastic Foundation, 16–18 July 2003, Paper 67.

[6] O'Donnell, W.J., The Additional Deflection of a Cantilever Due to the Elasticity of the Support, *ASME Journal of Applied Mechanics*, **27**, 461 (1960).

[7] *Mil-Hdbk 17-3F,* vol. 3, chap. 5, Fig 5.7.2.h, 17 June 2002.

[8] *Mil-Hdbk 17-3F,* vol. 3, chap. 5, Fig 5.7.2.f, 17 June 2002.

[9] Hart-Smith, L.J., The Key to Designing Efficient Bolted Composite Joints, *Composites*, **25**, 835–837 (1994).

[10] Hart-Smith, L.J., Bolted Joints in Graphite/Epoxy Composites, *McDonnell-Douglas Corporation, Report A600503*, 1976.

[11] Ramkumar, R.L., Saether, E.S. and Cheng, D., Design Guide for Bolted Joints in Composite Structures, *Air Force Wright Aeronautical Report AFWAL-TR-86-3035*, 1986.

9

Skin–Stiffened Structure

The individual constituents, skin and stiffeners, have been examined in previous chapters. Based on that discussion, the behaviour and the design of a stiffened panel such as the one shown in Figure 9.1 can be summarized as follows: The skin takes pressure loads via in-plane stretching (membrane action) and shear loads. It also takes compression loads up to buckling. Beyond buckling, extra care must be exercised to account for the skin deformations and additional failure modes, not examined so far, such as the skin–stiffener separation, which is discussed in this chapter. The stiffeners take bending and compression loads. It is readily apparent that the robustness and efficiency of a design will strongly depend on how one can 'sequence' the various failure modes so benign failures occur first and load can be shared by the rest of the structure, and how one can eliminate certain failure modes without unduly increasing the weight of the entire panel.

In this chapter, some aspects that manifest themselves at the component level, with both skin and stiffeners present, are examined. This includes modelling aspects such as smearing of stiffness properties and additional failure modes such as the skin–stiffener separation.

9.1 Smearing of Stiffness Properties (Equivalent Stiffness)

If the number of stiffeners is sufficiently large and/or their spacing is sufficiently narrow, accurate results for the overall panel performance can be obtained by smearing the skin and stiffener properties in combined, equivalent stiffness, expressions. This can be done for both in-plane (membrane) and out-of-plane (bending) properties.

9.1.1 Equivalent Membrane Stiffnesses

A composite stiffened panel is shown in Figure 9.2. The stiffener spacing is d_s and the width of the panel is b_p.

As can be seen from Equation (8.13), the equivalent in-plane stiffness of the skin–stiffener combination is the sum of the individual stiffnesses of skin and stiffeners. This means that the

Design and Analysis of Composite Structures: With Applications to Aerospace Structures, Second Edition. Christos Kassapoglou.
© 2013 John Wiley & Sons, Ltd. Published 2013 by John Wiley & Sons, Ltd.

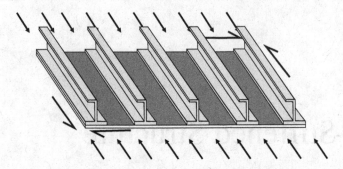

Figure 9.1   Stiffened panel under combined loads

A matrix for the entire panel, which is the membrane stiffness per unit width, is given by the sum of the corresponding terms for skin and stiffeners considered separately:

$$(A_{ij})_{eq} = (A_{ij})_{skin} + (A_{ij})_{stiffeners} \tag{9.1}$$

with the subscript ij denoting the ij element of the A matrix. Also, again from Equation (8.13),

$$(A_{ij})_{stiffeners} = n_s (A_{ij})_{singlestiff} \tag{9.2}$$

where n_s is the number of stiffeners.

Determining the number of stiffeners involves some approximation caused by the presence or lack of stiffeners at the panel edges. If there are stiffeners right at the panel edges as in Figure 9.2, the number of stiffeners is given by

$$n_s = \text{int}\left[\frac{b_p}{d_s}\right] + 1 \tag{9.3}$$

where int[...] denotes the integer that is obtained when the quantity in brackets is rounded down to the nearest integer.

If the stiffener spacing is sufficiently small, the second term in the right-hand side of Equation (9.3) can be neglected and the number of stiffeners approximated by

$$n_s \approx \frac{b_p}{d_s} \tag{9.4}$$

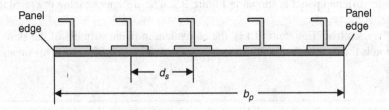

Figure 9.2   Section cut of a composite stiffened panel

If there are no stiffeners at the panel edges, that is, the skin overhangs on either side of the panel, b_p in Equation (9.3) must be reduced by the total amount of overhang. Again, for sufficiently large b_p and/or small d_s Equation (9.4) can be used. Note that Equation (9.4) is typically a rational number as the division b_p/d_s is an integer only for judiciously chosen values of b_p and d_s. But, for the purpose of stiffness estimation, using the rational number obtained from Equation (9.4) without round-off is a reasonable approximation.

Now the A_{ij} term for a single stiffener can be estimated by averaging the corresponding membrane stiffness of the stiffener over the skin width b_p. For the case of A_{11} this gives

$$(A_{11})_{\text{singlestiff}} = \frac{(EA)_{\text{stiff}}}{b_p} \tag{9.5}$$

Placing Equations (9.2), (9.4) and (9.5) into Equation (9.1) and recognizing that a one-dimensional stiffener has negligible contribution to stiffnesses other than the one parallel to its own axis gives the A matrix terms:

$$
\begin{aligned}
(A_{11})_{\text{eq}} &\approx (A_{11})_{\text{skin}} + \frac{(EA)_{\text{stiff}}}{d_s} \\
(A_{12})_{\text{eq}} &\approx (A_{12})_{\text{skin}} \\
(A_{22})_{\text{eq}} &\approx (A_{22})_{\text{skin}} \\
(A_{66})_{\text{eq}} &\approx (A_{66})_{\text{skin}}
\end{aligned}
\tag{9.6}
$$

9.1.2 Equivalent Bending Stiffnesses

The derivation of the bending stiffnesses proceeds in a similar fashion. Based on Equation (8.20) the bending stiffnesses per unit width can be written as

$$\left(D_{ij}\right)_{\text{eq}} = \left(D_{ij}\right)_{\text{skin}} + \left(D_{ij}\right)_{\text{stiffeners}} \tag{9.7}$$

with

$$\left(D_{ij}\right)_{\text{stiffeners}} = n_s \left(D_{ij}\right)_{\text{singlestiff}} \tag{9.8}$$

The bending stiffness D_{11} for a single stiffener can be determined by smearing its contribution over the entire width b_p:

$$(D_{11})_{\text{singlestiff}} = \frac{(EI)_{\text{stiff}}}{b_p} \tag{9.9}$$

While there are no contributions to the D_{12} and D_{22} terms because the bending stiffness contribution from the stiffeners is negligible in these directions, the contribution to D_{66} requires a detailed derivation.

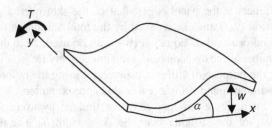

Figure 9.3   Laminate under applied torque

Consider the situation shown in Figure 9.3 where a laminate deforms under an applied torque.

The angle α is given by

$$\alpha = \frac{\partial w}{\partial x} \tag{9.10}$$

From torsion theory [1], the rate of change of angle α as a function of y is given by

$$\frac{d\alpha}{dy} = \frac{T}{GJ} \tag{9.11}$$

where T is the applied torque, G the shear modulus and J the torsional rigidity. Combining Equations (9.10) and (9.11) gives

$$\frac{d\alpha}{dy} = \frac{\partial^2 w}{\partial x \partial y} = \frac{T}{GJ} \tag{9.12}$$

Now from classical laminated-plate theory (see Equation 3.49 and assuming no coupling is present) the torque per unit width M_{xy} is given by

$$M_{xy} = -2D_{66}\frac{\partial^2 w}{\partial x \partial y} \tag{9.13}$$

Since now

$$\frac{T}{b_p} = -M_{xy} \tag{9.14}$$

Equations (9.12), (9.13) and (9.14) can be combined and applied to a single stiffener to give

$$(D_{66})_{\text{singlestiff}} = \frac{(GJ)_{\text{stiff}}}{2b_p} \tag{9.15}$$

This equation is analogous to Equation (9.9), but there is a factor of 2 in the denominator. Summing up the contributions of all stiffeners and using Equation (9.4), the contribution of all stiffeners to D_{66} for the combined skin–stiffener configuration is

$$D_{66} = n_s \, (D_{66})_{\text{singlestiff}} = \frac{(GJ)_{\text{stiff}}}{2d_s} \tag{9.16}$$

Finally, combining Equations (9.7) to (9.9) and (9.16) gives the final approximate expressions for the bending stiffnesses of a stiffened panel:

$$\begin{aligned}
(D_{11})_{\text{eq}} &\approx (D_{11})_{\text{skin}} + \frac{(EI)_{\text{stiff}}}{d_s} \\
(D_{12})_{\text{eq}} &\approx (D_{12})_{\text{skin}} \\
(D_{22})_{\text{eq}} &\approx (D_{22})_{\text{skin}} \\
(D_{66})_{\text{eq}} &\approx (D_{66})_{\text{skin}} + \frac{(GJ)_{\text{stiff}}}{2d_s}
\end{aligned} \tag{9.17}$$

which are analogous to Equations (9.6) in the previous section.

If the stiffeners have an open cross-section (such as L, C, Z, T, I, J) the torsional rigidity J in the last of Equations (9.17) is negligibly small and the stiffener contribution (second term in that equation) can be neglected altogether. If the stiffeners have a closed cross-section (such as a hat stiffener) the second term in the last of Equations (9.17) is significant and cannot be neglected.

In addition to the approximation introduced by Equation (9.4) when the number of stiffeners is small, there is an approximation in Equations (9.17) introduced by the fact that stretching–bending coupling terms (B matrix contribution) were neglected. The skin–stiffener cross-section in Figure 9.2 is asymmetric and there is a contribution to the bending stiffnesses coming from the axial stiffnesses of the stiffeners and skin. These are analogous to the B matrix terms of an asymmetric laminate and they become more significant as the stiffeners become bigger (e.g. greater web heights). Only in a situation where the stiffeners are mirrored to the other side of the skin, giving a symmetric configuration with respect to the skin mid-plane, will these coupling terms be exactly zero and no additional correction terms needed in Equation (9.17).

9.2 Failure Modes of a Stiffened Panel

Failure modes that are specific to the individual constituents, skin and stiffeners, were examined in previous chapters. Here, a summary of all failure modes, including those pertaining to the interaction between the skin and stiffeners, is given. The most important failure modes are presented in Figure 9.4. Of these, the material strength failure modes (either for the stiffener or for the skin) are typically covered by a first-ply failure analysis (see Chapter 4) supported and modified by test results. They were also briefly invoked in the discussion on crippling (Section 8.5) and skin post-buckling (see Sections 7.1 and 7.2). Flange crippling was examined in Section 8.5. Inter-rivet buckling was discussed in Section 8.7. Panel buckling failure modes were discussed in Chapter 6 for plates and Section 8.3 for beams. Whichever buckling mode

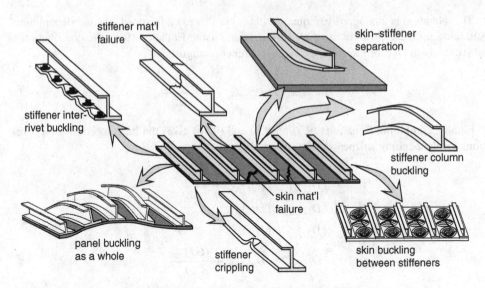

stiffener mat'l failure

skin–stiffener separation

stiffener inter-rivet buckling

stiffener column buckling

skin mat'l failure

panel buckling as a whole

stiffener crippling

skin buckling between stiffeners

Figure 9.4   Failure modes of a stiffened panel

occurs first, overall buckling of the panel or buckling of the skin between stiffeners is a function of the relative stiffnesses and geometry of skin and stiffeners, and conditions ensuring precedence of one buckling mode over another are discussed in this chapter. Finally, the skin–stiffener separation mode was briefly mentioned in association with post-buckling (at the start of Chapter 7) and will be examined in detail in this chapter.

As might be expected, unless explicitly designed for this, failure modes do not occur simultaneously. In certain situations, designing so that some (or all) failure modes occur at the same time gives the most efficient design in the sense that no component is over-designed. This is not always true. It assumes that different components are used to take different types of loading and fail in different failure modes that are independent of one another. In general, formal optimization shows that the lightest designs are not always the ones where the critical failure modes occur simultaneously.

In view of this, knowing when one failure mode may switch to another is critical. In addition, sequencing failure modes so they occur in a predetermined sequence is also very useful for the creation of robust designs. For example, relatively benign failure modes such as crippling (as opposed to column buckling) and local buckling between stiffeners (as opposed to overall buckling) contribute to creating a damage-tolerant design, in the sense that catastrophic failure is delayed and some load sharing with components that have not yet failed occurs. One such case of finding when one failure mode changes to another was examined in Section 8.7 where the condition for switching from crippling to inter-rivet buckling was determined.

9.2.1   Local Buckling (between Stiffeners) versus Overall Panel Buckling (the Panel Breaker Condition)

As mentioned in the previous section, confining the buckling mode between stiffeners is preferable. In general it leads to lighter designs and keeps the overall panel from buckling, which, typically, leads to catastrophic failure. From a qualitative point of view, as the bending stiffness of the stiffeners increases, it becomes harder for them to bend. Then under compressive loading, for example, if the stiffeners are sufficiently stiff, the skin between the stiffeners will buckle first. The stiffeners remain straight and act as 'panel breakers'. For a given stiffener stiffness, this behaviour can be assured if the stiffener spacing is sufficiently wide. Then, even for relatively soft stiffeners, the skin between them will buckle first. This means that the panel breaker condition will involve both the stiffness and spacing of the stiffeners compared with the skin stiffness and its overall dimensions.

There are two main scenarios to quantify this sequence of events. In the first, a non-buckling design is all that is required (no post-buckling capability). In such a case, the stiffeners must have properties such that the buckling load of the panel as a whole equals the buckling load of the skin between stiffeners. The two failure modes, local and global buckling, occur simultaneously. In the second scenario, the skin is allowed to buckle. This means that the stiffeners must stay intact and not bend, until the skin reaches the desired post-buckling load and fails.

9.2.1.1 Global Buckling = Local Buckling (Compression Loading)

It is important to recognize that the total applied force F_{TOT} is distributed between the skin and stiffeners according to their respective in-plane EA stiffnesses. This was expressed by Equation (8.12). With reference to Figure 9.5, the membrane stiffness EA of the skin is approximated by bA_{11}. Note that a more accurate expression would be $b(A_{11} - A_{12}^2/A_{22})$, as is indicated by Equations (8.5) and (8.6), but the second term is neglected here, assuming the skin has at least 40% $0°$ plies aligned with the load so that $A_{12} \ll A_{11}$. If this requirement is not satisfied, the equations that follow can be modified accordingly.

Using Equation (8.12), the force acting on the skin alone can be determined as

$$F_{skin} = \frac{bA_{11}}{bA_{11} + b\dfrac{EA}{d_s}} F_{TOT} = \frac{A_{11}}{A_{11} + \dfrac{EA}{d_s}} F_{TOT} \qquad (9.18)$$

It should be emphasized that F_{TOT} here refers to the total force on both skin and stiffeners. This is different than Equations (8.8), (8.9), (8.10), (8.11) and (8.12) where F_{TOT} referred to the total force on the stiffener alone.

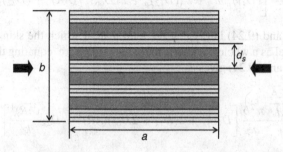

Figure 9.5 Stiffened skin under compression

Then, the force per unit length acting on the skin alone is

$$N_{xskin} = \frac{F_{skin}}{b} \qquad (9.19)$$

If the skin between stiffeners is assumed to be simply supported, its buckling load can be obtained from Equation (6.7):

$$N_{xskin} = \frac{\pi^2}{a^2}\left[D_{11}k^2 + 2(D_{12} + 2D_{66})(\overline{AR})^2 + D_{22}\frac{(\overline{AR})^4}{k^2}\right] \qquad (9.20)$$

where k is the number of half-waves into which the skin buckles, D_{ij} are skin bending stiffnesses and $\overline{AR}$ is the aspect ratio a/d_s.

Combining Equations (9.18) to (9.20) and solving for the total force F_{TOT} gives

$$F_{TOT} = \frac{A_{11} + \dfrac{EA}{d_s}}{A_{11}} \frac{\pi^2 b}{a^2}\left[D_{11}k^2 + 2(D_{12} + 2D_{66})(\overline{AR})^2 + D_{22}\frac{(\overline{AR})^4}{k^2}\right] \qquad (9.21)$$

where A_{11} is the skin membrane stiffness and EA is the stiffener membrane stiffness.

Assuming now that the panel as a whole is simply supported all around its boundary, its buckling load will also be given by Equation (6.7):

$$N_{panel} = \frac{\pi^2}{a^2}\left[(D_{11})_p m^2 + 2\left((D_{12})_p + 2(D_{66})_p\right)(AR)^2 + (D_{22})_p\frac{(AR)^4}{m^2}\right] \qquad (9.22)$$

where m is the number of half-waves into which the entire panel would buckle (note that m for the overall panel and k for the skin between the stiffeners can be different) and the subscript p denotes the entire panel. The aspect ratio AR of the panel is a/b. The bending stiffnesses $(D_{ij})_p$ for the panel are given by Equation (9.17).

The force per unit length N_{panel} is given by

$$N_{panel} = \frac{F_{TOT}}{b} \qquad (9.23)$$

Combining Equations (9.22) and (9.23) and solving for F_{TOT} gives

$$F_{TOT} = \frac{b\pi^2}{a^2}\left[(D_{11})_p m^2 + 2\left((D_{12})_p + 2(D_{66})_p\right)(AR)^2 + (D_{22})_p\frac{(AR)^4}{m^2}\right] \qquad (9.24)$$

Equations (9.21) and (9.24) imply that the total load at which the skin between stiffeners buckles and the panel as a whole buckles is the same. Therefore, equating the right-hand sides of Equations (9.21) and (9.24) gives

$$\frac{A_{11} + \dfrac{EA}{d_s}}{A_{11}} \frac{\pi^2 b}{a^2}\left[D_{11}k^2 + 2(D_{12} + 2D_{66})(\overline{AR})^2 + D_{22}\frac{(\overline{AR})^4}{k^2}\right]$$

$$= \frac{b\pi^2}{a^2}\left[(D_{11})_p m^2 + 2\left((D_{12})_p + 2(D_{66})_p\right)(AR)^2 + (D_{22})_p\frac{(AR)^4}{m^2}\right] \qquad (9.25)$$

Equation (9.25) can be simplified by cancelling out common factors and using Equation (9.17) to express the panel bending stiffnesses EI. Here, it will be assumed that the stiffener has an open cross-section so its GJ is very small and does not contribute to the panel D_{66} value. Then, Equation (9.25) reads

$$\frac{A_{11} + \dfrac{EA}{d_s}}{A_{11}} \left[D_{11}k^2 + 2(D_{12} + 2D_{66})(\overline{AR})^2 + D_{22}\frac{(\overline{AR})^4}{k^2} \right]$$

$$= \left[\left(D_{11} + \frac{EI}{d_s} \right) m^2 + 2(D_{12} + 2D_{66})(AR)^2 + D_{22}\frac{(AR)^4}{m^2} \right] \quad (9.26)$$

Note that, in Equation (9.26), EA and EI both refer to stiffener quantities.

Further simplification is possible if the values of k and m are approximated. As mentioned in Section 6.2, k and m are the integer values of half-waves that minimize the corresponding buckling loads (skin between stiffeners or entire panel). If k and m were continuous variables (instead of only taking integer values) differentiating the corresponding buckling expressions and setting the result equal to zero would give the values to use (see the derivation of Equation 8.45). Since they are integers, the rational expression resulting from differentiation would have to be rounded up or down to the nearest adjacent integer that minimizes the buckling load.

Differentiating the right-hand side of Equation (9.20) with respect to k and setting the result equal to zero gives

$$\frac{dN_{xskin}}{dk} = 0 \Rightarrow k^* = \left(\frac{D_{22}}{D_{11}} \right)^{1/4} \left(\frac{a}{d_s} \right) \quad (9.27)$$

where k^* denotes the continuous variable ($k = k^*$ when the right-hand side of Equation 9.27 is an integer).

Using Equation (9.27), the value of k is given by either

$$k = \text{int}[k^*]$$

or

$$k = \text{int}[k^*] + 1$$

whichever of the two minimizes the right-hand side of Equation (9.20). The symbol int[x] denotes the integer obtained if x is rounded down to the next integer.

Similarly for m,

$$m^* = \left(\frac{D_{22}}{D_{11} + \dfrac{EI}{d_s}} \right)^{1/4} \left(\frac{a}{b} \right) \quad (9.28)$$

By necessity, if either of k^* or m^* is less than 1, the corresponding value of k or m will be set equal to 1.

Now for typical applications of panels under compression, the quantity $D_{11} + \mathrm{EI}/d_s$ is greater than D_{22} because of the contribution of the stiffeners EI/d_s and the tendency to align fibres with the load direction which would give $D_{11} \geq D_{22}$. So unless $a/b \gg 1$ the quantity in the right-hand side of Equation (9.28) is less than 1 and m will be equal to 1.

To proceed, set $k = k^*$ as obtained from Equation (9.27). Then, substituting for k and m in Equation (9.26)

$$D_{11} + \frac{(\mathrm{EI})_{\mathrm{stiff}}}{d_s} + 2\,[D_{12} + 2D_{66}]\,(AR)^2 + D_{22}(AR)^4$$

$$= \underbrace{\frac{\left(A_{11} + \dfrac{EA}{d_s}\right)}{A_{11}}}_{\lambda} \left[D_{11}\sqrt{\frac{D_{22}}{D_{11}}}\,\overline{AR}^2 + 2\,[D_{12} + 2D_{66}]\,(\overline{AR})^2 + D_{22}\frac{(\overline{AR})^4}{\sqrt{\dfrac{D_{22}}{D_{11}}}\,\overline{AR}^2} \right] \tag{9.29}$$

Denoting by λ the term multiplying the quantity in brackets on the right-hand side, solving for the stiffener EI, and dropping the subscript 'stiff', for convenience, gives the final expression:

$$\mathrm{EI} = D_{11}d_s \left[\sqrt{\frac{D_{22}}{D_{11}}}\left(2\lambda\overline{AR}^2 - \sqrt{\frac{D_{22}}{D_{11}}}\,(AR)^4\right) + \frac{2\,(D_{12} + 2D_{66})}{D_{11}}\left(\lambda\overline{AR}^2 - (AR)^2\right) - 1 \right] \tag{9.30}$$

Equation (9.30) gives the minimum bending stiffness EI that the stiffeners must have in order for buckling of the skin between stiffeners to occur at the same time as overall buckling of the stiffened panel. If EI is greater than the right-hand side of Equation (9.30), the skin between stiffeners buckles first.

9.2.1.2 Stiffener Buckling = PB × Buckling of Skin between Stiffeners (Compression Loading)

This scenario covers the case where skin is allowed to post-buckle. It is assumed that the skin is loaded over the b_{eff} portion, which was determined in Section 7.1. The stiffener must stay straight all the way up to the load that fails the skin. That load is given by the buckling load of the skin between stiffeners multiplied by the post-buckling ratio PB.

In general, when the skin has buckled the compressive load on it is not constant (see Figure 7.8 for example) and the skin strains are not constant across its width. If the skin is replaced by the b_{eff} portion shown in Figure 9.6, then the skin load is constant over b_{eff} and the strain, given by inverting Equation (8.4), is also constant. Thus, strain compatibility can be applied.

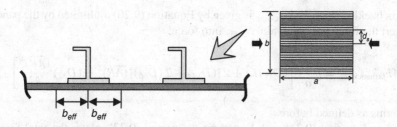

Figure 9.6 Effective skin in the vicinity of a stiffener in a post-buckled stiffened panel

Considering a $2b_{\text{eff}}$ portion of skin and its corresponding stiffener as shown in Figure 9.6 and using Equation (8.12), the individual forces on skin and stiffener can be found to be

$$F_{\text{skin}} = \frac{A_{11}\dfrac{b}{d_s}2b_{\text{eff}}}{2A_{11}\dfrac{b}{d_s}b_{\text{eff}} + \text{EA}\dfrac{b}{d_s}}F_{\text{TOT}} \Rightarrow F_{\text{skin}} = \frac{2A_{11}b_{\text{eff}}}{2A_{11}b_{\text{eff}} + \text{EA}}F_{\text{TOT}} \tag{9.31}$$

$$F_{\text{stiffeners}} = \frac{\text{EA}}{2A_{11}b_{\text{eff}} + \text{EA}}F_{\text{TOT}} \tag{9.32}$$

where EA is the membrane stiffness of each stiffener ($=$ membrane modulus $\times$ cross-sectional area)

For a single stiffener, dividing the right-hand side of Equation (9.32) by the number of stiffeners given by Equation (9.4) gives

$$F_{\text{stiff}} = \frac{d_s}{b}\frac{\text{EA}}{2A_{11}b_{\text{eff}} + \text{EA}}F_{\text{TOT}} \tag{9.33}$$

Now, the column buckling load of a stiffener is, for simply supported ends, given by Equation (8.21), repeated here for convenience:

$$F_{\text{stiff}b} = \frac{\pi^2\text{EI}}{a^2} \tag{9.34}$$

Equating the right-hand sides of Equations (9.33) and (9.34) relates the load in each stiffener to the buckling load of that stiffener and can be solved for the total force on the panel F_{TOT}:

$$\frac{d_s}{b}\frac{\text{EA}}{2A_{11}b_{\text{eff}} + \text{EA}}F_{\text{TOT}} = \frac{\pi^2\text{EI}}{a^2} \Rightarrow F_{\text{TOT}} = \frac{\pi^2\text{EI}}{a^2}\frac{b}{d_s}\frac{2A_{11}b_{\text{eff}} + \text{EA}}{\text{EA}} \tag{9.35}$$

Now it is postulated that final failure occurs when the required post-buckling ratio PB is reached. At that point, the force in the skin F_{skin} must equal the buckling load of the skin between the stiffeners multiplied by PB:

$$F_{\text{skin}} = F_{\text{skinbuckling}}(PB) \tag{9.36}$$

The skin buckling load $F_{\text{skinbuckling}}$ is given by Equation (9.20) multiplied by the panel width b to convert the force per unit width $N_{x\text{skin}}$ into force:

$$F_{\text{skinbuckling}} = b\frac{\pi^2}{a^2}\left[(D_{11})k^2 + 2\left[(D_{12}) + 2(D_{66})\right](\overline{AR})^2 + (D_{22})\frac{(\overline{AR})^4}{k^2}\right] \quad (9.37)$$

with all terms as defined before.

Combining Equation (9.31) with Equations (9.36) and (9.37) relates the total force to the skin buckling load:

$$F_{\text{TOT}} = \frac{2A_{11}b_{\text{eff}} + \text{EA}}{2A_{11}b_{\text{eff}}}b\frac{\pi^2}{a^2}\left[(D_{11})k^2 + 2\left[(D_{12}) + 2(D_{66})\right](\overline{AR})^2 + (D_{22})\frac{(\overline{AR})^4}{k^2}\right](PB) \quad (9.38)$$

Equations (9.35) and (9.38) can now be combined to yield the condition for column buckling of the stiffeners occurring when the final PB is reached:

$$\frac{\text{EI}}{d_s\text{EA}} = \frac{(PB)}{2A_{11}b_{\text{eff}}}\left[D_{11}k^2 + 2\left[D_{12} + 2D_{66}\right](\overline{AR})^2 + D_{22}\frac{(\overline{AR})^4}{k^2}\right] \quad (9.39)$$

which relates stiffener properties on the left-hand side with skin properties on the right-hand side.

Equation (9.39) can be further manipulated using the definition of the parameter λ that was introduced when Equation (9.29) was derived. Using that definition, it can be shown that

$$\lambda = \frac{A_{11} + \dfrac{\text{EA}}{d_s}}{A_{11}} \Rightarrow \lambda A_{11} - A_{11} = \frac{\text{EA}}{d_s} \Rightarrow A_{11}(\lambda - 1) = \frac{\text{EA}}{d_s} \Rightarrow \frac{\text{EA}}{A_{11}} = (\lambda - 1)d_s \quad (9.40)$$

Introducing this result in Equation (9.39) and solving for the stiffener bending stiffness EI results in the expression

$$\text{EI} = (\lambda - 1)(PB)d_s\frac{d_s}{2b_{\text{eff}}}\left[D_{11}k^2 + 2\left[D_{12} + 2D_{66}\right](\overline{AR})^2 + D_{22}\frac{(\overline{AR})^4}{k^2}\right] \quad (9.41)$$

Also note that using the expression for b_{eff}, Equation (7.15), derived earlier (assuming the boundary conditions for the skin between stiffeners reasonably approximate those used in Section 7.1) the following can be shown:

$$\frac{d_s}{b_{\text{eff}}} = 2\left[1 + 2\left(1 + \frac{A_{12}}{A_{11}}\right)\left(1 - \frac{1}{(PB)}\right)\frac{A_{11}}{A_{11} + 3A_{22}}\right] \quad (9.42)$$

which can be substituted in Equation (9.41).

Equation (9.41) determines the minimum bending stiffness of the stiffeners so that they do not buckle until the final failure load in the post-buckling regime is reached. This would guarantee that the stiffeners will stay straight and thus act as panel breakers, all the way to the

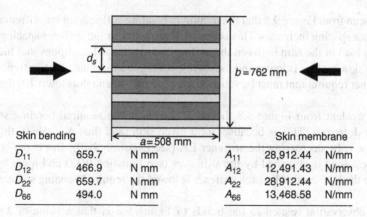

Skin bending			Skin membrane		
D_{11}	659.7	N mm	A_{11}	28,912.44	N/mm
D_{12}	466.9	N mm	A_{12}	12,491.43	N/mm
D_{22}	659.7	N mm	A_{22}	28,912.44	N/mm
D_{66}	494.0	N mm	A_{66}	13,468.58	N/mm

Figure 9.7   Example of stiffened panel under compression

failure load of the panel. It should be noted that EI in Equation (9.41) is related to the stiffener EA through the parameter λ so the two are not entirely independent and some iterations may be needed between stiffener geometry and layup to arrive at the required bending stiffness.

9.2.1.3   Example

As an application of the two conditions (9.30) and (9.41) consider a skin panel with dimensions $a = 508$ mm and $b = 762$ mm loaded in compression parallel to dimension a as shown in Figure 9.7. The skin layup is $[(\pm 45)/(0/90)/(\pm 45)]$ with stiffness properties given in Figure 9.7. The stiffener spacing, geometry and layup are unknown. The minimum EI for the stiffeners must be determined subject to conditions (9.30) and (9.41).

The procedure is as follows: First a value of the parameter λ is selected. Then, for that value, the stiffener spacing d_s is varied between 75 and 300 mm. For each value of d_s the aspect ratio $(\overline{AR})$ is calculated and the corresponding buckling load of the skin between the stiffeners is determined using Equation (9.20). Finally, the minimum EI required is determined using Equations (9.30) and (9.41). When using Equation (9.41) a PB ratio of 5 is assumed. The results are shown in Figure 9.8 for two different λ values.

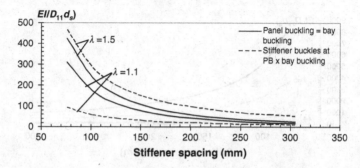

Figure 9.8   Normalized minimum bending stiffness required for stiffeners of a stiffened panel under compression (PB $= 5$)

It can be seen from Figure 9.8 that the minimum bending stiffness for the stiffener decreases as the stiffener spacing increases. This is due to the fact that, as the stiffener spacing increases, the buckling load of the skin between the stiffeners decreases. This implies that the total load at which the skin buckles is lower and the corresponding load applied to the stiffeners is lower. So the stiffener requirement must be satisfied for a lower load and thus lower bending stiffness is needed.

It is also evident from Figure 9.8 that as λ decreases, the required bending stiffness for the stiffener decreases. This is because, for a given skin (and thus A_{11} value), the only way to decrease λ is by decreasing the stiffener EA (see Equation 9.40). But if the stiffener EA decreases, less load is absorbed by the stiffeners (see Equation 9.32) and more by the skin. Again, since the load carried by each stiffener is lower, the required bending stiffness will also be lower.

The last observation related to the trends of Figure 9.8 is that a value of λ is reached beyond which the condition that the stiffeners buckle at PB $\times$ bay buckling dominates. This is demonstrated by the fact that the continuous curve is above the dashed curve for $\lambda = 1.1$, but below the dashed curve for $\lambda = 1.5$. This means that, for a given stiffener spacing, there is a λ value between 1.1 and 1.5 at which the design driver switches from Equation (9.30) to Equation (9.41). This means that one should check both conditions, (9.30) and (9.41), and use the one that gives the more conservative results. If no post-buckling is allowed in the design, only condition (9.30) should be used.

Selecting the more critical of the two conditions, for each value of λ, results in the curves of Figure 9.9. The lowest three curves ($\lambda = 1.1$, 1.5, 3.0) correspond to Equation (9.30) dominating the design and the highest two curves ($\lambda = 5.0$, 10.) correspond to Equation (9.40) dominating the design.

9.2.2 Skin–Stiffener Separation

As load is transferred between the skin and stiffeners, out-of-plane loads develop at their interface or at the flange edges. These loads develop even when the applied loads are in the plane of the skin and they can lead to separation of the stiffeners from the skin. There are two main mechanisms for the development of these out-of-plane stresses. The first is associated with the presence of any stress-free edges such as the flange edges [2–7]. Load present in

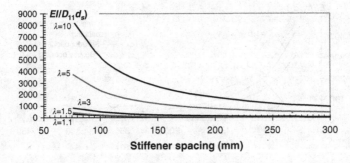

Figure 9.9   Normalized minimum bending stiffness of stiffeners for various λ values (PB = 5)

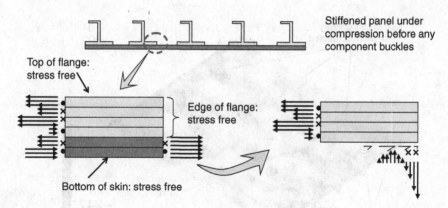

Figure 9.10 Development of out-of-plane stresses at the interface between the flange and skin prior to buckling

one component such as the flange has to transfer to the other as the free edge is approached. The local stiffness mismatch caused by the presence of the free edge (and the differences in stacking sequence between flange and skin) creates out-of-plane stresses which are the main culprit for separation of the flange from the skin. This is shown in Figure 9.10.

Away from the flange edge, closer to the location where the stiffener web meets the flange, a two-dimensional state of stress develops that can be determined with the use of classical laminated-plate theory once the local loads are known. A section including the flange edge and the skin below can be cut off and placed in equilibrium (bottom left part of Figure 9.10). If then the flange alone is sectioned off (bottom right of Figure 9.10), it is not in equilibrium unless out-of-plane shear and normal stresses develop at the interface between the flange and the skin. This is the only location where this can happen since the top of the flange and the right edge of the flange are, by definition, stress free. Zooming to the sectioned detail at the bottom right of Figure 9.10 gives the situation shown in Figure 9.11.

For the purposes of discussion, the skin and stiffener in Figures 9.10 and 9.11 are assumed to be under compression. A local coordinate system is established in Figure 9.11 to facilitate

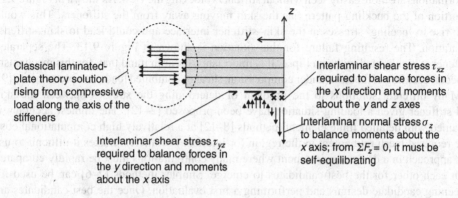

Figure 9.11 Free-body diagram of the flange

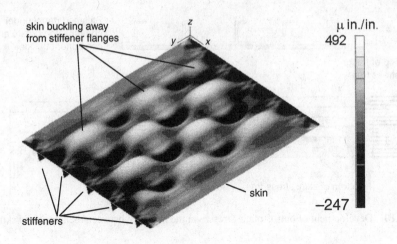

skin buckling away
from stiffener flanges

μ in./in.
492

skin

−247

stiffeners

Figure 9.12   Post-buckled shape of blade-stiffened panel under compression

the discussion. As shown in Figure 9.11, to maintain force equilibrium in the y direction, an interlaminar shear stress τ_{yz} must develop at the flange–skin interface (bottom of the flange in Figure 9.11). Also, the presence of in-plane shear stresses at the left end of the flange, as predicted by the classical laminated-plate theory, leads to a net force in the x direction. In order to balance that force, an interlaminar shear stress τ_{xz} must also develop at the flange–skin interface. Finally, to balance the moments (about the bottom right corner of the flange, say), an out-of-plane normal stress σ_z must develop at the flange–skin interface. However, since there is no net force in the z direction in Figure 9.11, the σ_z stress must be self-equilibrating. Thus, it will be tensile over a portion of the region over which it acts and compressive over the remaining portion. This is why σ_z is shown as both tensile and compressive in Figure 9.11.

The second mechanism that gives rise to these separation stresses is associated with the skin deformation after buckling in a post-buckling situation. This is shown in Figure 9.12 based on an example from [8].

The stiffened panel of Figure 9.12 is shown with the stiffeners at the bottom so the skin deformations are more easily seen without stiffeners blocking the view. As shown in Figure 9.12 a portion of the buckling pattern has the skin moving away from the stiffeners. This would give rise to 'peeling' stresses at the skin–stiffener interface and could lead to skin–stiffener separation. The resulting failure for this situation is shown in Figure 9.13. The separated stiffener is highlighted with an ellipse. It is important to bear in mind that, for this mechanism to occur, buckling of the skin (under compression, shear or combined loads) is a prerequisite [9].

Many different solutions to the problem of determining the stresses between the skin and stiffener given a loading situation have been proposed [2–12]. The highest accuracy is obtained with detailed finite element methods [8–12] at a relatively high computational cost. The required mesh refinement in the region (or interface) of interest makes it difficult to use this approach in a design environment where many configurations must be rapidly compared with each other for the best candidates to emerge. Simpler methods [2–6] can be used for screening candidate designs and performing a first evaluation. Once the best candidates are selected, more detailed analysis methods using finite elements can be used for more accurate predictions.

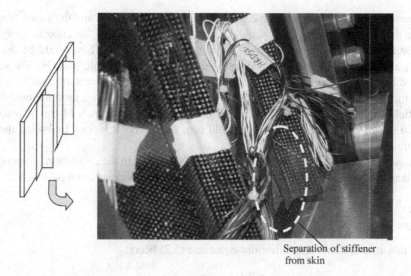

Separation of stiffener
from skin

Figure 9.13 Skin–stiffener separation failure mode (see Plate 22 for the colour figure)

In general, solutions that calculate the stresses at the skin–stiffener interface assume a perfect bond between the stiffener flange and the skin and require the use of some out-of-plane failure criterion [13, 14] to determine when delamination starts. This can be quite conservative as a delamination starting at the skin–stiffener interface rarely grows in an unstable fashion to cause final failure. To model the presence of a delamination and to determine when it will grow, methods based on energy release rate calculations [10–12] are very useful. In what follows, only the problem of determining the out-of-plane stresses in a pristine structure will be presented.

The approach is adapted from [3, 4] and can be applied to any situation for which the loads away from the flange edge are known, irrespective of whether the structure is post-buckled or not. The situation is shown in Figure 9.14. The flange end is isolated from the rest of the stiffener and a portion of the skin below it is shown.

Two coordinate systems are used in Figure 9.14, one for the flange and one for the skin. They have a common origin at the end of the flange where it interfaces with the skin. The z axis (out-of-plane) for the flange is going up while it is going down for the skin. The y

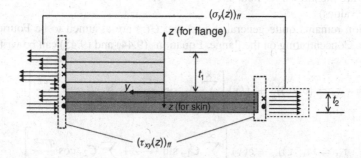

Figure 9.14 Geometry and coordinate systems for the skin–stiffener separation problem

axis is moving away from the flange edge towards the stiffener web and the x axis is aligned with the axis of the stiffener. The stresses away from the origin of the coordinate systems, at the far left and right end in Figure 9.14, are assumed known. They would be the result of classical laminated-plate theory and/or other two-dimensional solutions. For the solution discussed here, these far-field stresses are assumed to be only in-plane stresses. This means that the three-dimensional stresses that arise in the vicinity of the flange termination die out in the far field so the known solution at the left and right ends of Figure 9.14 can be recovered. Note that out-of-plane stresses at the far field can, if present, be added following the same procedure as the one outlined below.

It is assumed that the structure shown in Figure 9.14 is long in the x direction (perpendicular to the plane of the figure) so no quantity depends on the x coordinate, that is,

$$\frac{\partial(\ldots)}{\partial x} = 0$$

With this assumption, the equilibrium equations (5.2) become

$$\frac{\partial \tau_{xy}}{\partial y} + \frac{\partial \tau_{xz}}{\partial z} = 0$$

$$\frac{\partial \sigma_y}{\partial y} + \frac{\partial \tau_{yz}}{\partial z} = 0 \qquad\qquad (9.43\text{a–c})$$

$$\frac{\partial \tau_{yz}}{\partial y} + \frac{\partial \sigma_z}{\partial z} = 0$$

This means that the first equilibrium equation, (9.43a), uncouples from the other two. This, in turn, suggests that if one of the stresses τ_{xy} or τ_{xz} were somehow known, Equation (9.43a) could be used to determine the other. Similarly, if one of the stresses σ_y, τ_{yz} or σ_z were known, the other two could be determined from Equations (9.43b) and (9.43c).

Pursuing this line of thought, a fairly general assumption is made for τ_{xy} and σ_y in the form

$$\sigma_y = \left(\sigma_y(z)\right)_{ff} + f(y)F(z) \qquad\qquad (9.44)$$

$$\tau_{xy} = \left(\tau_{xy}(z)\right)_{ff} + g(y)G(z) \qquad\qquad (9.45)$$

where $f(y)$, $g(y)$, $F(z)$ and $G(z)$ are unknown functions of the respective coordinates and $(\sigma_y(z))_{ff}$ and $(\tau_{xy}(z))_{ff}$ are the known far-field stresses away from the flange edge (i.e. at large positive or negative y values).

The solution remains quite general if $F(z)$ and $G(z)$ are assumed to be Fourier sine and cosine series. Concentrating on the flange, Equations (9.44) and (9.45) can be written as

$$\sigma_y = \left(\sigma_y(z)\right)_{ff} + f(y)\left[\sum_{m=1}^{\infty} A_m \sin\frac{m\pi z}{t_1} + \sum_{n=1}^{\infty} B_n \cos\frac{n\pi z}{t_1}\right] \qquad (9.44\text{a})$$

$$\tau_{xy} = \left(\tau_{xy}(z)\right)_{ff} + g(y)\left[\sum_{p=1}^{\infty} C_{1p} \sin\frac{p\pi z}{t_1} + \sum_{q=1}^{\infty} C_{2q} \cos\frac{q\pi z}{t_1}\right] \qquad (9.45\text{a})$$

Here a simplification is introduced by truncating the infinite series in Equations (9.44a) and (9.45a) after the first term. This will give good solutions in terms of trends and will simplify the algebra considerably. Additional terms may be included for more accurate solutions. Then the two stresses in the flange have the form

$$\sigma_y = \left(\sigma_y(z)\right)_{ff} + f(y)\left[A_1 \sin \frac{\pi z}{t_1} + B_1 \cos \frac{\pi z}{t_1}\right] \tag{9.44b}$$

$$\tau_{xy} = \left(\tau_{xy}(z)\right)_{ff} + g(y)\left[C_1 \sin \frac{\pi z}{t_1} + C_2 \cos \frac{\pi z}{t_1}\right] \tag{9.45b}$$

where, for simplicity, we set $C_1 = C_{11}$ and $C_2 = C_{21}$. The coefficients A_1, B_1, C_1 and C_2 and the functions $f(y)$ and $g(y)$ are unknown at this point.

Now use Equation (9.44b) to substitute in Equation (9.43b) to obtain

$$\frac{\partial \tau_{yz}}{\partial z} = -f'\left(A_1 \sin \frac{\pi z}{t_1} + B_1 \cos \frac{\pi z}{t_1}\right) \tag{9.46}$$

where $f' = df/dy$.

Integrating Equation (9.46) with respect to z gives

$$\tau_{yz} = -f'\left(-A_1 \frac{t_1}{\pi} \cos \frac{\pi z}{t_1} + B_1 \frac{t_1}{\pi} \sin \frac{\pi z}{t_1}\right) + P_1(y) \tag{9.47}$$

where $P_1(y)$ is an unknown function of y.

Now the top of the flange is stress free, which means that $\tau_{yz}(z = t_1) = 0$, or,

$$-f'\left(A_1 \frac{t_1}{\pi}\right) + P_1(y) = 0 \Rightarrow P_1(y) = f'\left(A_1 \frac{t_1}{\pi}\right) \tag{9.48}$$

and $P_1(y)$ is determined. This would give the following expression for τ_{yz}:

$$\tau_{yz} = f'\left(A_1 \frac{t_1}{\pi}\left(1 + \cos \frac{\pi z}{t_1}\right) - B_1 \frac{t_1}{\pi} \sin \frac{\pi z}{t_1}\right) \tag{9.49}$$

In a similar fashion, Equation (9.49) can be used to substitute for τ_{yz} in Equation (9.43c) to obtain

$$\frac{\partial \sigma_z}{\partial z} = -f''\left(A_1 \frac{t_1}{\pi}\left(1 + \cos \frac{\pi z}{t_1}\right) - B_1 \frac{t_1}{\pi} \sin \frac{\pi z}{t_1}\right) \tag{9.50}$$

which can be integrated with respect to z to give

$$\sigma_z = -f''\left(A_1 \frac{t_1}{\pi}\left(z + \frac{t_1}{\pi} \sin \frac{\pi z}{t_1}\right) + B_1 \left(\frac{t_1}{\pi}\right)^2 \cos \frac{\pi z}{t_1}\right) + P_2(y) \tag{9.51}$$

where $f'' = d^2f/dy^2$ and $P_2(y)$ is an unknown function.

Again, the requirement that the top of the flange be stress free leads to $\sigma_z(z = t_1) = 0$ or

$$
-f''\left(A_1\frac{t_1}{\pi}(t_1) + B_1\left(\frac{t_1}{\pi}\right)^2\cos\pi\right) + P_2(y) = 0 \Rightarrow P_2(y) = f''\left(A_1\frac{t_1^2}{\pi} - B_1\left(\frac{t_1}{\pi}\right)^2\right)
$$

$$(9.52)$$

With $P_2(y)$ known, the final expression for σ_z can be obtained:

$$
\sigma_z = f''\left(A_1\frac{t_1}{\pi}\left(t_1 - z - \frac{t_1}{\pi}\sin\frac{\pi z}{t_1}\right) - B_1\left(\frac{t_1}{\pi}\right)^2\left(1 + \cos\frac{\pi z}{t_1}\right)\right)
$$

$$(9.53)$$

In a completely analogous fashion, placing Equation (9.45b) into Equation (9.43a) and solving for τ_{xz} gives

$$
\tau_{xz} = g'\left[C_1\frac{t_1}{\pi}\left(1 + \cos\frac{\pi z}{t_1}\right) - C_2\frac{t_1}{\pi}\sin\frac{\pi z}{t_1}\right]
$$

$$(9.54)$$

Equations (9.44b), (9.45b), (9.49), (9.51) and (9.54) determine the stresses σ_y, τ_{xy}, τ_{yz}, σ_z and τ_{xz} to within some unknown constants and the two unknown functions f and g and their derivatives. At this point the stress σ_x is unknown. It does not appear in the equilibrium equations (9.43a–c) and other means must be invoked for its determination.

By inverting the stress–strain equations (5.4), the following relations are obtained:

$$
\begin{Bmatrix} \varepsilon_x \\ \varepsilon_y \\ \varepsilon_z \\ \gamma_{yz} \\ \gamma_{xz} \\ \gamma_{xy} \end{Bmatrix} = \begin{bmatrix} S_{11} & S_{12} & S_{13} & 0 & 0 & S_{16} \\ S_{12} & S_{22} & S_{23} & 0 & 0 & S_{26} \\ S_{13} & S_{23} & S_{33} & 0 & 0 & S_{36} \\ 0 & 0 & 0 & S_{44} & S_{45} & 0 \\ 0 & 0 & 0 & S_{45} & S_{55} & 0 \\ S_{16} & S_{26} & S_{36} & 0 & 0 & S_{66} \end{bmatrix} \begin{Bmatrix} \sigma_x \\ \sigma_y \\ \sigma_z \\ \tau_{yz} \\ \tau_{xz} \\ \tau_{xy} \end{Bmatrix}
$$

$$(9.55)$$

where S_{ij} $(i,j = 1\text{–}6)$ are compliances for the flange as a whole. They can be computed as thickness-averaged sums of the corresponding compliances for the individual plies, and the compliances of the individual plies are obtained using standard tensor transformation equations (Equations 3.8–3.10).

In addition, the strain compatibility relations (5.10) and (5.12) can be rewritten as

$$
\frac{\partial^2\gamma_{xy}}{\partial x\partial y} = \frac{\partial^2\varepsilon_x}{\partial y^2} + \frac{\partial^2\varepsilon_y}{\partial x^2}
$$

$$(9.56)$$

$$
\frac{\partial^2\gamma_{xz}}{\partial x\partial z} = \frac{\partial^2\varepsilon_x}{\partial z^2} + \frac{\partial^2\varepsilon_z}{\partial x^2}
$$

$$(9.57)$$

As was mentioned earlier, there is no dependence on the x coordinate so all derivatives with respect to x are zero and Equations (9.56) and (9.57) simplify to

$$0 = \frac{\partial^2 \varepsilon_x}{\partial y^2} \tag{9.58}$$

$$0 = \frac{\partial^2 \varepsilon_x}{\partial z^2} \tag{9.59}$$

The first of Equations (9.55) can be combined with Equations (9.58) and (9.59) to give

$$\frac{\partial^2}{\partial y^2} \left[S_{11}\sigma_x + S_{12}\sigma_y + S_{13}\sigma_z + S_{16}\tau_{xy} \right] = 0 \tag{9.60}$$

$$\frac{\partial^2}{\partial z^2} \left[S_{11}\sigma_x + S_{12}\sigma_y + S_{13}\sigma_z + S_{16}\tau_{xy} \right] = 0 \tag{9.61}$$

The quantity in brackets is the same for both Equations (9.60) and (9.61). The only way these two equations can be compatible with each other is if the quantity in brackets has the following form:

$$S_{11}\sigma_x + S_{12}\sigma_y + S_{13}\sigma_z + S_{16}\tau_{xy} = yG_1(z) + G_2(z) \tag{9.62}$$

with $G_1(z)$ and $G_2(z)$ unknown functions of z.

Using Equation (9.62) to substitute in Equation (9.61) gives

$$y\frac{d^2 G_1(z)}{dz^2} + \frac{d^2 G_2(z)}{dz^2} = 0$$

from which

$$G_1(z) = k_o + k_1 z$$

$$G_2(z) = k_3 + k_4 z$$

with k_o, k_1, k_3 and k_4 unknown constants.

Using this result and Equation (9.62) to solve for σ_x gives

$$\sigma_x = K_o + K_1 y + K_2 z + K_3 yz - \frac{S_{12}}{S_{11}}\sigma_y - \frac{S_{13}}{S_{11}}\sigma_z - \frac{S_{16}}{S_{11}}\tau_{xy} \tag{9.63}$$

with K_o, K_1, K_2 and K_3 new unknown constants (combinations of $k_o - k_4$).

Equation (9.63) determines σ_x as a function of the other stresses σ_y, σ_z and τ_{xy} which were determined earlier. The unknown coefficients $K_o - K_3$ are determined from matching the far-field solution. That is, Equation (9.63) is evaluated for large values of y and compared with the known solution there.

At this point, all stresses in the flange have been determined to within some unknown coefficients and two unknown functions $f(y)$ and $g(y)$. The stresses in the skin for $y > 0$ are determined in a completely analogous fashion. One additional set of conditions is imposed here, namely, stress continuity at the flange–skin interface. By using an overbar to denote skin quantities, these conditions have the form

$$\tau_{xz}(z = 0) = -\overline{\tau_{xz}}(z = 0)$$
$$\tau_{yz}(z = 0) = -\overline{\tau_{yz}}(z = 0) \tag{9.64}$$
$$\sigma_z(z = 0) = \overline{\sigma_z}(z = 0)$$

Note that a minus sign is needed in front of the shear stresses on the right-hand side to account for the orientation of the coordinate systems in Figure 9.14. As a result of Equations (9.64), the unknown functions $f(y)$ and $g(y)$ have to be the same for both flange and skin. In addition, the coefficients A_1, B_1, etc. for the skin, corresponding to the coefficients present in Equations (9.44b), (9.45b), (9.49), (9.51) and (9.54) for the flange, are also determined from Equations (9.64).

In order to determine the unknown functions $f(y)$ and $g(y)$, the principle of minimum complementary energy (see Section 5.4) is invoked. This means that the quantity

$$\Pi_C = \frac{1}{2} \iiint \underline{\sigma}^T \underline{S} \underline{\sigma} \, dy dx dz + \frac{1}{2} \iiint \overline{\underline{\sigma}}^T \overline{\underline{S} \underline{\sigma}} \, dy dx dz - \iint \underline{T}^T \underline{u}^* dy dz \tag{9.65}$$

must be minimized. Underscores denote vectors and matrices. Overbars, as already mentioned, denote skin quantities. Specifically,

$$\sigma^T = \begin{bmatrix} \sigma_x & \sigma_y & \sigma_z & \tau_{yz} & \tau_{xz} & \tau_{xy} \end{bmatrix}$$

and $\underline{S}$ was given in Equation (9.55) above.

The last term of Equation (9.65) is the work term. It consists of the tractions $\underline{T}$ multiplying the prescribed displacements $\underline{u}^*$.

The stress expressions already determined are used to substitute in Equation (9.65). The x and z integrations can be carried out without difficulty because they only involve either powers or sines and cosines of the variables. Thus, after x and z integration, an expression for the energy is obtained in the form

$$\Pi_C = \frac{1}{2} \int H \left(\frac{d^2 f}{dy^2}, \frac{df}{dy}, f, \frac{dg}{dy}, g, y \right) dy \tag{9.66}$$

The problem has thus been recast as one in which the functions $f(y)$ and $g(y)$ must be determined such that the integral in the right-hand side of Equation (9.66) is minimized. This can be done by using the calculus of variations [15]. The general form of the Euler equations for f and g is as follows:

$$\frac{d^2}{dy^2} \left(\frac{\partial H}{\partial f''} \right) - \frac{d}{dy} \left(\frac{\partial H}{\partial f'} \right) + \frac{\partial H}{\partial f} = 0 \tag{9.67}$$

$$\frac{\partial H}{\partial g} - \frac{d}{dy} \left[\frac{\partial H}{\partial g'} \right] = 0 \tag{9.68}$$

Using the detailed expression of H to substitute in Equations (9.67) and (9.68) yields the two equations:

$$\frac{d^4 f}{dy^4} + R_1 \frac{d^2 f}{dy^2} + R_2 f + R_3 \frac{d^2 g}{dy^2} + R_4 g = 0 \tag{9.67a}$$

$$\frac{d^2 g}{dy^2} + R_5 g + R_6 \frac{d^2 f}{dy^2} + R_7 f = 0 \tag{9.68a}$$

where R_1–R_7 are constants obtained from the x and z integrations implied by Equation (9.65) and can be found in [2].

It should be noted that Equations (9.67a) and (9.68a) are given as homogeneous equations. In fact, Equations (9.67a) and (9.68a) would yield a nonhomogeneous term, that is, the right-hand side of Equations (9.67a) and (9.68a) is, in general, nonzero. However, it can be shown [2] that the nonhomogeneous terms affect only the far-field behaviour of the stresses which is already incorporated in the stress expressions. So the nonhomogeneous part of the solution can be neglected without loss of generality.

The two Equations (9.67a) and (9.68a) are coupled ordinary differential equations with constant coefficients. Following standard procedures the solutions can be written using exponentials:

$$f(y) = S_{1f} e^{-\varphi_1 y} + S_{2f} e^{-\varphi_2 y} + S_{3f} e^{-\varphi_3 y} \tag{9.69}$$

$$g(y) = S_{1g} e^{-\varphi_1 y} + S_{2g} e^{-\varphi_2 y} + S_{3g} e^{-\varphi_3 y} \tag{9.70}$$

where the exponents φ_i are solutions to

$$\varphi^6 + (R_1 + R_5 - R_3 R_6)\varphi^4 + (R_1 R_5 + R_2 - R_3 R_7 - R_4 R_6)\varphi^2 + R_2 R_5 - R_4 R_7 = 0 \tag{9.71}$$

There are, in general, six solutions to Equation (9.71), but because only even powers of φ are present, they will appear in positive and negative pairs. Positive values of φ (or φ values with positive real parts) imply that f and g grow indefinitely as y increases, which implies that the interlaminar stresses containing f and g and their derivatives will tend to infinity for large values of y. This, however, is unacceptable as the interlaminar stresses must go to zero for large values of y in order for the far-field solution to be recovered. This is the reason for the negative exponents in Equations (9.69) and (9.70). It is assumed that φ_1, φ_2 and φ_3 correspond to the positive solutions of Equation (9.71) (or those with positive real parts) so that the expressions for f and g are in terms of decaying exponentials.

One additional comment pertaining to the limits of the integral in Equation (9.66) is in order. The lower limit is zero, the edge of the flange. The upper limit is any large but finite value of y, corresponding to a point where the far-field stresses are recovered. Of course, if the flange of the stiffener is very narrow, the negative exponentials in Equations (9.69) and (9.70) have not died out and those two expressions for f and g must be modified to include the remaining three φ solutions of (9.71) which correspond to increasing exponentials. As already alluded to, some of the solutions to Equation (9.71) may be complex, in which case, expressions (9.69) and (9.70) will include complex conjugates.

Proceeding with the solution to the system of the two Equations (9.67a) and (9.68a) it can be shown that

$$\frac{S_{if}}{S_{ig}} = -\frac{\varphi_i^2 + R_5}{R_6\varphi_i^2 + R_7} \tag{9.72}$$

which relates the coefficients in function $f(y)$ to those in function $g(y)$.

At this point in the solution, the following unknowns remain: S_{1f}, S_{2f}, S_{3f}, B_1 and C_2 in the flange and $\overline{C}_2$ in the skin. To determine these, the remaining boundary conditions, namely that the flange edge is stress free, are imposed:

$$\sigma_y(y = 0) = 0$$

$$\tau_{xy}(y = 0) = 0 \tag{9.73}$$

$$\tau_{yz}(y = 0) = 0$$

which, using the expressions for the stresses and the solution to the governing equations for f and g, become

$$\left(S_{1f} + S_{2f} + S_{3f}\right)\left(A_1 \sin\frac{\pi z}{t_1} + B_1 \cos\frac{\pi z}{t_1}\right) + \left(\sigma_y(z)\right)_{ff} = 0$$

$$\left(S_{1g} + S_{2g} + S_{3g}\right)\left(C_1 \sin\frac{\pi z}{t_1} + C_2 \cos\frac{\pi z}{t_1}\right) + \left(\tau_{xy}(z)\right)_{ff} = 0 \tag{9.74a–c}$$

$$\left(\varphi_1 S_{1f} + \varphi_2 S_{2f} + \varphi_3 S_{3f}\right)\left(A_1\frac{t_1}{\pi}\left(1 + \cos\frac{\pi z}{t_1}\right) - B_1\frac{t_1}{\pi}\sin\frac{\pi z}{t_1}\right) = 0$$

Since Equations (9.74a–c) involve sines and cosines of the variable z, the far-field stresses $(\sigma_y(z))_{ff}$ and $(\tau_{xy}(z))_{ff}$ are expanded in Fourier series and the first terms used to match the corresponding terms in Equations (9.74a–c). This introduces an additional approximation in the solution, but the results are still accurate enough to give reliable trends of the behaviour.

Once Equations (9.74a–c) are solved, all unknown constants in the stress expressions are determined except for $\overline{C}_2$ in the skin. This, again, is determined by energy minimization

$$\frac{\partial \Pi_c}{\partial \overline{C}_2} = 0$$

which yields a linear equation for $\overline{C}_2$. The details of the algebra can be found in [2].

The predictions of the method presented here have been compared with finite element results [2–4] and shown to be in good to excellent agreement. Discrepancies and reasons for them are discussed in the references. In what follows, the solution will be used to generate trend curves and discuss the implications for design.

To gain insight on how different parameters affect the tendency of a stiffener to peel away from a skin, a typical flange and skin portion of a stiffened panel is isolated in Figure 9.15. The applied loading is simplified to an applied moment M. This could be the result of bending

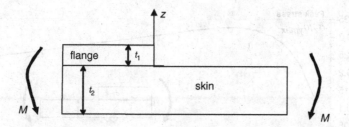

Figure 9.15 Skin–flange configuration under bending load

loads (e.g. pressure) applied on the panel or even post-buckling where local in-plane axial and/or shear loads are neglected.

The basic material properties are as follows:

$$E_x = 137.9 \text{ GPa} \qquad E_z = 11.03 \text{ GPa}$$
$$E_y = 11.03 \text{ GPa} \qquad G_{xz} = 4.826 \text{ GPa}$$
$$G_{xy} = 4.826 \text{ GPa} \qquad G_{yz} = 3.447 \text{ GPa}$$
$$\nu_{xy} = 0.29 \qquad \nu_{xz} = 0.29$$
$$t_{ply} = 0.152 \text{ mm} \qquad \nu_{yz} = 0.4$$

Note that, for this type of problem where out-of-plane stresses are involved, the out-of-plane stiffness properties E_z, G_{xz}, G_{yz}, ν_{xz} and ν_{yz} of the basic ply are also needed.

For the first set of results, the skin and flange layup are assumed to be the same [45/−45/−45/45]n in order to eliminate stiffness mismatch due to differences in moduli in the flange and skin. Only the thicknesses of skin and flange are allowed to vary by specifying different values of n. The normal stress σ_z at the interface between the skin and flange is plotted against distance from the flange edge for different values of the ratio t_1/t_2 in Figure 9.16. It is normalized with the maximum tensile value of σ_y at the far field which is given by

$$\sigma_{ymax} = \frac{6M}{(t_1 + t_2)^2}$$

with M the applied moment per unit of stiffener length.

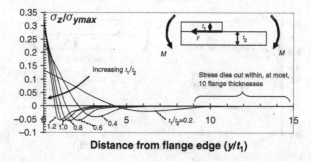

Figure 9.16 Normal stress as a function of distance from flange end for various thickness ratios

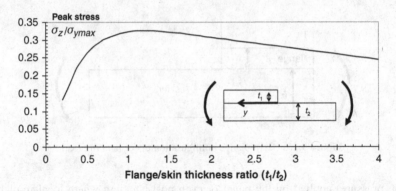

Figure 9.17   Peak normal stress at the flange edge as a function of flange to skin thickness ratio

As expected from the qualitative discussion in association with Figure 9.11 at the beginning of this section, the normal stress has a maximum value at the edge of the flange and then reduces rapidly to negative values and decays to zero. The higher the value of t_1/t_2 the more rapidly the stress decays to zero. The distance over which the normal stress decays to zero does not exceed 10 flange thicknesses as is shown in Figure 9.16.

It is also apparent from Figure 9.16 that starting from low t_1/t_2 values and going up, the peak stress at the flange edge increases. However, this trend is not monotonic. As is shown in Figure 9.17, for the same skin and flange layups as in Figure 9.16, the highest peaks are reached for t_1/t_2 values between 1 and 1.5 and then decrease again. This means that flange thicknesses close to the skin thickness should be avoided because they maximize the normal stresses at the interface.

The discussion so far has attempted to isolate the effect of geometry by keeping the layup of the flange and skin the same. Now, the thicknesses of the flange and skin are fixed to the same value and the layup is varied. This is shown in Figure 9.18 where the case of $[45/-45]_s$ flange and skin is compared with the extreme case of an all $0°$ flange, $[0_4]$, on an all $90°$ skin, $[90_4]$. This is the most extreme case because it has the largest stiffness mismatch between the skin and flange. Note that the 0 direction is taken to be parallel to the y axis for this example.

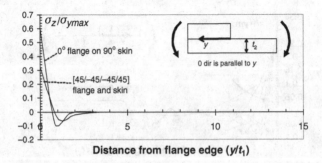

Figure 9.18   Dependence of interface normal stress on layup

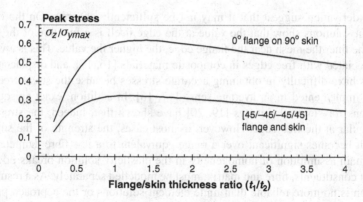

Figure 9.19 Peak normal stress as a function of thickness ratio (t_1/t_2) for two different layup configurations

The [0_4] flange on [90_4] skin has a peak normal stress that is twice as high as the peak stress when skin and flange are both [45/–45]$_s$. And, because the peak value is higher, the stress decays to zero faster than in the case of [45/–45]$_s$ skin and flange. An attempt to combine the effect of layup and geometry is shown in Figure 9.19 where the peak stress is plotted as a function of the ratio t_1/t_2 for the two layups of Figure 9.18.

Again, the highest peaks occur for t_1/t_2 ratios between 1 and 1.5. In addition, the 0° flange on 90° skin has much higher peaks (as much as two times higher) than a situation where the flange and skin have the same layup. Only for t_1/t_2 values lower than 0.3 do the two cases approach each other but even for $t_1/t_2 = 0.2$ the 0° flange on 90° skin has 40% higher peak normal stress at the flange edge.

The results presented in Figures 9.16–9.19 were based on a case where only a bending moment M was applied. Similar results are obtained for other types of loading (but see Exercise 9.5 for some important differences). The trends can be summarized into the following recommendations or guidelines for design:

1. The interlaminar stresses die out within 10–15 flange thicknesses away from the flanged edge.
2. The interlaminar normal stress peaks at the flange edge and decays rapidly to zero.
3. The flange thickness must be either less than the skin thickness or at least 1.5 times greater to minimize the peak stress at the flange edge.
4. The closer the stiffness of the flange to that of the skin, the lower the interlaminar stresses at the flange–skin interface. Since the thicknesses have to be different, this suggests a situation where the flange and skin have the same repeating base layup and only the number of times it repeats in one is different than in the other.

These implications for design rely heavily on the peak value of the normal stress at the flange–skin interface. This value occurs at the edge of the flange. It is important to note that the exact value at that location is not easy to determine. While the method presented in this section gives a well-defined value for the peak stress, the approximations and assumptions made in the derivation suggest that it may not be sufficiently accurate. On the other hand, finite element solutions show that the value at the edge itself is a function of the mesh size. In general, the finer the mesh near the flange edge, the higher the value. This is a well-known problem associated with free edges in composite materials [16, 17], and displacement-based formulations have difficulty in obtaining accurate stresses because the stress-free boundary condition is implemented in an average sense [17, 18]. In addition, exact anisotropic elasticity solutions for simple laminates [19, 20] have shown that, indeed, the stresses are, in general, singular at the free edge. However, in most cases, the strength of the singularity is so low that it becomes significant over a range equivalent to a few fibre diameters. In such a case, the main assumption of homogeneity in the elasticity solution breaks down and the two different constituents, fibre and matrix, must be modelled separately. As a result, the elasticity solution is no more reliable than finite element solutions or the approach presented in this section.

From a design perspective, any method that gives accurate stresses near the free edge of the flange (but not necessarily the flange edge itself) can be used to differentiate between design candidates. Configurations with higher peak values at the edge are expected to have inferior performance. Predicting the exact load at which a delamination will start requires the use of some failure criterion [13, 14] and some test results to adjust any discrepancies between the value predicted at the edge by the analysis method used and the actual test value.

The design guidelines presented in this section can now be used to revisit and upgrade the design of the stiffener cross-section that was last reviewed in Figure 8.36. The revised design is shown in Figure 9.20. The two main differences from Figure 8.36 are the stepped flange next to the skin where plies are dropped with the distance between drops at least 10 times the height of the dropped ply (or plies if more than one ply is dropped at the same location) and the requirement that the skin and flange thicknesses be different ($t_{\text{flange}}/t_{\text{skin}} <1$ or >1.5) with

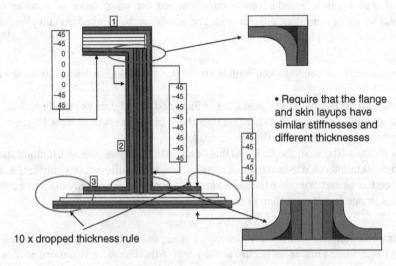

Figure 9.20 Stiffener cross-section design incorporating guidelines from this section

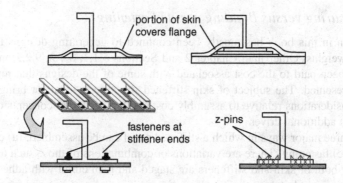

Figure 9.21   Different options for delaying skin–stiffener separation

layups that are, if possible, multiples of the same base layup to keep the respective in-plane stiffnesses as close as possible.

It is recognized, of course, that by using a stepped flange next to the skin the manufacturing cost increases significantly. A tradeoff between the increased cost associated with the stepped flange and the improvement in performance (and thus decreased weight) is necessary in such cases. Alternatives to the stepped flange are shown in Figure 9.21. While they all improve the performance of the skin-stiffened panel, they all carry a significant cost penalty with them.

9.3   Additional Considerations for Stiffened Panels

9.3.1   'Pinching' of Skin

With shear loads present, the skin between stiffeners goes into diagonal tension (see Section 7.2). Resolving the shear load in biaxial tension and compression as was done in Section 7.2 results in the situation shown in Figure 9.22. Only the base flanges of the stiffeners are shown in this figure for clarity.

Of particular importance during testing of such configurations are the compression regions at the top left and bottom right of each bay (the term bay here refers to the section between stiffeners). Locally there, an originally rectangular piece of skin deforms such that the angle at the corners of interest is less than 90° (Figure 9.22). If the applied load is sufficiently high, this can cause the skin to fail in compression. This 'pinching' phenomenon is more pronounced on components isolated from the surrounding structure during, for example, testing of individual panels. In a complete structure such as a fuselage or wing skin, the conditions at the edges of the panel in Figure 9.22 are different from when it is isolated in a test fixture and the compliance of the adjacent structure relieves this phenomenon.

Pinching of skin at the corners may lead to a premature failure when testing in the laboratory. This can happen even before the skin buckles and it can be exacerbated by local eccentricities that may introduce additional bending moments in the region where the skin is under compression. For this reason special fixtures and specimen geometries have been designed to eliminate this problem [21].

9.3.2   Co-curing versus Bonding versus Fastening

The discussion in this book has mostly been confined to generating designs that meet the loads at low weights. Other than Chapter 2 and Sections 5.1.1, 5.1.2, 9.2.2 and 13.4, little attention has been paid to the cost associated with some of the designs that result from the approaches presented. The subject of skin-stiffened structure is ideal for bringing up some additional considerations relative to assembly cost and how different concepts can be traded with cost as an additional driver.

There are three major ways in which a stiffened panel can be assembled: (a) co-curing, (b) bonding and (c) fastening. There are variations or combinations of those such as co-bonding (where one or both of skin and stiffeners are staged and then cured with adhesive present), bonding and fastening, but these three major options are a good starting point for cost tradeoffs.

To discuss these approaches some basic aspects and experience-based conclusions should be laid down first:

1. In general, the larger the part the lower the cost per unit weight. There are economies of scale, elimination of secondary process steps (such as off-line part preparation), and, most importantly, elimination of assembly time required to put together the final product if it is made in smaller pieces instead of one larger part.
2. Eliminating additional cure cycles reduces the cost. Ideally, one should have one cure cycle (or cure at room temperature if materials and loads permit).
3. The higher the complexity of the part being made the higher the cost.
4. If the process requires additional inspection to assure structural integrity and to check that tolerances are met, the cost increases.
5. The risk associated with something going wrong during fabrication adds to the cost. This risk is higher for parts of greater complexity and size. Thus, complex parts have a higher rework and scrap rate than simpler parts, which add to their cost.
6. Automated processes are more accurate and reliable and have lower recurring cost, but can be limited by how complex a part they can make. Also, the nonrecurring cost associated with acquiring the equipment (e.g. an automated fibre/tow placement machine) can be high and not easily justifiable for relatively short production runs.

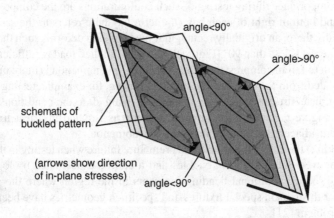

Figure 9.22   Local tension and compression regions in skin under shear

In co-curing, the skin and stiffeners are cured at the same time. This requires detailed (and costly) tooling to accurately locate the stiffeners during cure and to ensure uniform pressure everywhere. So the nonrecurring cost associated with tooling is relatively high. The recurring cost (labour hours) per unit weight is relatively low according to item (1) above. On the other hand, the risk of something going wrong during cure of the combined skin and stiffeners is relatively high, which, according to item (5) above, adds to the cost.

In a bonded configuration, skin and stiffeners are made separately which minimizes the risk. But bonding requires the extra assembly step of skin and stiffeners thus adding to the cost. In addition, there are currently no reliable consistent nondestructive inspection methods to verify that the bond is everywhere effective and meets minimum strength requirements. This means that additional process steps ensuring proper surface preparation of the surfaces to be bonded, full coverage with adhesive, cleanliness and avoidance of contamination, etc. have to be in place to guarantee a good bondline. This adds to the cost. In some cases, to protect against defective bondlines that were missed during fabrication and inspection, it is required to demonstrate that the structure can meet limit load with a significant portion of the bondline ineffective, which adds to the weight of the structure.

Fastening of the stiffeners to the skin eliminates the problems associated with bonding and improves the post-buckling performance. However, the extra assembly time associated with fastening is a significant cost increase. In addition, the use of fasteners typically increases the weight.

It can be seen from the previous discussion that each of the three approaches has advantages and disadvantages, and deciding which approach to follow is a function of the amount of risk to be undertaken and which process steps a particular facility is more comfortable with and more efficient in. It is possible, for example, if the assembly process steps are streamlined and automated, that the cost associated with them is only a small fraction of the total and can lead to overall cost savings compared with a co-cured configuration [22].

Exercises

9.1   Assume that the stiffener used in the application in Section 8.5.3.1 is to be used on a square skin panel of dimensions 508 × 508 mm loaded in compression. The skin layup is $[45/{-}45]_s$ (same material as the stiffener). Determine the largest stiffener spacing (and thus the spacing that minimizes the number of stiffeners and therefore the fabrication cost) such that the overall buckling load equals the buckling load of the skin portion between the stiffeners.

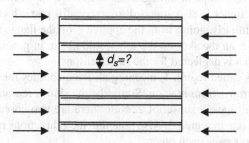

9.2   Consider the stiffener terminating somewhere on a skin as shown in the figure.

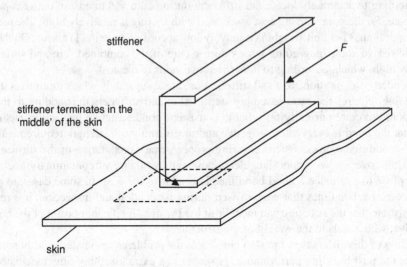

Further isolating the flange and skin portions included in the dashed rectangle above and viewing them from the side (enlarged):

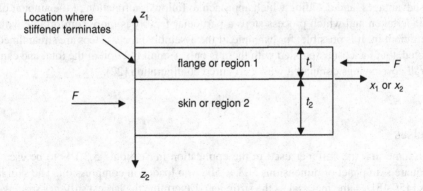

Note the following:

(a)   The load F applied to the stiffener in the first figure, when transmitted to the flange in the second figure, would, in general, also exert a moment. To simplify the calculations this moment is neglected. There are applications where this is valid (when the stiffener is also attached to a very stiff part which absorbs the bending moment). Similarly, going from the applied F on the flange in the second figure to the reacting F on the skin on the left would again, in general, result in a bending moment which is neglected for the same reasons.

(b)   The flange is called region 1, meaning all the quantities relating to the flange will have a subscript 1 from now on. Similarly, the skin is region 2 and all skin-related quantities will have a subscript 2. Note there are two coordinate systems, one in region 1 and one in region 2. Also note that the z directions in these two coordinate systems are opposite each other.

(c)   The left edge of the flange in the second figure above, where the stiffener terminates, has no loads applied and therefore stresses are zero there. All stresses in the flange transfer to the skin as that edge is approached.

(d)   The flange and skin are assumed to be wide in the y direction (perpendicular to the plane of the figure) and, therefore, there is no dependence on y_1 or y_2.

It is assumed that far from the terminating flange edge (i.e. for high values of x_1) the classical laminated-plate theory is recovered according to which the average stress in the flange σ_1 is given by

$$\sigma_1 = \frac{F}{t_1 b_1} = A_1$$

where b_1 is the flange width.

It is now assumed that the normal stress σ_{x1} in the flange is given by the following expression:

$$\sigma_{x1} = f\left(\sin\frac{\pi z}{2t_1} + D_1 \sin\frac{3\pi z}{2t_1}\right) + A_1$$

where f is an unknown function of x_1 and D_1 is an unknown constant.

Note that this expression can be viewed as consisting of the first few terms of a Fourier series. The requirement, of course, is that f, for large x_1, tends to zero so $\sigma_{x1} \to A_1$ away from the flange termination.

Use the equilibrium equations to determine the stresses τ_{xz1} and σ_{z1} (all other stresses are zero or close to zero in the flange). To do this you will also need to make sure that $\tau_{xz1} = 0$ and $\sigma_{z1} = 0$ at the top of the flange. Also, write the corresponding stress expressions for region 2 (by analogy with region 1).

Require that τ_{xz} and σ_z are continuous at the flange–skin interface (i.e. the value of τ_{xz} obtained at the flange–skin interface coming from the flange equals the value of τ_{xz} obtained at the same location coming from the skin, and similarly for σ_z) and obtain the values of D_1 and D_2. (Watch out for the definition of the coordinate systems and the z direction.)

9.3   Continuing from Exercise 9.2, the function f is still unknown. To determine it you need to minimize the energy in the flange–skin system. The energy has the form

$$\Pi = U - W$$

where the internal potential energy U is the sum of U_1 and U_2 and W is the external work. It turns out that W does not contribute to the solution so neglect it.

Now U_1 is given by

$$U_1 = \frac{1}{2}\iiint \left[S_{11}\sigma_x^2 + 2S_{13}\sigma_x\sigma_z + S_{33}\sigma_z^2 + S_{55}\tau_{xz}^2\right]dxdydz$$

where all quantities in the right-hand side should have an additional subscript 1 to denote the flange (omitted here for simplicity) and the quantities S_{ij} are compliances (obtained from inverting the stiffness matrix for the entire flange).

The y integration only yields a constant b_1, the width of the flange, since there is no dependence on y.

Without substituting for D_1 (and D_2) use your expressions for the stresses obtained in Exercise 9.2 and perform the z integrations in the expression for U_1. Note that, because these involve sines and cosines, quite a few of the integrals are zero and you can derive simple expressions for those that are not.

Write the analogous expression for U_2 and perform the z integration. Note that the y integration in the skin will also yield a constant multiplier of b_1 because we are assuming that the skin–flange portion we have isolated for analysis is of width b_1 both for the flange and the skin.

If you substitute in the expression for $\Pi = U_1 + U_2$ you will now have a long integral with respect to x only (y and z integrations already performed) which will be a function of f and its first two derivatives.

9.4 Continuing from Exercise 9.3, using the calculus of variations write down the Euler equation for this problem and derive the governing equation for f so that the energy is minimized. Write the governing equation in the form

$$R_1 \left(d^n f / dx^n \right) + R_2 \left(d^{n-1} f / dx^{n-1} \right) + \cdots = 0$$

Start with R_1 with the highest derivative of f, and neglect the constant term (because it does not contribute to the solution). Write down the expressions for $R_1, R_2, \ldots$.
Solve the equation obtained writing down the solution in the form

$$f = C_1 e^{m_1 x} + C_2 e^{m_2 x} + \cdots$$

Also write down the expression for m_i.

9.5 (Continuing from Exercise 9.4) At this point, the fact that none of the stresses can increase with increasing x is invoked and only exponentials with negative real parts are used. In addition, the boundary conditions that require $\tau_{xz1} = 0$ and $\sigma_{x1} = 0$ at $x = 0$ are invoked (the second one approximately only) and the constants $C_1, C_2, \ldots$ are determined. After that is done, one notices that at $x_1 = 0$ at the flange–skin interface, the interlaminar shear stresses τ_{xz1} and τ_{xz2} are zero and only the normal stress σ_{z1} or σ_{z2} is nonzero. Therefore, the only stress that contributes to delamination is σ_{z1} (or σ_{z2} which is the same at the flange–skin interface by stress continuity).

Substituting in the stress expressions and evaluating σ_{z1} at $x = z = 0$ shows that

$$\sigma_{z\text{crit}} = \frac{-8A_1}{\pi} \left(1 - \frac{t_2}{3t_1} \right) \sqrt{ \frac{ (S_{11})_1 \left(1 + \dfrac{D_1^2}{2} \right) + (S_{11})_2 \dfrac{t_2}{t_1} \left(1 + \dfrac{D_2^2}{2} \right) }{ 2 (S_{33})_1 \left[T_1 + \left(\dfrac{t_2}{t_1} \right)^5 T_2 \right] } }$$

where the compliances for each laminate (flange or skin) can be obtained by averaging the compliances over all plies (springs in series):

$$\frac{t_1}{(S_{ij})_1} = \sum \frac{t_i}{(S_{ij})_{ithply}}$$

with an analogous expression for region 2, and the compliances for each ply of orientation θ are given by the following expressions where quantities with superscript 0 refer to a 0° ply of the material being used:

$$S_{11} = S_{11}^0 \cos^4 \theta + \left(2S_{12}^0 + S_{66}^0\right) \sin^2 \theta \cos^2 \theta + S_{22}^0 \sin^4 \theta$$

$$S_{13} = S_{13}^0 \cos^2 \theta + S_{23}^0 \sin^2 \theta$$

$$S_{33} = S_{33}^0$$

$$S_{55} = S_{55}^0 \cos^2 \theta + S_{44}^0 \sin^2 \theta$$

$$S_{11}^0 = \frac{1}{E_{11}}$$

$$S_{12}^0 = -\frac{\nu_{12}}{E_{11}}$$

$$S_{66}^0 = \frac{1}{G_{12}}$$

$$S_{22}^0 = \frac{1}{E_{22}}$$

$$S_{13}^0 = -\frac{\nu_{13}}{E_{11}}$$

$$S_{23}^0 = -\frac{\nu_{23}}{E_{22}}$$

$$S_{33}^0 = \frac{1}{E_{33}}$$

$$S_{44}^0 = \frac{1}{G_{23}}$$

$$S_{55}^0 = \frac{1}{G_{13}}$$

$$T_1 = \frac{3}{2} - \frac{4}{\pi} + \frac{D_1^2}{81}\left(\frac{3}{2} - \frac{4}{3\pi}\right) + \frac{2D_1}{9}\left(1 - \frac{8}{3\pi}\right)$$

$$T_2 = \frac{3}{2} - \frac{4}{\pi} + \frac{D_2^2}{81}\left(\frac{3}{2} - \frac{4}{3\pi}\right) + \frac{2D_2}{9}\left(1 - \frac{8}{3\pi}\right)$$

A material is made available with the following properties:

$$E_{11} = 137.88 \, \text{GPa}$$

$$E_{22} = E_{33} = 11.72 \, \text{GPa}$$

$$G_{12} = G_{13} = 4.825 \, \text{GPa}$$

$$G_{23} = 4.0 \, \text{GPa}$$

$$\nu_{12} = \nu_{13} = 0.3$$

$$\nu_{23} = 0.45$$

$$t_{\text{ply}} = 0.1524 \, \text{mm}$$

$$X^t = 2068 \, \text{MPa}$$

$$X^c = 1379 \, \text{MPa}$$

$$Y^t = 68.94 \, \text{MPa}$$

$$Y^c = 310.2 \, \text{MPa}$$

$$S = 124.1 \, \text{MPa}$$

It is also given that the skin at the location of interest has the following layup: $[45/-45/0/90]_{s4}$.

9.6 (a) If the skin is not to exceed 4500 µs, determine the value of F that barely fails the skin. (b) Assume that the flange is made from one of the three layups: $[0/90]_{s^n}$ or $[45/-45]_{s^m}$ or $[45/-45/0/90]_{s^p}$. Select a value of n, m and p and plot the value of $\sigma_{z\text{crit}}$ as a function of t_2/t_1. Which value(s) of t_2/t_1 would you recommend to use in this case and why? (c) Use your results in (a) and (b) to select the lightest flange layup you should use.

References

[1] Megson, T.H.G., *Aircraft Structures for Engineering Students*, Elsevier, Oxford, UK, 2007, Chapter 3.

[2] Kassapoglou, C., Stress Determination at Skin-Stiffener Interfaces of Composite Stiffened Panels under Generalized Loading, *Journal of Reinforced Plastics and Composites*, **13**, 1555–1572 (1994).

[3] Kassapoglou, C., Calculation of Stresses at Skin-Stiffener Interfaces of Composite Stiffened Panels under Shear Loads, *International Journal of Solids and Structures*, **30**, 1491–1503 (1993).

[4] Kassapoglou, C. and DiNicola, A.J., Efficient Stress Solutions at Skin Stiffener Interfaces of Composite Stiffened Panels, *AIAA Journal*, **30**, 1833–1839 (1992).

[5] Hyer, M.W. and Cohen, D., Calculation of Stresses and Forces between the Skin and Stiffener in Composite Panels, *AIAA Paper 87-0731, 28th SDM Conference*, Monterey, CA, April 1987.

[6] Cohen, D. and Hyer, M.W., Influence of Geometric Non-linearities on Skin-Stiffener Interface Stresses, *AIAA Paper 88-2217, 29th SDM Conference*, Williamsburg, VA, April 1988.

[7] Volpert, Y. and Gottesman, T., Skin-Stiffener Interface Stresses in Tapered Composite Panel, *Composite Structures*, **33**, 1–6 (1995).

[8] Baker, D.J. and Kassapoglou, C., Post Buckled Composite Panels for Helicopter Fuselages: Design, Analysis, Fabrication and Testing. *Presented at the American Helicopter Society International Structures Specialists' Meeting, Hampton Roads Chapter, Structures Specialists Meeting*, Williamsburg, VA, 30 October–1 November 2001.

[9] Wiggenraad, J.F.M. and Bauld, N.R., Global/Local Interlaminar Stress Analysis of a Grid-Stiffened Composite Panel, *Journal of Reinforced Plastics and Composites*, **12**, 237–253 (1993).

[10] Minguet, P.J. and O'Brien, T.K., Analysis of Test Methods for Characterizing Skin/Stringer Debonding Failures in Reinforced Composite Panels, in *Composite Materials: Testing and Design ASTM STP 1274*, R.B. Deo and C.R. Saff (eds), American Society for Testing and Materials, Philadelphia, PA, 1996, pp 105–124.

[11] Wang, J.T. and Raju, I.S., Strain Energy Release Rate Formulae for Skin-Stiffener Debond Modeled with Plate Elements, *Engineering Fracture Mechanics*, **54**, 211–228 (1996).

[12] Wang, J.T., Raju, I.S. and Sleight, D.W., Composite Skin-Stiffener Debond Analyses, Using Fracture Mechanics Approach with Shell Elements, *Composites Engineering*, **5**, 277–296 (1995).

[13] Brewer, J.C. and Lagace, P.A., Quadratic Stress Criterion for Initiation of Delamination, *Journal of Composite Materials*, **22**, 1141–1155 (1988).

[14] Fenske, M.T., and Vizzini, A.J., The Inclusion of In-Plane Stresses in Delamination Criteria, *Journal of Composite Materials*, **35**, 1325–1342 (2001).

[15] Hildebrandt, F.B., *Advanced Calculus for Applications*, Prentice Hall, Englewood Cliffs, NJ, 1976, Section 7.8.

[16] Wang, A.S.D. and Crossman, F.W., Some New Results on Edge Effect in Symmetric Composite Laminates, *Journal of Composite Materials*, **11**, 92–106 (1977).

[17] Bauld, N.R., Goree, J.G. and Tzeng, L.-S., A Comparison of Finite-Difference and Finite Element Methods for Calculating Free Edge Stresses in Composites, *Composites and Structures*, **20**, 897–914 (1985).

[18] Spilker, R.L. and Chou, S.C., Edge Effects in Symmetric Composite Laminates: Importance of Satisfying the Traction-Free-Edge Condition, *Journal of Composite Materials*, **14**, 2–20 (1980).

[19] Wang, S.S. and Choi, I., Boundary-Layer Effects in Composite Laminates, Part I – Free Edge Stress Solutions and Basic Characteristics, *ASME Transactions Applied Mechanics*, **49**, 541–548 (1982).

[20] Wang, S.S. and Choi, I., Boundary-Layer Effects in Composite Laminates, Part II – Free Edge Stress Solutions and Basic Characteristics, *ASME Transactions Applied Mechanics*, **49**, 549–560 (1982).

[21] Farley, G.L. and Baker, D.J., In-Plane Shear Test of Thin Panels, *Experimental Mechanics*, **23**, 81–88 (1983).

[22] Apostolopoulos, P. and Kassapoglou, C., Cost Minimization of Composite Laminated Structures – Optimum Part Size as a Function of Learning Curve Effects and Assembly, *Journal of Composite Materials*, **36**(4), 501–518 (2002).

10

Sandwich Structure

A sandwich structure (Figure 10.1) typically consists of two facesheets separated by lightweight core. Usually, the facesheets are bonded to the core with the use of adhesive but, under certain circumstances (for example, using X-cor® or K-cor® [1, 2]), it is possible to eliminate the use of adhesive.

Composite laminates make up the facesheets. There is a variety of materials and configurations used for the core depending on the application and the desired properties: foam, honeycomb, low-density foaming aluminium, etc. Most core materials, in particular honeycombs, are anisotropic. They have different stiffnesses and strengths in different directions. In general, the purpose of the core is to increase the bending stiffness of the sandwich by moving material away from the neutral axis of the cross-section. The stiffness (and strength) of the core are, typically, much lower than those of the facesheets. As a result, for general loading situations such as that shown in Figure 10.1, with applied bending moment M and in-plane axial and shear loads N and V, respectively, all the load is taken by the facesheets. The bending moment (per unit width) is resolved into a force couple where one facesheet is loaded by a positive force per unit width N_m and the other by an equal and opposite force per unit width $-N_m$. The magnitude of that force is such that the force couple generates a moment equal to M:

$$N_m = \frac{M}{t_c + t_f}$$

The axial load N and the shear load V are divided equally between the two facesheets. The core must still have minimum strength and stiffness in certain directions so that

i. the sandwich does not collapse under pressure during cure;
ii. load can be transferred between facesheets; and
iii. core ramp-downs, where the core gradually transitions to monolithic laminate for attachment to adjacent structure, do not fail prematurely.

With reference to Figure 10.2, aside from the core thickness t_c which determines the overall bending stiffnesses of the sandwich, the most important core properties are the transverse

---

Design and Analysis of Composite Structures: With Applications to Aerospace Structures, Second Edition. Christos Kassapoglou.
© 2013 John Wiley & Sons, Ltd. Published 2013 by John Wiley & Sons, Ltd.

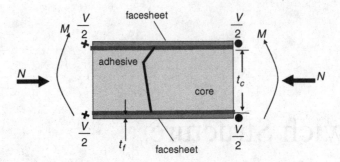

Figure 10.1   Sandwich configuration

shear stiffnesses G_{xz} and G_{yz}, the corresponding transverse shear strengths Z_{xz} and Z_{yz}, the out-of-plane Young's modulus E_c and the corresponding (flatwise) tension and compression strengths Z^t and Z^c, respectively. Finally, for the case of honeycomb core such as that shown in Figure 10.2, the core cell size s plays a role in some of the failure modes. A thorough investigation of sandwich structures with isotropic facesheets can be found in reference [3].

10.1   Sandwich Bending Stiffnesses

A sandwich can be treated as a laminate where the core is just another ply with negligible stiffness and strength properties and thickness equal to the core thickness. Standard classical laminated-plate theory (see Section 3.3) can be used to determine the corresponding A, B and D matrices. The presence of the core does not change the A matrix, but will affect the B (if the total layup is asymmetric) and D matrices significantly. This can be seen by applying

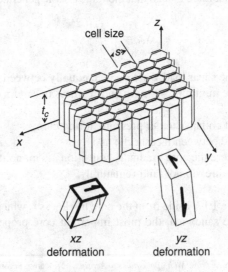

Figure 10.2   Honeycomb core geometry

Equation (8.14) to obtain the D matrix of a sandwich. Rewriting Equation (8.14) it can be shown that for a sandwich with identical facesheets

$$D_{ij} = 2(D_{ij})_f + 2(A_{ij})_f \left(\frac{t_c + t_f}{2} \right)^2 \tag{10.1}$$

where the multiplicative factors of 2 appearing in the right-hand side account for the presence of two facesheets. The first term in the right-hand side of Equation (10.1) is the same as the EI term in Equation (8.14) with the stiffness E incorporated in the corresponding D_{ij} term. The second term is the product of the stiffness and the distance from the neutral axis present in Equation (8.14) with the modulus this time lumped in the A_{ij} term.

To see the effect of the core on increasing the bending stiffness of a sandwich, consider two facesheets of layup $(\pm 45)/(0/90)/(\pm 45)$ separated by a core of varying core thickness. The individual A and D matrices for each facesheet are given by

A_{11}	28,912.44	N/mm		D_{11}	659.7	N mm
A_{12}	12,491.43	N/mm		D_{12}	466.9	N mm
A_{22}	28,912.44	N/mm		D_{22}	659.7	N mm
A_{66}	13,468.58	N/mm		D_{66}	494.0	N mm

Using Equation (10.1), the ratio of D_{11} for the entire sandwich divided by D_{11} for each facesheet can be determined as a function of varying core thickness. The result is shown in Figure 10.3.

It can be seen from Figure 10.3 that even very small core thicknesses (5 mm) result in a 1000-fold increase of the bending stiffness. The range of typical core thicknesses used in many applications is also shown in Figure 10.3, indicating that, for such applications, the core increases the bending stiffness anywhere between 4000 and 15,000 times.

This kind of improvement at a relatively small increase in weight, due to the presence of the core and adhesive, makes sandwich structure ideal for many stability-driven applications where high buckling loads are important. In fact, judicious selection of facesheet material and layup and core material and thickness would result in sandwich being the most weight-efficient structure if it were not for a variety of new failure modes associated with such configurations.

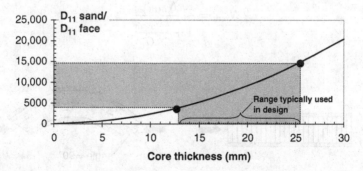

Figure 10.3 Variation of sandwich bending stiffness as a function of core thickness

Each of the components, facesheet, adhesive or core, can fail and there are more than one failure modes for each component. Some of these failure modes are quite limiting and tend to drive the design. The result is that sandwich is not always more efficient than the alternative(s) such as a skin-stiffened structure. It depends on the geometry, loading and design philosophy (e.g. whether post-buckling is allowed and at how high a post-buckling ratio). The most important of these failure modes are examined in subsequent sections.

In addition, it should be pointed out that in many applications where sandwich is selected, ramp-downs are used at the edges of the sandwich for attachment to adjacent structure. Ramp-downs are briefly discussed in Section 10.6.1. They tend to increase cost and introduce additional failure modes that must be checked to make sure they do not lead to premature failure of the entire structure.

10.2 Buckling of Sandwich Structure

Buckling is one of the critical failure modes for sandwich structure in particular for relatively large panels. The reason is that it is hard to design against all possible failure modes in the post-buckling regime and, as a result, buckling is usually considered to coincide with final failure.

10.2.1 Buckling of Sandwich under Compression

The procedure to determine the buckling load of a sandwich structure is very similar to that presented in Chapter 6 for monolithic laminates, with one important difference. The presence of the core makes the effects of transverse shear very important. If they are not properly accounted for, the predicted buckling load is very unconservative (higher than the case where transverse shear effects are accounted for).

In a uniform thickness plate where transverse shear effects are significant the Kirchhoff hypothesis is no longer valid. Plane sections remain plane, but are no longer perpendicular to the plate mid-plane. This is shown schematically in Figure 10.4.

The sandwich under compression is treated as a wide beam. Following the derivation in [4] for isotropic beams, the buckling load is given by

$$N_{x\text{crit}} = \frac{N_{E\text{crit}}}{1 + \dfrac{k N_{E\text{crit}}}{t_c G_c}} \qquad (10.2)$$

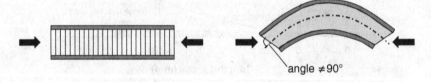

angle $\neq 90°$

Figure 10.4 Bending of a sandwich panel under compression

In Equation (10.2), $N_{E\text{crit}}$ is the buckling load N_o of the sandwich if transverse shear effects are neglected, given, for simply supported edges, by Equation (6.7) repeated below for convenience:

$$N_{E\text{crit}} = \frac{\pi^2}{a^2}\left[D_{11}m^2 + 2(D_{12} + 2D_{66})(AR)^2 + D_{22}\frac{(AR)^4}{m^2} \right] \tag{10.7}$$

Also, t_c and G_c are the core thickness and transverse shear stiffness, respectively. It is important to note that G_c is the value of the shear modulus (typically G_{xz} or G_{yz}) aligned with the loading direction.

Finally, k in Equation (10.2) is the shear correction factor. The shear correction factor is introduced to reconcile the inconsistency between the derived and assumed transverse shear stress distributions through the thickness of the plate. Engineering bending theory leads to a quadratic distribution of shear stress through the thickness while first-order shear deformation theory assumes the shear strain (and thus the shear stress) is independent of the through-thickness coordinate. This inconsistency [5] is reconciled by requiring that the work done following either formulation is the same. This leads to an expression for the transverse shear force (per unit width) of the form [5],

$$Q_x = kG_{xz}h\gamma = \frac{5}{6}G_{xz}h\gamma$$

where the shear correction factor is $k = 5/6$ in this case, h is the plate thickness and γ is the transverse shear strain.

As mentioned earlier, most sandwich structures use core with very small shear stiffness G_{xz} compared with that of the facesheets. As a result, the shear stress through the thickness is very nearly uniform. This is consistent with the fact that bending stresses are not linearly distributed through the thickness of the core because, as was mentioned earlier, bending moments are transmitted through a sandwich as a force couple. Thus, there is (almost) no inconsistency between bending theory and first-order shear deformation theory and $k \approx 1$. Thus,

$$N_{x\text{crit}} = \frac{N_{E\text{crit}}}{1 + \dfrac{N_{E\text{crit}}}{t_c G_c}} \tag{10.3}$$

An example would help illustrate the importance of transverse shear in a sandwich. The same facesheet properties as those in Section 10.1 and Figure 10.3 are used here. The core material is assumed to have a shear stiffness $G_{xz} = 42.1$ N/mm^2. Equation (10.3) is used to calculate the buckling load of a square sandwich panel of side 508 mm and compare it with Equation (6.7), that is, without accounting for transverse shear. This is done for different core thicknesses and the results are shown in Figure 10.5.

It can be seen that once the core thickness exceeds 5 mm, the buckling load including transverse shear effects diverges drastically from the buckling load without transverse shear effects. If transverse shear effects are included the buckling load is always lower. Even for a core thickness of 3 mm the two buckling loads (with and without transverse shear effects) differ by 21%.

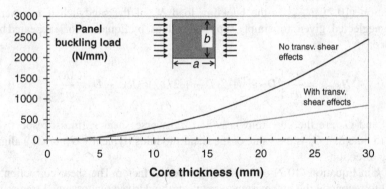

Figure 10.5 Buckling load of a sandwich with and without transverse shear effects as a function of core thickness

10.2.2 Buckling of Sandwich under Shear

The situation is shown in Figure 10.6. The form of the equation predicting buckling of a simply supported sandwich under shear is the same as Equation (10.3):

$$N_{xycrit} = \frac{N_{xyc}}{1 + \dfrac{N_{xyc}}{t_c G_{45}}} \tag{10.4}$$

where N_{xyc} is the shear buckling load without transverse shear effects (see, e.g. Sections 6.3–6.5) and G_{45} is the core shear modulus in the 45° direction, as shown in Figure 10.6.

The shear modulus G_{45} is used because it is aligned with the primary load direction. Since a pure shear loading is equivalent to biaxial loading with tension in one direction and compression in the other (see, e.g. Section 7.2), the tendency for buckled half-waves to form is along the 45° line in Figure 10.6 (the direction of maximum compression) and G_{45}, the core shear stiffness in that direction, opposes that tendency.

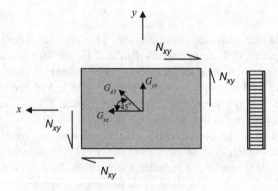

Figure 10.6 Sandwich panel under shear load

To determine G_{45} standard tensor transformation equations are used (Section 3.2, Equation 3.8). The result is

$$G_{45} = \sin^2 45 G_{yz} + \cos^2 45 G_{xz} = \frac{G_{yz} + G_{xz}}{2} \tag{10.5}$$

Using this result to substitute in Equation (10.4) and rearranging lead to

$$N_{xycrit} = \frac{(G_{xz} + G_{yz})\, t_c}{\dfrac{(G_{xz} + G_{yz})\, t_c}{N_{xyc}} + 2} \tag{10.6}$$

10.2.3   Buckling of Sandwich under Combined Loading

For combined loading situations the same interaction curves as those presented in Sections 6.5 and 6.6 can be used, provided the individual buckling loads are corrected for transverse shear effects as presented in the two previous sections.

10.3   Sandwich Wrinkling

Wrinkling is a local buckling phenomenon where the facesheet of a sandwich buckles over a characteristic half-wavelength ℓ, which is unrelated to the overall length or width of the panel. There are three possible modes, symmetric, antisymmetric and mixed-mode wrinkling. These are shown schematically in Figure 10.7 for applied compression, but can also occur under applied shear or combined loads.

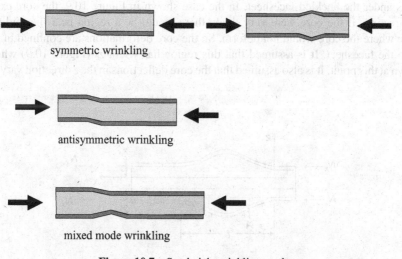

symmetric wrinkling

antisymmetric wrinkling

mixed mode wrinkling

Figure 10.7   Sandwich wrinkling modes

Figure 10.8   Wrinkling failure of sandwich tested in compression

10.3.1   Sandwich Wrinkling under Compression

The symmetric wrinkling case is examined here in detail (see also [6]). A sandwich compression specimen, which failed in wrinkling, is shown in Figure 10.8.

The deformed shape after the facesheet has buckled in the wrinkling mode is idealized in Figure 10.9. This shape extends through the width of the sandwich (perpendicular to the plane of Figure 10.9). It is assumed that the sandwich is very long in the y direction. It is also assumed that at the edges of the buckled shape, at $x = 0$ and $x = \ell$, the boundary conditions on the facesheet are those of simple support, that is, $w = 0$ there.

One important aspect of the formulation is modelling the behaviour of the core. Assuming perfect bonding between core and facesheet, it is obvious from Figure 10.9 that the core deforms under the buckled facesheet. In the case shown in Figure 10.9, the core extends in the z direction. If the core were very thick, there would be a region near the mid-plane of the core where the core would not deform. So the core deformations are confined in a region close to the facesheet. It is assumed that this region has width z_c (Figure 10.9) where z_c is unknown at this point. It is also assumed that the core deflections in the z direction vary linearly with z.

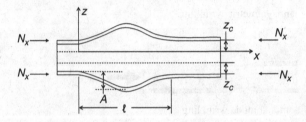

Figure 10.9   Deformed configuration of sandwich undergoing symmetric wrinkling

Combining the assumption of simply supported ends for the deformed facesheet and linear variation of deflection for the core, the following expression for w is introduced:

$$w = A\frac{z}{z_c} \sin\frac{\pi x}{\ell} \tag{10.7}$$

Equation (10.7) satisfies the requirement that $w = 0$ at $x = 0$ and $x = \ell$. It also satisfies the linear variation of w as a function of z with $w = 0$ at $z = 0$ (i.e. at the interface where core deformations seize to be significant) and reproducing the facesheet sinusoidal deformation at $z = z_c$, which is the intersection of the core with the facesheet.

The wrinkling load is determined by energy minimization. During wrinkling, energy is stored in bending the facesheet and extending the core. So the energy expression has the form

$$\Pi_c = 2U_f + U_c - 2W \tag{10.8}$$

where U_f is the energy in each facesheet and U_c is the energy stored in the core. W is the work done by the applied load on one end of the sandwich. The factors of 2 in this equation account for the presence of two (identical) facesheets.

Neglecting deformations u and v in the plane of the facesheet, the strain and stress in the facesheet can be obtained from Equations (5.6), (5.7) and (5.4):

$$\varepsilon_x = -z\frac{\partial^2 w}{\partial x^2}$$
$$\sigma_x = E_f\varepsilon_x \tag{10.9}$$

where E_f is the facesheet membrane modulus obtained from Equation (8.6).

Then,

$$\sigma_x\varepsilon_x = E_f z^2 \left(\frac{\partial^2 w}{\partial x^2}\right)^2$$

Thus, the facesheet energy can be written as

$$U_f = \frac{1}{2}\iint E_f z^2\left[\left(\frac{\partial^2 w}{\partial x^2}\right)^2\right]_{z=z_c} dzdx = \frac{1}{2}E_f\bar{I}\int_0^\ell\left[\left(\frac{\partial^2 w}{\partial x^2}\right)^2\right]_{z=z_c} dx$$

$$= \frac{(E\bar{I})_f}{2}\int_0^\ell\left[\left(\frac{\partial^2 w}{\partial x^2}\right)^2\right]_{z=z_c} dx \tag{10.10}$$

where $\bar{I}$ is the moment of inertia of the facesheet per unit width b. It should be noted that this expression can also be obtained from Equation (5.62) assuming a symmetric facesheet and noticing that only the D_{11} term contributes with $D_{11} = EI/b$, where b is the width of the facesheet perpendicular to the plane of Figure 10.9.

The strain–displacement and stress–strain equations (5.9) and (5.4) applied to the core give

$$\varepsilon_z = \frac{\partial w}{\partial z}$$

$$\gamma_{xz} = \frac{\partial w}{\partial x} + \frac{\partial u}{\partial z}$$

$$\sigma_z = E_c \varepsilon_z$$

$$\tau_{xz} = G_{xz} \gamma_{xz}$$

where E_c is the core modulus in the z direction and G_{xz} is the transverse shear modulus of the core for shearing in the xz plane (see Figure 10.9).

As already mentioned, the u deflection of the core is negligible. Then, the above equations can be combined to

$$\sigma_z \varepsilon_z = E_c \left(\frac{\partial w}{\partial z} \right)^2$$

$$\tau_{xz} \gamma_{xz} = G_{xz} \left(\frac{\partial w}{\partial x} \right)^2$$

which, in turn, can be substituted for in the core energy expression:

$$U_c = \frac{1}{2} \int_0^\ell 2 \int_0^{z_c} (E_c \sigma_z \varepsilon_z + G_{xz} \tau_{xz} \gamma_{xz}) \, dzdx = \int_0^\ell \int_0^{z_c} \left(E_c \left(\frac{\partial w}{\partial z} \right)^2 + G_{xz} \left(\frac{\partial w}{\partial x} \right)^2 \right) dzdx$$

$$(10.11)$$

Note the factor of 2 in front of the second integral, which is introduced to account for the fact that there are two portions of the core (one above and one below the mid-plane) both of thickness z_c that contribute to the core energy.

The external work done by the applied load N_x per facesheet per unit width is given by

$$W = N_x \delta$$

$$\delta = \ell - \int_0^\ell dx$$

with δ the deflection at the edge of the sandwich portion considered, that is, at $x = 0$ and $x = \ell$.

Considering the deformed shape of the facesheet shown in Figure 10.10, the deflection δ can be calculated using Pythagoras' theorem and assuming small deflections w.

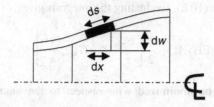

Figure 10.10 Deformed facesheet and local geometry

By Pythagoras' theorem,

$$(dx)^2 + (dw)^2 = (ds)^2 \Rightarrow dx = ds\sqrt{1 - \left(\frac{dw}{ds}\right)^2}$$

The quantity involving the square root can be expanded into the first two terms of a Taylor series (valid for small $(dw/ds)^2$) to give

$$\sqrt{1 - \left(\frac{dw}{ds}\right)^2} \approx 1 - \frac{1}{2}\left(\frac{dw}{ds}\right)^2 \quad \text{for small} \left(\frac{dw}{ds}\right)^2$$

and for small deflections w

$$\frac{dw}{ds} \approx \frac{dw}{dx}$$

Therefore, substituting in the expression for δ,

$$W = \frac{N_x}{2}\int_0^\ell \left(\frac{\partial w}{\partial x}\right)^2\bigg|_{z=z_c} dx$$

At this point, the relevant w derivatives present in the energy expression are evaluated

$$\left(\frac{\partial w}{\partial x}\right)^2 = \frac{A^2 z^2}{z_c^2}\frac{\pi^2}{2\ell^2}\left(1 + \cos\frac{2\pi x}{\ell}\right)$$

$$\left(\frac{\partial w}{\partial z}\right)^2 = \frac{A^2}{z_c^2}\frac{1}{2}\left(1 - \cos\frac{2\pi x}{\ell}\right)$$

$$\left(\frac{\partial^2 w}{\partial x^2}\right)^2 = \frac{A^2 z^2}{z_c^2}\frac{\pi^4}{2\ell^4}\left(1 - \cos\frac{2\pi x}{\ell}\right)$$

and substituted in Equation (10.8). Evaluating the integrals gives

$$\Pi_c = \frac{\pi^4}{2\ell^3}(E\bar{I})_f A^2 + \frac{1}{2}\left[\frac{E_c\ell}{z_c} + \frac{1}{3}G_{xz}z_c\frac{\pi^2}{\ell}\right]A^2 - N_x\frac{A^2\pi^2}{2\ell} \tag{10.12}$$

Equation (10.12) must be minimized with respect to the unknown amplitude A. This implies

$$\frac{\partial \Pi_c}{\partial A} = 0 \Rightarrow 2A\left[\frac{\pi^4}{2\ell^3}(E\bar{I})_f + \frac{1}{2}\left[\frac{E_c\ell}{z_c} + \frac{1}{3}G_{xz}z_c\frac{\pi^2}{\ell}\right] - N_x\frac{\pi^2}{2\ell}\right] = 0 \tag{10.13}$$

For nonzero values of A, the quantity in brackets must be zero. This gives a condition for the wrinkling load N_x. Denoting the wrinkling load by N_{xwr}:

$$N_{xwr} = \frac{\pi^2(E\bar{I})_f}{\ell^2} + \frac{E_c\ell^2}{\pi^2 z_c} + G_{xz}\frac{z_c}{3} \tag{10.14}$$

Examining Equation (10.14) it can be seen that the first term of the right-hand side is the buckling load of a beam column (per unit width). The second term is the contribution to the buckling load of a beam by an elastic foundation when the stiffness of the foundation k equals E_c/z_c. This can be readily seen by comparing this term with Equation (8.33) when $m = 1$. The third term is the contribution of the elastic foundation when it consists of torsional instead of extensional springs.

This expression for the wrinkling load is still in terms of two unknowns: ℓ the half-wavelength during wrinkling and z_c the portion of the core undergoing deformations during wrinkling. Each of them is determined by noticing that if N_x starts increasing from zero, then wrinkling will occur at the lowest possible value that Equation (10.14) allows as a function of ℓ and z_c. Therefore, minimizing N_{xwr} with respect to ℓ gives:

$$\frac{\partial N_{xwr}}{\partial \ell} = 0 \Rightarrow \ell = \pi\left(\frac{(E\bar{I})_f}{E_c}z_c\right)^{1/4} \tag{10.15}$$

which gives a condition relating ℓ and z_c. Using it to eliminate ℓ from Equation (10.14) results in

$$N_{xwr} = \frac{2\sqrt{E_c(E\bar{I})_f}}{\sqrt{z_c}} + \frac{G_{xz}z_c}{3} \tag{10.16}$$

which is only in terms of z_c. Differentiating now with respect to z_c and setting the result equal to 0 gives

$$\frac{\partial N_{xwr}}{\partial z_c} = 0 \Rightarrow z_c = 3^{2/3}\left(\frac{E_c(E\bar{I})_f}{G_{xz}^2}\right)^{1/3} \tag{10.17}$$

Now the moment of inertia per unit width

$$\bar{I} = \frac{t_f^3}{12}$$

is used to substitute in Equation (10.17) to obtain the final expression for z_c:

$$z_c = \frac{3^{2/3}}{12^{1/3}} t_f \left(\frac{E_c E_f}{G_{xz}^2} \right)^{1/3} = 0.91 t_f \left(\frac{E_c E_f}{G_{xz}^2} \right)^{1/3} \tag{10.18}$$

This expression can now be used to substitute in Equation (10.15) to get the final value for the half-wavelength:

$$\ell = \frac{\pi 3^{1/6}}{12^{1/3}} t_f \left(\frac{E_f}{\sqrt{E_c G_{xz}}} \right)^{1/3} = 1.648 t_f \left(\frac{E_f}{\sqrt{E_c G_{xz}}} \right)^{1/3} \tag{10.19}$$

Finally, Equation (10.18) can be used to substitute in Equation (10.16) to obtain the wrinkling load

$$N_{xwr} = 0.91 t_f \left(E_f E_c G_{xz} \right)^{1/3} \tag{10.20}$$

Equation (10.20) has been derived in many different ways [6, 7]. In fact, depending on the assumptions, the form of the equation remains the same and only the coefficient in the right-hand side changes [7].

It is important to keep in mind that the derivation so far has assumed that the core was sufficiently thick that the portion z_c of the core undergoing deformations is less than or equal to half the core thickness $t_c/2$. If z_c given by Equation (10.18) is greater than half the core thickness, then the entire core deforms during wrinkling and

$$z_c = \frac{t_c}{2} \tag{10.18a}$$

With this new value of z_c new values of ℓ and N_{xwr} must be calculated. Following the same procedure as before, z_c is substituted for in Equation (10.15) to get

$$\ell = \frac{\pi}{24^{1/4}} \left(\frac{E_f}{E_c} t_f^3 t_c \right)^{1/4} \quad \text{for} \quad z_c = \frac{t_c}{2} \tag{10.19a}$$

Then, the new value of z_c is substituted in Equation (10.16) to obtain

$$N_{xwr} = 0.816 \sqrt{\frac{E_f E_c t_f^3}{t_c} + G_{xz} \frac{t_c}{6}} \tag{10.20a}$$

The condition for the full depth of the core being involved in the wrinkling deformations can be obtained from Equation (10.18). If the right-hand side of Equation (10.18) is greater than $t_c/2$, then the entire core thickness deforms. Therefore, if

$$t_c < 1.817 t_f \left(\frac{E_f E_c}{G_{xz}^2} \right)^{1/3} \tag{10.21}$$

the core deforms in its entirety, and Equations (10.18a), (10.19a) and (10.20a) are valid. Otherwise, only a portion z_c of the core deforms and Equations (10.18), (10.19) and (10.20) are valid.

It should be noted that, according to Equation (10.20a), as the core thickness increases the wrinkling load decreases. So it would be expected that wrinkling would become the primary failure mode beyond a certain core thickness. However, this is only true as long as Equation (10.21) is satisfied. Once the core thickness exceeds the right-hand side of Equation (10.21) the governing equation is Equation (10.20), which is independent of the core thickness.

For antisymmetric wrinkling, an analogous approach to that presented above, but with a different expression for the w deflection of the core in order to satisfy the different boundary conditions, leads to the following results [6,8]:

$$N_{xwr} = 0.51 t_f \left(E_f E_c G_{xz} \right)^{1/3} + \frac{G_{xz} t_c}{3} \tag{10.22}$$

$$\ell = 2.15 t_f \left(\frac{E_f^2}{E_c G_{xz}} \right)^{1/6} \tag{10.23}$$

$$z_c = \frac{3}{2} t_f \left(\frac{E_f E_c}{G_{xz}^2} \right)^{1/3} \tag{10.24}$$

for sufficiently thick core, that is, when

$$t_c \geq 3 t_f \left(\frac{E_f E_c}{G_{xz}^2} \right)^{1/3} \tag{10.25}$$

or when the entire core thickness undergoes deformations (core is relatively thin),

$$N_{xwr} = 0.59 t_f^{3/2} \sqrt{\frac{E_f E_c}{t_c}} + 0.378 G_{xz} t_c \tag{10.22a}$$

$$\ell = 1.67 t_f \left(\frac{E_f t_c}{E_c t_f} \right)^{1/4} \tag{10.23a}$$

valid when

$$t_c < 3 t_f \left(\frac{E_f E_c}{G_{xz}^2} \right)^{1/3} \tag{10.25a}$$

Table 10.1 Analytical predictions versus FE results for sandwich wrinkling

E_c (MPa)	G_{xz} (MPa)	N_{xwr}/t_f (MPa) present	N_{xwr}/t_f (MPa) FE	Δ (%)	ℓ (mm) present	ℓ (mm) FE	Δ (%)
133	42	646	658	−1.8	11.3	11.4	−0.9
266	42	842	1033	−18.5	9.5	8.9	+6.7
133	84	808	821	−1.6	10.6	13.2	−19.7

In practice, one would have to evaluate both symmetric and antisymmetric wrinkling loads for a given application and use the lowest of the two. However, it can be demonstrated (see Exercise 10.1) that only for very thin cores is antisymmetric wrinkling possible. For typical core thicknesses, symmetric wrinkling is the mode of failure.

Comparison of the predictions for symmetric and antisymmetric wrinkling with experimental results is difficult to do with the equations presented so far. The main reason is that sandwich structure is most often fabricated by co-curing core, adhesive and facesheets all at once. As a result of this process the facesheets are not perfectly flat, but have some waviness. This waviness is not included in the analysis presented so far. Only if the facesheets are pre-cured separately and then bonded on a perfectly flat core will the waviness be (mostly) eliminated.

For this reason, the predictions presented so far are compared with finite element models in which the facesheets are perfectly flat. Such a comparison can be found in reference [9] and its conclusions are summarized here.

A sandwich with 25.4 mm honeycomb core and facesheets made with four plain weave fabric plies and layup $[(\pm 45)/(0/90)_2/(\pm 45)]$ was modelled under compression using finite elements. The facesheet stiffness E_f was 64 GPa. Since the core was thick, only symmetric wrinkling predictions were used. The pertinent core properties and a comparison or prediction from Equations (10.19) and (10.20) are shown in Table 10.1 for three different core materials with the same facesheet.

It can be seen that the predictions for the wrinkling stress N_{xwr}/t_f and the corresponding half-wavelength ℓ are in good agreement with finite elements with the highest discrepancy being less than 20%. It should be noted that the predicted wrinkling loads are always less than the finite element result. Also, the greatest discrepancy in wrinkling stress (case 2) does not correspond to the case with the greatest discrepancy in the half-wavelength (case 3). The discrepancies are attributed to a combination of finite element modelling issues related to proper load introduction and boundary conditions, and to the fact that Equation (10.7) is an approximation, especially considering the assumed linear variation with out-of-plane coordinate z.

The discussion so far has not explicitly accounted for the fact that the facesheets are made of composite materials. Only by substituting the appropriate value for the facesheet in-plane stiffness E_f in the equations derived does one account for composite laminates. For more accurate values for E_f in case of composite facesheets, the following expression can be used [7]:

$$E_f = \frac{12(1 - \nu_{xy}\nu_{yx})D_{11f}}{t_f^3} \tag{10.26}$$

where ν_{xy}, ν_{yx} and D_{11f} are Poisson's ratios and bending stiffness of the facesheet.

Other models explicitly incorporating composite layups can be found in the literature. For example, symmetric wrinkling was determined by Pearce and Webber [10] as

$$N_{xwr} = \frac{\pi^2}{a^2} \left[(D_{11})_f \, m^2 + 2 \left((D_{12})_f + 2 (D_{66})_f \right) \left(\frac{a}{b} \right)^2 + \frac{(D_{22})_f}{m^2} \left(\frac{a}{b} \right)^4 \right] + \frac{2E_c a^2}{m^2 \pi^2 t_c} \quad (10.27)$$

Comparing Equation (10.27) with Equation (8.33) suggests that the wrinkling load consists of two parts, the buckling load of the facesheet and a contribution from the core acting as an elastic foundation with spring constant $k = 2E_c/t_c$. Indeed, the first part of Equation (10.27) is identical to the buckling load of a simply supported plate under compression given by Equation (6.7). Finally, comparing Equation (10.27) with the wrinkling expression (10.14) derived earlier without explicitly incorporating the fact that the facesheet is composite, it is seen that the first two terms of Equation (10.14) have a one-to-one correspondence with the two terms of Equation (10.27). The first term corresponds to buckling of the facesheet and the second to the core acting as an elastic foundation and storing energy in deformation in the z direction. However, Equation (10.14) has an additional term dependent on the core shear stiffness which represents core shear deformations. This term is not present in Equation (10.27). It is, therefore, expected that Equation (10.27) may not be as accurate when the core shear deformations are appreciable.

As already alluded to, the fact that sandwich structures are usually co-cured results in facesheet waviness, which may significantly affect the performance of the sandwich and limit the usefulness of the design equations presented so far. One attempt to include the effect of waviness can be found in [9]. A typical cross-section of a $[(\pm 45)/(0/90)/(\pm 45)]$ facesheet on honeycomb core is shown in Figure 10.11 (taken from [9]).

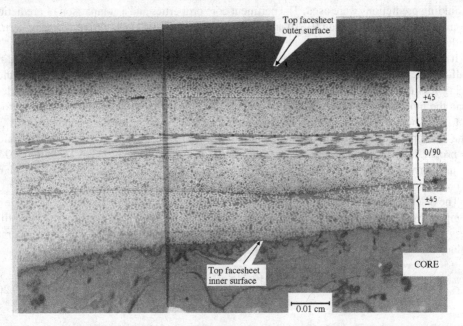

Figure 10.11   Sandwich cross-section showing facesheet and portion of the core (200× magnification from [9])

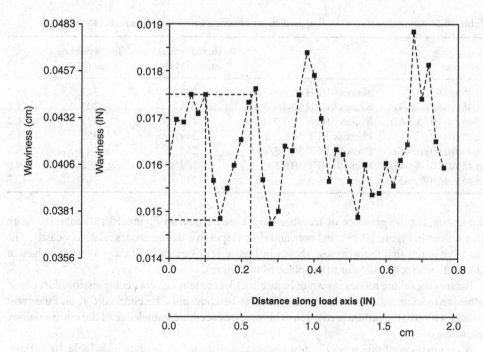

Figure 10.12   Waviness of outer surface of facesheet of Figure 10.11 (from [9])

Using Figure 10.11, the waviness of the facesheet at that section cut through the specimen was measured and plotted in Figure 10.12 (taken from [9]). It is evident from Figure 10.12 that the waviness can be significant and its amplitude can approach one-quarter to one-third of the facesheet thickness t_f (=0.5717 mm in this case).

Even though there is an element of randomness in the waviness of Figure 10.12, a main sinusoidal component of a specific amplitude and wavelength can be estimated. Assuming that component is present everywhere in the facesheet, a new model of facesheet deformations under compression accounting for the waviness can be created [9]. This model assumes a sinusoidal shape of the facesheet shown schematically in Figure 10.13. This model permits

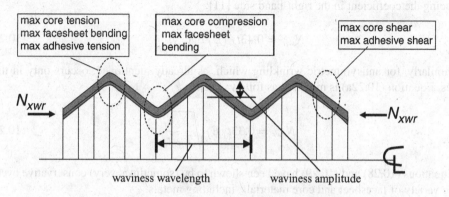

Figure 10.13   Sandwich with waviness

Table 10.2 Comparison of wrinkling predictions obtained with a waviness model to test results

Facesheet	Core	Predicted wrinkling stress (MPa)	Test wrinkling stress (MPa)	Δ (%)
(±45)/(0/90)	Nomex®HRH 10-1/8-3.0	295	313	−5.8
(±45)/(0/90)/(±45)	Nomex®HRH 10-1/8-3.0	264	297	−11.2
(±45)/(0/90)$_2$/(±45)	Nomex®HRH 10-1/8-3.0	426	337	+26.4
(±45)/(0/90)	Phenolic HFT 3/16-3.0	344	350	−1.8
(±45)/(0/90)/(±45)	Phenolic HFT 3/16-3.0	255	349	−26.9
(±45)/(0/90)$_2$/(±45)	Phenolic HFT 3/16-3.0	309	382	−19.0
(±45)/(0/90)/(±45)	Korex® 1/8-3.0	246	365	−32.7

accounting for the presence of facesheet (light grey colour in Figure 10.13), adhesive (dark grey colour in Figure 10.13), and core and their respective failure modes relatively easily. The model assumes that the waviness shown in Figure 10.13 extends all the way to the edges of the sandwich (perpendicular to the plane of the figure).

Each of the failure modes shown in Figure 10.13, core tension, core compression, core shear, adhesive tension, adhesive shear and facesheet bending must be checked for, and the most critical will give the failure prediction. This requires accurate knowledge of the corresponding allowables.

A comparison of this model to test results, taken from [9], is shown in Table 10.2. Here, three different facesheet layups and three different cores were used. The predictions range from excellent to barely acceptable (for the last case in Table 10.2). The main reasons for the discrepancies from test results are

(a) not so accurate knowledge of all allowables for the failure modes mentioned above; and
(b) in all cases, one amplitude and waviness were used obtained from Figure 10.12, which is not sufficiently accurate for some of the cores and facesheet layups in Table 10.2.

Still, using a waviness model is promising and, combined with accurate allowables, can yield very reliable predictions.

To account for the effect of (usually unknown) waviness and other complicating factors, it is customary to knockdown the predictions of Equation (10.20) for symmetric wrinkling by reducing the coefficient in the right-hand side [11]:

$$N_{xwr} = 0.43 t_f \left(E_f E_c G_{xz} \right)^{1/3} \tag{10.28}$$

Similarly, for antisymmetric wrinkling which, as already mentioned, occurs only in thin cores, Equation (10.22a) is modified as follows [12]:

$$N_{xwr} = 0.33 t_f E_f \sqrt{\frac{E_c}{E_f} \frac{t_f}{t_c}} \tag{10.29}$$

Equations (10.28) and (10.29) have been shown to be (sometimes very) conservative over a wide variety of facesheet and core materials, including metals.

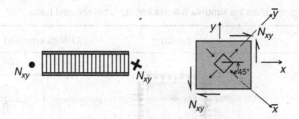

Figure 10.14   Sandwich under shear with shear load resolved into compression and tension loads

What has been presented so far is only a small portion of sandwich wrinkling modelling approaches. There are many more models, each with its own range of applicability. A discussion of the accuracy of the various models and their applicability can be found in [13].

10.3.2   Sandwich Wrinkling under Shear

Wrinkling of a sandwich structure can also occur under applied shear load. Since pure shear can be resolved in compression in one direction and tension in the other, the compression portion can cause wrinkling of the sandwich along a line at 45° to the applied shear load (see Figure 10.14). A conservative way to estimate the wrinkling load under shear is to analyse the sandwich as loaded under compression along the 45° line and neglect the tension load. The reason is that in biaxial loading situations with compression and tension, the tension load tends to stabilize the structure and the buckling load is higher than if only compression were applied. This was demonstrated in Figure 6.3 where the buckling load was higher for compression and tension than for biaxial compression.

Therefore, conservatively, the equations derived in the previous section for wrinkling under compression can also be used here, provided the relevant quantities E_f, E_c and G_{xz} are rotated to the direction of applied compression. Of these, the core Young's modulus in the z direction E_c remains unaffected. The facesheet modulus E_f is rotated by 45° by simply rotating the stacking sequence by that angle and calculating the corresponding membrane modulus of the resulting laminate in that direction using Equation (8.6). The core shear modulus G_{45} also changes if the core is not isotropic in its plane. The corresponding transformation was given by Equation (10.5). So, in the coordinate system $\bar{x}\bar{y}$, with compression parallel to the $\bar{x}$ axis, the rotated core shear stiffnesses are

$$G_{\bar{x}z} = \sin^2\theta\, G_{yz} + \cos^2\theta\, G_{xz} = \frac{G_{yz} + G_{xz}}{2} \quad \text{for} \quad \theta = -45°$$

$$G_{\bar{y}z} = \cos^2\theta\, G_{yz} + \sin^2\theta\, G_{xz} = \frac{G_{yz} + G_{xz}}{2} \quad \text{for} \quad \theta = -45°$$

(10.5a)

10.3.3   Sandwich Wrinkling under Combined Loads

Wrinkling under combined loads is analysed using interaction curves [7,14], which are very similar to the interaction curves for buckling of monolithic plates presented in Sections 6.5 and 6.6. A summary of the most common cases is given in Table 10.3. To use Table 10.3 the following ratios are defined:

Table 10.3 Interaction curves for sandwich wrinkling under combined loads

Case	Loading	Design equation
Biaxial compression		$N_x = \dfrac{N_{xwr}}{\left(1 + \left(\dfrac{N_y}{N_x}\right)^3\right)^{1/3}}$
Compression in x direction and tension in y direction		$N_x = N_{xwr}$
Combined compression and shear		$R_c + R_s^2 = 1$
Biaxial compression and shear		$R_c + R_s^2 = 1$ $R_c = N_x/N_{xwr}$ $R_s = N_{xy}/N_{xywr}$ N_{xwr} is wrinkling load in major core direction when biaxial compression acts alone (1st case in this table)
Compression in x direction, tension in y direction, and shear		$R_c + R_s^2 = 1$ $R_c = N_x/N_{xwr}$ $R_s = N_{xy}/N_{xywr}$ N_{xwr} is wrinkling load in x direction when compression acts alone

For compression alone,

$$R_c = N_x/N_{xwr} \tag{10.30}$$

where N_{xwr} is the wrinkling load under compression.
   For shear alone,

$$R_s = N_{xy}/N_{xywr} \tag{10.31}$$

where N_{xywr} is the wrinkling load under shear.

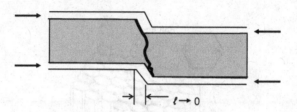

Figure 10.15 Sandwich crimping

10.4 Sandwich Crimping

This failure mode is shown in Figure 10.15. It occurs when the core shear stiffness is very low and is quite sensitive to the presence of eccentricities (e.g. when the core thickness is not uniform or if there is an abrupt change in the facesheet thickness when many plies are dropped).

This is a failure mode that is similar to antisymmetric wrinkling with, essentially, zero wavelength ($\ell \to 0$).

10.4.1 Sandwich Crimping under Compression

If the wavelength ℓ of the buckling mode tends to zero, the corresponding buckling load tends to infinity since the buckling load is proportional to $1/\ell^2$. Then, the basic buckling equation for a sandwich under compression, Equation (10.3), can be used:

$$N_{x\text{crit}} = \frac{N_{E\text{crit}}}{1 + \dfrac{N_{E\text{crit}}}{t_c G_c}} \tag{10.3}$$

letting $N_{E\text{crit}}$ tend to infinity.

It can be seen that Equation (10.3) is of the form ∞/∞ as $N_{E\text{crit}} \to \infty$, so l'Hopital's rule can be used to determine the limit of $N_{x\text{crit}}$. Differentiating numerator and denominator with respect to $N_{E\text{crit}}$ and, subsequently letting $N_{E\text{crit}}$ tend to infinity, the crimping load $N_{x\text{crit}}$ is shown to be

$$N_{x\text{crit}} = t_c G_c \tag{10.32}$$

where G_c is either G_{xz} or G_{yz}, whichever is aligned with the direction of the load.

10.4.2 Sandwich Crimping under Shear

A semi-empirical formula is used in this case, which is analogous to Equation (10.32):

$$N_{xy\text{crim}} = t_c \sqrt{G_{xz} G_{yz}} \tag{10.33}$$

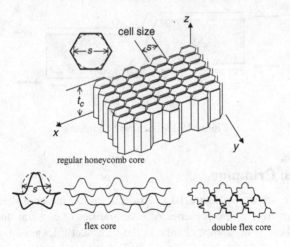

Figure 10.16   Cores susceptible to dimpling

10.5   Sandwich Intracellular Buckling (Dimpling) under Compression

This is a failure mode specific to honeycomb or other open-cell cores. Such representative cores are shown in Figure 10.16. Flex and double-flex core are used in structures with single or compound curvature to allow the sandwich to conform to the curved shape and eliminate anticlastic curvature effects.

When cores as those shown in Figure 10.16 are used, if the cell size is big enough it is possible for the unsupported facesheet between the cell walls to buckle. To analyse this intracellular buckling or dimpling mode requires developing a buckling solution for a composite facesheet with hexagonal or highly irregular (for flex or double-flex cores in Figure 10.16) boundaries. The complexity of such solution is prohibitive. Instead, a one-dimensional column-buckling type solution combining Equations (8.56) and (10.26) with a semi-empirical factor is used:

$$N_{x\text{dim}} = 2 \frac{E_f t_f^3}{1 - \nu_{xy} \nu_{yx}} \frac{1}{s^2} \tag{10.34a}$$

or

$$N_{x\text{dim}} = 24 \frac{D_{11f}}{s^2} \tag{10.34b}$$

where s is the core cell size shown in Figure 10.16.

10.6   Attaching Sandwich Structures

As mentioned in Section 10.1, sandwich has superior bending stiffness properties and would be the preferred design configuration had it not been for several failure modes such as wrinkling or crimping that limit its performance. Another problem that limits the usage of sandwich structure is the difficulty of attaching it to adjacent structure with adequate load transfer at the attachment region without undue increase in weight and cost. While it is relatively easy

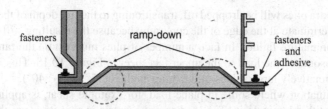

Figure 10.17 Attaching sandwich structure through the use of ramp-downs

to attach sandwich structure when the applied loads are low, it is quite a challenge to do so for highly loaded structure. Some considerations and options are discussed in the following two sections.

10.6.1 Core Ramp-Down Regions

One of the most common methods of providing adequate means of attachment is through the use of a ramp-down. By eliminating the core at the attachment region one does not have to deal with the fact that core has low compression and shear strengths which would compromise the strength of an attachment. As seen in Figure 10.17, the attachment can be through fasteners or adhesive (or both) connecting monolithic laminates at the edge of the ramp-down region.

The monolithic laminate created by eliminating the core (consisting of the two facesheets) may not be sufficient if the loads are high, and local reinforcement may be necessary to transfer bearing and shear loads (Figure 10.18). This creates the additional problem of deciding how

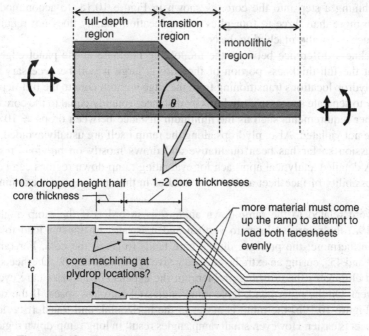

Figure 10.18 Transitioning from monolithic laminate to full-depth core

and where the extra plies will be dropped off, transitioning to the full depth of the core. Clearly, they cannot all terminate at the edge of the core ramp because the resulting stiffness mismatch would lead to premature failure. In fact, a number of plies must go up the ramp to stiffen it and thus attract some of the load to the upper facesheet in Figure 10.18. This is of particular importance for relatively large values of the ramp angle θ ($15° < \theta < 40°$).

Consider a situation where some in-plane load, for example shear, is applied at the edge where the monolithic laminate is. The loading is eccentric in that it acts at the neutral axis of the monolithic laminate which is offset from the neutral axis of the full-depth core. This causes a twisting moment. At low loads, the entire shear load stays in the flat facesheet, the bottom facesheet in Figure 10.18. At higher loads, bending of the full-depth core is more pronounced, and a significant fraction of the applied load starts to get transmitted up the ramp into the ramped facesheet (top facesheet in Figure 10.18). This is the reason plies must go up the ramp and into the upper facesheet, to transfer that load without failure. For typical designs of ramped sandwich under shear, 60% of the load stays in the straight (bottom) facesheet and 40% is transmitted up the ramp to the top facesheet.

Since the monolithic region is typically designed for bearing strength (at least when fasteners are used to connect to adjacent structure) and the facesheets away from the ramped regions are designed for buckling and notched strength requirements, the thickness of the monolithic region is, usually, significantly higher than the sum of the thicknesses of the two facesheets. This poses the problem of smoothly transitioning the thicker monolithic laminate to the thinner facesheets. A typical transition with some guidelines is shown in Figure 10.18. Successive plydrops are separated by at least 10 times the thickness of the dropped plies. This is in agreement with the findings of Figure 9.16. At the same time, again consistent with the results of Section 9.2.2, dropping a large number of plies should be avoided because of the high normal and shear stresses created. In addition, dropping many plies at the same location may require machining a step into the core, as shown in Figure 10.18, to accommodate them. Dropping no more than three to four plies at one location usually does not require special provisions such as locally machining the core.

If the thickness difference between the monolithic laminate at the panel edges and the facesheets at the full-thickness portion of the core is large it will be necessary to have a number of plydrop locations transitioning from the edge without core to the full-depth core. It is customary to separate successive plydrops by distances roughly equal to the core thickness, provided other requirements such as the minimum distance between drops = 10× plydrop thickness are not violated. Also, plydrops along the ramp itself are usually avoided.

The discussion so far has been qualitative and draws mostly on previous results from Chapter 9. A detailed analytical approach for evaluating ramp-down regions can be found in [15]. The possibility of facesheet and/or core failures in the ramp region is examined in that reference.

A final word related to the ramp-down angle θ is in order. If the ramp angle is large (Figure 10.19a) the ramp is closer to vertical and it is hard to transfer load to the upper facesheet. Furthermore, the pressure during cure tends to crush the core. For ramp angles between 30° and 45°, curing an extra layer of adhesive prior to curing the facesheets stabilizes the core and eliminates the crushing problem (at the expense of an extra cure cycle for the adhesive layer). On the other hand, large ramp angles take up less space. If the ramp angle is shallow (Figure 10.19b) the ramp is closer to the horizontal, and transferring load to the upper facesheet is easier. However, small ramp angles result in long ramp-down regions which means significant portions of the sandwich panel have core with lower thickness and thus

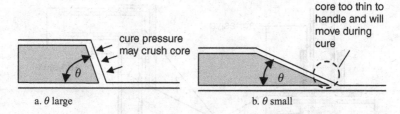

Figure 10.19 Steep versus shallow ramp-down regions

lower bending stiffness. This can cause stability problems. In addition, for small ramp angles, the end of the core, where the monolithic laminate starts, ends up being very thin. This causes handling problems during fabrication and it is hard to keep the core edge from moving and/or getting crushed under pressure during cure.

The optimum ramp angle will depend on which of the factors mentioned above are favoured for a given design and by specific factory practices. Lightly loaded ramps and situations with limited available space tend to favour larger values of θ, while more highly loaded applications will favour lower values of θ which approach a more even distribution of load between the facesheets, provided the local loss of bending stiffness is not prohibitive. For applied bending loads on small panels, where the ramped portions on either side of the panel are a significant fraction of the total panel size, it can be shown that the optimum value of θ is 12–18° (depending on panel size) [15]. This is a result of two opposing tendencies. For large values of θ the core shear and normal stresses are high, and lead to failure. For low values of θ, the core stresses are low, but the deformations are high due to reduced bending stiffness, and they cause failure. The best compromise is reached at intermediate θ values.

10.6.2 Alternatives to Core Ramp-Down

While using a ramp-down has certain advantages, especially for highly loaded situations, it does not come without a price, in particular because of the additional failure modes (core compression or shear) in the ramp-down region that require detailed analysis. Alternatives have been used for a long time and are based on the experience with metal cores [16].

These methods make use of inserts and bushings that span the full depth of the core so there is no need for ramp-down (Figure 10.20). Locally, the core may be densified with

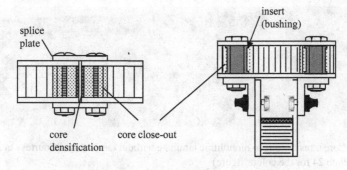

Figure 10.20 Attachments of sandwich parts without ramp-downs

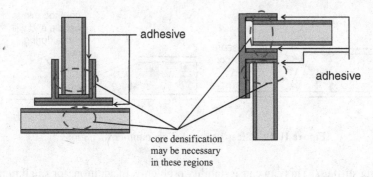

Figure 10.21 Alternate means of joining sandwich structures (see Plate 23 for the colour figure)

higher-density core or other material that has the required compression and shear stiffness to meet the loads exerted by clamp-up of bolts or other localized loads.

In addition to these, bonded configurations making use of special purpose joints such as the 'pi' and 'F' joints shown in Figure 10.21 can be used. Again, local densification of the core may be necessary. Due to the difficulty in accurately controlling the final thickness of the parts to be connected, and the width of the opening of the 'pi' or 'F' joint (mainly due to spring-back after cure), paste adhesive is used. Controlling the thickness of the bondline and making sure it is within the required range (too thin leads to early failure, too thick causes eccentricities that lead to high bending-induced loads) is the major challenge for these configurations. In addition, the lack of a reliable nondestructive inspection (NDI) technique, that determines whether the bond has the required strength or not, may either force the designer to use fasteners or to build into the configuration sufficient strength and alternate load paths so that, if a significant portion of the bond is compromised, the remainder can still meet limit load requirements. Despite these issues, bonded joints similar to those in Figure 10.21 have been used successfully in airframe structures (see, e.g. [17, 18]).

Finally, for relatively thin cores, transitioning to a thick monolithic laminate that forms an 'F' joint is also a possibility and has been used successfully (Figure 10.22).

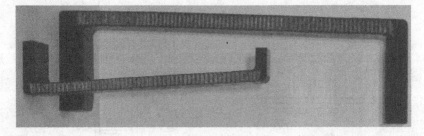

Figure 10.22 Core transitioning to monolithic laminate without ramp-down (Courtesy of Aurora Flight Sciences) (see Plate 24 for the colour figure)

Exercises

10.1 Prove that for antisymmetric wrinkling to occur, the core thickness must satisfy the following relation:

$$t_c < 1.047 t f \left(\frac{E_f E_c}{G_{xz}^2} \right)^{1/3}$$

10.2a Prove that for a simply supported square composite panel for which $D_{11} = D_{22}$, the number of half-waves m into which the panel buckles under compression is always 1. What should the condition be between D_{11} and D_{22} for the square panel to buckle in two half-waves?

10.2b Assume a layup consisting of n plies of the same material all at the same orientation (not necessarily $0°$). Let E be the Young's modulus of a single ply at that orientation, G the corresponding shear modulus, and v_{12}, v_{21} the two Poisson's ratios. Derive analytical expressions for $A_{11}, A_{12}, A_{22}, A_{66}, D_{11}, D_{12}, D_{22}, D_{66}$ as functions of E, G, v_{12}, v_{21}, and the thickness h of the laminates (still having all plies with the same fibre orientation).

10.2c A simply supported square sandwich panel of dimension a and core thickness t_c is under a compression load N_a (units: force/width). Use the results of Exercises 10.2a and 10.2b to express the buckling load N_{crit} as a function of E, G, v_{12}, v_{21}, h and t_c. Assume now that the material used is plain weave fabric for which $E_x = E_y$ and simplify the expression you derived (v_{12}, v_{21} are replaced by a single Poisson's ratio v).

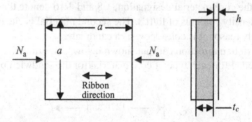

10.2d In certain circumstances, optimizing a structure that is likely to fail in more than one failure modes with corresponding loads 'reasonably' close to each other is equivalent to making sure that all failure modes occur simultaneously since this guarantees that the structure is not over-designed (and thus heavier than it needs to be) for any of the failure modes. This is not true in general, but it is true in quite a few cases. Assuming that the wrinkling failure and the buckling failure for the simply supported sandwich of Exercise 10.2c above occur at the same time, derive an expression for the facesheet thickness h (independent of t_c) and the core thickness t_c (which will be a function of h).

10.2e Let $a = 381$ mm, $N_a = 175$ N/mm. For the facesheet material assume that $E_x = E_y = 68.94$ GPa, $G_{xy} = 4.826$ GPa, $v_{xy} = 0.05$ and $t_{ply} = 0.1905$ mm. For the core assume $E_c = 133.05$ MPa (out-of-plane stiffness) and $G_{xz} = 42.05$ MPa (shear stiffness in the ribbon direction). If the ribbon direction is aligned with the loading N_a and the facesheet consists exclusively of (±45) plies, use the results of Exercise 10.2d to determine the minimum number of plies and minimum facesheet thickness needed. Is this the minimum weight configuration (i.e. is there another pair of values of h, t_c that gives lower weight)? For the optimum solution you found are the crimping and intracellular buckling requirements also satisfied? (For the latter assume a core cell size of 6.35 mm.)

10.3a A skin panel has dimensions 1270×1016 mm and is loaded in compression (along the long dimension of the panel) with $N_x = 121.45$ N/mm. A sandwich design is proposed for this application. The skin layup has been fixed to [45/–45/0/core/0/–45/45]. The facesheet material has the following properties:

$$E_x = 137.88\,\text{GPa}$$
$$E_y = 11.03\,\text{GPa}$$
$$G_{xy} = 4.826\,\text{GPa}$$
$$v_{xy} = 0.29$$
$$t_{ply} = 0.1524\,\text{mm}$$

Two Nomex honeycomb materials are proposed with the properties given below:

Material	E_c (MPa)	G_{xz} (MPa) (ribbon direction)	G_{yz} (MPa)
HRH-1/8-3.0	133.1	42.05	24.12
HRH-3/16-3.0	122.7	39.29	24.12

Note that in the core material designation, 1/8 and 3/16 denote the cell size (in inches!) and 3.0 the density (in units of lb/ft^3). The second material is cheaper than the first and this is the only reason it is considered as a candidate.

Given the ribbon direction call-out shown below, determine the minimum core thickness needed for each type of core material for the sandwich panel not to fail.

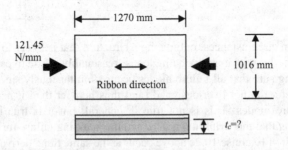

10.3b The manufacturing personnel in the factory who will fabricate this panel are very sloppy and careless. The engineer designing the panel is concerned that they will misorient the core in the panel and the ribbon direction will be perpendicular to the load. What is the minimum core thickness needed in this case (which will cover all possible errors in core placement during fabrication)? (Do this problem *only* for the first of the two core materials).

10.4 (May be done in conjunction with Exercise 8.6 which has the exact same requirements for a stiffener.) Design a sandwich configuration to represent a composite stiffener under compression.

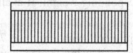

The sandwich design for each stiffener must fit within a 50 × 50 mm rectangle.

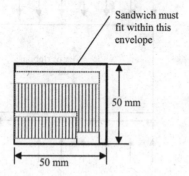

The applied load is 35,000 N (assume it is acting at the centre of gravity of the selected cross-section). The length ℓ of the stiffener is 550 mm.

Two composite materials are available and one core material with properties as follows:

Unidirectional tape graphite/epoxy	Plain weave fabric graphite/epoxy	Nomex core
$E_x = 131$ GPa	68.9 GPa	
$E_y = 11.4$ GPa	68.9 GPa	
$v_{xy} = 0.31$	0.05	
$G_{xy} = 5.17$ GPa	5.31 GPa	
$t_{ply} = 0.1524$ mm	0.1905 mm	
$\rho = 1611$ kg/m^3	1611 kg/m^3	48.2 kg/m^3

Also assume that the honeycomb core is attached to the facesheet (on either side) with an adhesive layer of density 0.147 kg/m^2. (Watch out for the units!) You are allowed to use any of the two graphite/epoxy materials or a combination thereof. Do not worry about any analysis for the adhesive, it is included here only for the weight calculation. Finally, assume a compression strain allowable (accounting for environment, damage and material scatter) of 4500 μs.

Determine the layup and core thickness of the sandwich, observing as many of the design rules as possible. (Note that the core is only allowed to take thickness values that are integral multiples of 3.175 mm; this is to save on machining costs.) Provide a simple sketch of the cross-section that shows the plies, layup, dimensions, etc. Calculate the corresponding weight. If available, compare with the answer from Exercise 8.6.

10.5 (May be done in conjunction with Exercise 7.6.) You are to design a composite panel under compressive load, using sandwich construction. The panel dimensions are 100×50 cm and the applied load is 1750 N/mm acting parallel to the 50 cm dimension.

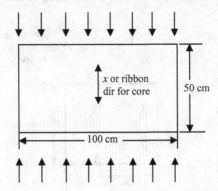

Two composite materials are available and one core material with properties as follows:

Unidirectional tape graphite/epoxy	Plain weave fabric graphite/epoxy	Nomex core
$E_x = 131$ GPa	68.9 GPa	
$E_y = 11.4$ GPa	68.9 GPa	$E_c = 133$ MPa
$\nu_{xy} = 0.31$	0.05	
$G_{xy} = 5.17$ GPa	5.31 GPa	$G_{xz} = 42.0$ MPa
$t_{ply} = 0.1524$ mm	0.1905 mm	Core cell size = 3.2 mm
$X^t = 2068$ MPa	1378.8 MPa	
$X^c = 1723$ MPa	1378.8 MPa	
$Y^t = 68.9$ MPa	1378.8 MPa	
$Y^c = 303.3$ MPa	1378.8 MPa	
$S = 124.1$ MPa	119.0 MPa	
$\rho = 1611$ kg/m^3	1611 kg/m^3	48.2 kg/m^3

Once you determine any strength values needed for any of the layups selected you are to assume the same knockdowns mentioned in Section 5.1.6 for environment, material scatter and damage, that is, any first-ply failure values should be reduced to the design (or allowable) values by multiplying them by $0.8 \times 0.65 \times 0.8 = 0.416$.

Determine the facesheet layup and core thickness for the sandwich panel not to buckle or fail in any other failure mode up to the applied load of 1750 N/mm. Also assume that the honeycomb core is attached to the facesheet (above and below) with an adhesive layer of density 0.147 kg/m². (Watch out for the units!) For the facesheet, you are allowed to use any of the two graphite/epoxy materials or a combination thereof. Do not worry about any analysis for the adhesive, it is included here only for the weight calculation. The core thickness cannot exceed 5 cm and cannot be any less than 6 mm. Each of the facesheets cannot be thinner than 0.57 mm.

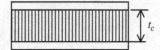

Determine the layup and core thickness of the sandwich, observing as many of the design rules as possible. Provide a simple sketch of the cross-section of the sandwich that shows the plies, layup, dimensions, etc. Calculate the corresponding weight and if available compare with the post-buckled design of Exercise 7.6.

10.6 A sandwich panel is 508 mm long and 203.2 mm wide. The applied loads are $N_x = 245.28$ N/mm and $N_{xy} = 133.15$ N/mm. A fabric material is available with properties:

$$
\begin{aligned}
E_x &= 68.9 \text{ GPa} & X^t &= 1034 \text{ MPa} \\
E_y &= 68.9 \text{ GPa} & X^c &= 689.4 \text{ MPa} \\
G_{xy} &= 5.17 \text{ GPa} & Y^t &= 1034 \text{ MPa} \\
\nu_{xy} &= 0.05 & Y^c &= 689.4 \text{ MPa} \\
t_{ply} &= 0.1905 \text{ mm} & S &= 124.1 \text{ MPa}
\end{aligned}
$$

Also, a core material with properties:

$$
\begin{aligned}
E_c &= 133.1 \text{ MPa} \\
G_{xz} &= 42.05 \text{ MPa} \\
G_{yz} &= 24.12 \text{ MPa}
\end{aligned}
$$

Using only three plies per facesheet and only in the orientations (± 45) and/or (0/90) such that each facesheet is symmetric, determine the lightest design. Do not do any crimping or dimpling checks. Also, do not use any knockdowns for material scatter, environments and damage.

10.7 A plain weave fabric composite material is available with the following properties:

$$
\begin{aligned}
E_x &= 68.9 \text{ GPa} \\
E_y &= 68.9 \text{ GPa} \\
\nu_{xy} &= 0.05 \\
G_{xy} &= 4.82 \text{ GPa} \\
t_{ply} &= 0.1905 \text{ mm}
\end{aligned}
$$

$$X^t = 1034 \text{ MPa}$$
$$X^c = 689.4 \text{ MPa}$$
$$Y^t = 82.73 \text{ MPa}$$
$$Y^c = 275.8 \text{ MPa}$$
$$S = 117.2 \text{ MPa}$$
$$\rho = 1636.5 \text{ kg/m}^3$$

Also, a core material is available with the following properties:

$$E_c = 221.3 \text{ MPa}$$
$$G_{xz} = 48.26 \text{ MPa}$$
$$G_{yz} = 24.13 \text{ MPa}$$
$$\text{Cell size} = 4.7625 \text{ mm}$$
$$\rho = 48.15 \text{ kg/m}^3$$

Making sure that the facesheet is symmetric and balanced, determine the lowest weight configurations for panels of dimensions $a = 1524$ mm and $b = 1016$ mm under shear loads 105, 157.5 and 210 N/mm. The panel is assumed to be simply supported all around.

10.8 If the lowest possible number of plies per facesheet in order to keep moisture out is 3 and the lowest available core thickness is 6.35 mm, determine the lowest weight configuration and the corresponding maximum load under compression given the materials from the previous exercise. The length a and width b are the same as in the previous exercise with a aligned with the x direction. If the applied load (along x) is lower than the maximum value determined here, is sandwich an efficient solution? Note that facesheets do not have to be symmetric in this problem.

References

[1] Carstensen, T., Cournoyer, D., Kunkel, E. and Magee, C., X-cor® Advanced Sandwich Core Material, *Proc. 33rd Int. SAMPE Conf.*, Seattle, WA, November 2001.

[2] Marasco, A.M., Cartié, D.D.R., Partridge, I.K. and Rezai, A., Mechanical Properties Balance in Novel Z-Pinned Sandwich Panels: Out-of-Plane Properties, *Composites Part A: Applied Science and Manufacturing*, **37**, 295–302 (2006).

[3] Plantema, F.J., *Sandwich Construction*, John Wiley & Sons, New York, 1966.

[4] Timoshenko, S.P. and Gere, J.M., *Theory of Elastic Stability*, McGraw-Hill, New York, 1961, p. 351.

[5] Whitney, J.M. and Pagano, N.J., Shear Deformation in Heterogeneous Anisotropic Plates, *Journal of Applied Mechanics*, **37**, 1031–1036 (1970).

[6] Hoff, N.J. and Mautner, S.E., The Buckling of Sandwich-Type Panels, *Journal of Aeronautical Sciences*, **12**, 285–297 (1945).

[7] Ley, R.P., Lin, W. and Mbanefo, U., Facesheet Wrinkling in Sandwich Structures, *NASA/CR-1999-208994* (1999).

[8] Vadakke, V. and Carlsson, L.A., Experimental Investigation of Compression Failure Mechanisms of Composite Faced Foam Core Sandwich Specimens, *Journal of Sandwich Structures and Materials*, **6**, 327–342 (2004).

[9] Kassapoglou, C., Fantle, S.C. and Chou, J.C., Wrinkling of Composite Sandwich Structures under Compression, *Journal of Composites Technology and Research*, **17**, 308–316 (1995).

[10] Pearce, T.R.A. and Webber, J.P.H., Buckling of Sandwich Panels with Laminated Face Plates, *Aeronautical Quarterly*, **23**, 148–160 (1972).

[11] Bruhn, E.F., *Analysis and Design of Flight Vehicle Structures*, S.R. Jacobs & Assoc, Indianapolis, IN, 1973, Section C12.10.3.

[12] Sullins, R.T., Smith, G.W. and Spier, D.D, Manual for Structural Stability Analysis of Sandwich Plates and Shells, *NASA CR 1457*, Section 2, (1969).

[13] Dobyns, A., Correlation of Sandwich Facesheet Wrinkling Test Results with Several Analysis Methods, *51st American Helicopter Society (AHS) Forum*, Fort Worth, TX, 1995, pp. 9–11.

[14] Birman, V. and Bert, C.W., Wrinkling of Composite Facing Sandwich Panels under Biaxial Loading, *Journal of Sandwich Structures and Materials*, **6**, 217–237 (2004).

[15] Kassapoglou, C., Stress Determination and Core Failure Analysis in Sandwich Rampdown Structures under Bending Loads, *in Fracture of Composites*, E. Armanios (ed.), TransTech Publications, Switzerland, 1996, pp. 307–326.

[16] Bruhn, E.F., *Analysis and Design of Flight Vehicle Structures*, S.R. Jacobs & Assoc., Indianapolis IN, 1973, Section C12.50.

[17] Wong, R., Sandwich Construction in the Starship, *Proc. 37th International SAMPE Symposium and Exhibition*, Anaheim CA, 9–12 March 1992, pp. 186–197.

[18] Jonas, P., and Aldag, T., Certification of Bonded Composite Structure, *SAE Technical Paper Series, Paper 871022*, 1987; also presented at *SAE General Aviation Aircraft Meeting and Exposition*, Wichita, KS, 28–30 April 1987.

11

Composite Fittings

A fitting is a part used to connect at least two other parts and transfer load effectively at the junction. Its main characteristic is that it, usually, transmits loads in multiple directions. Fittings are typically small compared to the parts they connect. Their relatively small size and the multi-directionality of applied loads make the designs of fittings challenging. While there are some standard fittings in terms of shape and purpose such as shear clips, lugs and bathtub fittings, in general, each fitting is a special-purpose part designed for use in one or very few locations. For this reason, the development of generic design and analysis methods can be difficult. Some typical fittings are shown in Figure 11.1.

11.1 Challenges in Creating Cost- and Weight-Efficient Composite Fittings

The directionality of composite properties may be a disadvantage when it comes to making composite fittings. Usually, one of the loads applied to the fitting is in the out-of-plane direction of a portion of the fitting. In general, the out-of-plane direction has no fibres. Only matrix, which has low strength, is present to carry loads. A typical example is shown, schematically, in Figure 11.2. (Note that this issue is discussed again in Chapter 12, Exercise 12.1.)

An out-of-plane load as shown in Figure 11.2 will load the matrix between plies in the flanges of the fitting. Making sure that there is no premature failure leads to increases in the surface area to minimize the out-of-plane load, introduction of additional load paths and use of reinforcements in the out-of-plane direction such as stitching and pinning.

The goal of any good design of a composite fitting is to minimize the out-of-plane loading situation that relies on the matrix to transfer the load. This, in the simplest solution, requires a combination of increased thickness and flange (or web) area and in the most complex the use of an out-of-plane reinforcement at the critical areas. In the first case the weight increases and the cost also increases through the use of more material and the need for careful planning for the tooling and cure cycle to ensure proper consolidation and avoid the creation of matrix cracks, adequate pressure everywhere in the part, appropriate heat-up and cool-down rates to keep the reaction from exotherming, etc. In the second case, the out-of-plane reinforcement reduces the in-plane strength and the extra fabrication steps add to the cost. The in-plane strength

Design and Analysis of Composite Structures: With Applications to Aerospace Structures, Second Edition. Christos Kassapoglou.
© 2013 John Wiley & Sons, Ltd. Published 2013 by John Wiley & Sons, Ltd.

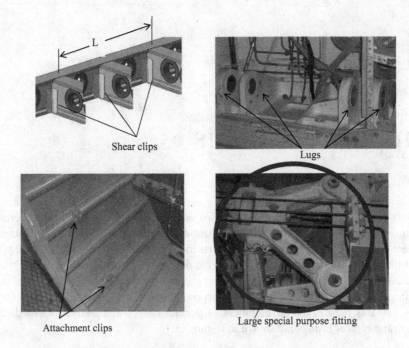

Figure 11.1   Various types of fittings (see Plate 25 for the colour figure)

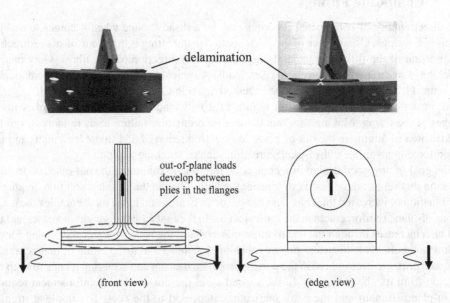

Figure 11.2   Tension fitting under out-of-plane load

reduction can be significant and must be accounted for. For example, 2.5–5% stitching may reduce the in-plane compression strength by 10–20% depending on the materials used.

11.2   Basic Fittings

The design and analysis of two types of common fittings, clips and lugs, are presented in this section. Clips and lugs may be used in isolation or may be parts of larger fittings. For example, the fitting of Figure 11.2 in the previous section combines a tension clip with a lug.

11.2.1   Clips

If two parts are aligned and need to be connected, a joint is, usually, used. If they are not aligned then clips are used. Clips are small fittings that usually connect two parts that are at an angle with each other (e.g. 90°). Clips provide an effective means for transferring load from one part to the next. There are two main types of clips: (1) tension clips and (2) shear clips. These are shown in Figure 11.3.

There are standard methods for the analysis of metallic tension and shear clips [1]. These methods are sometimes used for the analysis of composite clips with some modifications. In the next two sections, some basic analysis tools are presented for composite tension and shear clips.

11.2.1.1   Tension Clips

The situation is shown in Figure 11.4. A fastener is shown connecting the clip to the adjacent structure. Fasteners perform better as connecting media in this case because they are loaded in tension and have significantly higher strengths in tension than typical adhesives.

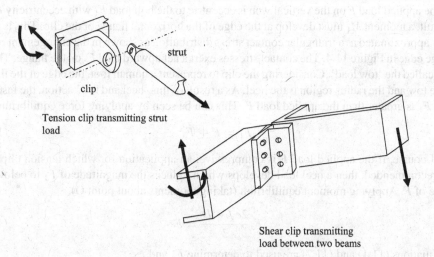

strut

clip

Tension clip transmitting strut
load

Shear clip transmitting
load between two beams

Figure 11.3   Tension and shear clips

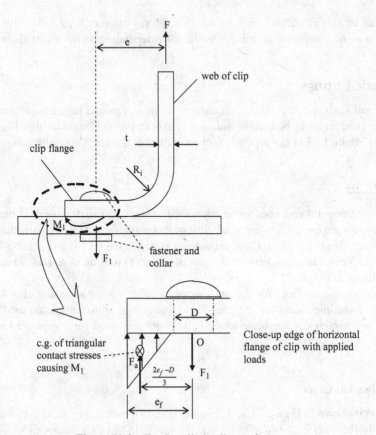

Figure 11.4 Tension clip loading and geometry

The applied load F on the vertical web is eccentric to the bolt load F_1 with eccentricity e. As a result, a moment M_1 must develop at the edge of the horizontal flange of the clip. This is typically approximated as a triangular contact stress distribution as shown in the enlargement of the flange edge in Figure 11.4. The contact stresses exert a net upward load F_e on the flange. This is also called the 'tow load'. Considering the clip to represent a human foot, the edge of the flange is the tow and the radius region is the heel. As a result of this 'heel and tow' action, the fastener load F_1 is greater than the applied load F. This can be seen by applying force equilibrium:

$$F_1 = F + F_a \tag{11.1}$$

Of course, if the applied load F is compressive, an application for which tension clips are not recommended, then a heel load develops which reduces the magnitude of F_1 to below the value of F. Applying moment equilibrium (taking moments about point O),

$$F_a \frac{2e_f - D}{3} = Fe \tag{11.2}$$

Equations (11.1) and (11.2) are used to determine F_a and F_1.

The analysis of the clip proceeds by starting from the top end of the web and analysing each portion of the clip until the edge of the horizontal flange is reached.

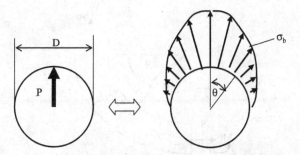

Figure 11.5 Bearing stresses caused by a fastener

The top of the web is checked for first-ply-failure using the methodology of Chapter 4. It should be noted that if the load F is not applied at the centre of the web as shown in Figure 11.4, then any additional loads present should be included in the analysis. For example, if there is additional eccentricity because the web is attached to a vertical part from one side only, then an additional bending moment may be applied to the clip web.

If the load is transmitted to the web via a fastener, similar to the situation shown in the flange, then the web must be checked for bearing failure. This is done by assuming that the fastener shank exerts a load P that results in a bearing stress on the contact surface between the fastener and web, which follows a cosine distribution as shown on the right of Figure 11.5:

$$\sigma_b = \sigma_{max} \cos \theta \tag{11.3}$$

The maximum bearing stress is determined by requiring that the total vertical load created by the stress distribution of Equation (11.3) equals the applied load P:

$$P = 2t \int_0^{\pi/2} \sigma_{max} \cos \theta \frac{D}{2} d\theta = t\sigma_{max} D \tag{11.4}$$

where D is the fastener diameter and t the web thickness. Solving for σ_{max},

$$\sigma_{max} = \frac{P}{Dt} \tag{11.5}$$

This maximum applied bearing stress must be compared to the bearing strength of the layup used in the web. Bearing strength values are obtained experimentally.

The next portion of the clip to be analysed is the radius region. It is important to recognize that the applied web load F, or per unit of clip width w (with w perpendicular to the page of Figure 11.4) $N_o = F/w$, would result in local axial N, shear S, and bending moment M at any θ location shown in Figure 11.6. These cause local stresses σ_r, σ_θ and $\tau_{r\theta}$. Of these, σ_r and $\tau_{r\theta}$ are interlaminar stresses that may cause delamination. This is a situation where the applied loads on the fitting result in stresses on the weakest link of the composite, the matrix.

The problem of the determination of the complete state of stress in a composite corner as that of Figure 11.6 under combined axial, shear and bending moment has been solved by Lehknitskii [2] and with a minor correction over Lehknitskii's solution, by Weber [3]. Here,

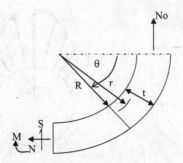

Figure 11.6   Isolated radius region of composite tension clip

only the axial load N_o is present. Following the solution in [3], the interlaminar stresses are
determined as

$$\sigma_r = \frac{N_o\,(2R_i + t)}{2b^2 g}\left[1 - \frac{1 - c^{\lambda+1}}{1 - c^{2\lambda}}\left(\frac{r}{b}\right)^{\lambda-1} - \frac{\left(1 - c^{\lambda-1}\right)c^{\lambda+1}}{1 - c^{2\lambda}}\left(\frac{b}{r}\right)^{\lambda+1}\right]$$

$$\tau_{r\theta} = N_o \sin\theta \frac{1}{g_1 r}\left[\left(\frac{r}{b}\right)^{\mu} + c^{\mu}\left(\frac{b}{r}\right)^{\mu} - 1 - c^{\mu}\right]$$

with

$$g = \frac{1 - c^2}{2} - \frac{\lambda\left(1 - c^{\lambda+1}\right)^2}{(\lambda + 1)\left(1 - c^{2\lambda}\right)} + \frac{\lambda c^2\left(1 - c^{\lambda-1}\right)^2}{(\lambda - 1)\left(1 - c^{2\lambda}\right)}$$

$$g_1 = \frac{2\left(1 - c^{\mu}\right)}{\mu} + \left(1 + c^{\mu}\right)\ln c$$

$$\lambda = \sqrt{\frac{\beta_{11}}{\beta_{22}}}$$

$$\mu = \sqrt{1 + \frac{\beta_{11} + 2\beta_{12} + \beta_{66}}{\beta_{22}}}$$

(11.6)

$$\beta_{11} = \frac{1}{E_r} - \frac{v_{xr}^2}{E_x}$$

$$\beta_{22} = \frac{1}{E_\theta} - \frac{v_{x\theta}^2}{E_x}$$

$$\beta_{12} = -\frac{v_{\theta r}}{E_\theta} - \frac{v_{xr}v_{x\theta}}{E_x}$$

$$\beta_{66} = \frac{1}{G_{r\theta}}$$

$$g_2 = \left(\frac{1 - c^2}{2}\right) - (c \ln c)^2$$

$$c = \frac{R}{R + t}$$

The in-plane stress σ_θ, not explicitly given in [3], is obtained from the stress equilibrium equation in cylindrical coordinates:

$$\frac{\partial \sigma_r}{\partial r} + \frac{1}{r}\frac{\partial \tau_{r\theta}}{\partial \theta} + \frac{\partial \tau_{rz}}{\partial z} + \frac{\sigma_r - \sigma_\theta}{r} = 0 \tag{11.7}$$

For a clip long in the z direction, perpendicular to the plane of Figure 11.4 or 11.6, there is no dependence on z. Then,

$$\frac{\partial \sigma_r}{\partial r} + \frac{\sigma_r}{r} + \frac{1}{r}\frac{\partial \tau_{r\theta}}{\partial \theta} = \frac{\sigma_\theta}{r} \tag{11.8}$$

or, multiplying through by r and rearranging

$$\frac{\partial}{\partial r}(r\sigma_r) + \frac{\partial \tau_{r\theta}}{\partial \theta} = \sigma_\theta \tag{11.9}$$

Using the first two of Equations (11.6), σ_θ is obtained as

$$\sigma_\theta = \frac{N_o(2R_i + t)}{2b^2 g}\left[1 - \lambda\frac{1 - c^{\lambda+1}}{1 - c^{2\lambda}}\left(\frac{r}{b}\right)^{\lambda-1} - (1-\lambda)\frac{\left(1 - c^{\lambda-1}\right)c^{\lambda+1}}{1 - c^{2\lambda}}\frac{b}{r^\lambda}^{\lambda+1}\right]$$
$$+ N_o \cos\theta\frac{1}{g_1 r}\left[\left(\frac{r}{b}\right)^\mu + c^\mu\left(\frac{b}{r}\right)^\mu - 1 - c^\mu\right] \tag{11.10}$$

The three stresses σ_r, σ_θ and $\tau_{r\theta}$ are used to check if there is failure anywhere in the radius region. Note that σ_θ is used to check for in-plane failure using a first-ply-failure criterion (see Chapter 4). Instead of σ_θ it is sometimes more convenient to use the combination of N and M at any θ location as direct inputs to a first-ply-failure analysis.

The remaining stresses σ_r and $\tau_{r\theta}$ are used to check for ply separation or delamination. There are several onset of delamination criteria [4, 5]. The criterion proposed by Brewer and Lagacé is more suited for interlaminar stresses developing at free edges so it would be more applicable near the two ends of the clip along the z axis and would require knowledge of σ_z, τ_{rz} and $\tau_{\theta z}$ stresses which were assumed negligible. For a complete analysis, these stresses, confined in a very small region at the specimen edges, should also be checked. The criterion by Chang and Springer [5] can be used directly with the σ_r and $\tau_{r\theta}$ stresses:

$$\left(\frac{\sigma_r}{Z_r}\right)^2 + \left(\frac{\tau_{r\theta}}{Z_{r\theta}}\right)^2 = 1 \tag{11.11}$$

where Z_r is the out-of-plane strength in the r direction and $Z_{r\theta}$ is the transverse shear strength in the $r\theta$ plane.

If σ_r is tensile, Z_r is the out-of-plane tension strength and if it is compressive, Z_r is the out-of-plane compression strength. Both Z_r and $Z_{r\theta}$ should be experimentally determined. A good approximation to Z_r can be obtained if the uni-directional ply transverse tension Y^t or compression Y^c strength is used [4]. For $Z_{r\theta}$, the short-beam shear strength value may be used but is not as accurate (within 20% for typical materials).

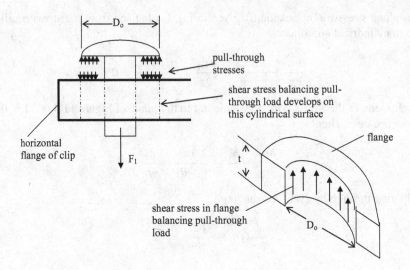

Figure 11.7   Transverse shear stresses balancing pull-through load

The last portion of the clip to be analysed is the horizontal flange in Figure 11.4. There are three areas and failure modes of interest: (a) out-of-plane failure of the flange under F_1 load on the fastener, (b) compressive failure of the flange bottom surface under the tow load (Figure 11.4), and (c) bending failure of the flange caused by the eccentric applied load F.

The out-of-plane failure of the flange is also referred to as 'pull-through' or 'push-through' failure caused by the fastener head being pulled through or pushed through (depending on one's perspective) the flange. This is a function of the contact area between the fastener head and the flange and the surface smoothness of the surfaces in contact. For this reason, reliable strength values are only obtained by testing. A conservative estimate of the pull-through strength can be obtained by assuming that the total fastener load F_1 in Equation (11.1) is balanced by transverse shear stresses in a cylindrical surface in the flange right below the outer edge of the fastener head as shown in Figure 11.7.

Conservatively then, pull-through failure occurs when

$$\frac{3F_1}{2\pi D_o t} = Z_{r\theta} \tag{11.12}$$

A couple of assumptions were used to derive Equation (11.12). The first was that, analogous to beam theory in isotropic materials, the shear stress shown in Figure 11.7 has a quadratic distribution through the flange thickness with maximum value 1.5 times the average. The second was that the transverse shear strength of the material is independent of θ and equals $Z_{r\theta}$. Both assumptions are approximations. Test results for typical carbon/epoxy thermoset materials show Equation (11.12) to give predictions as much as 50% lower than test results.

The compressive failure of the flange bottom surface under the triangular heel contact stresses does not occur in common aerospace composite materials. Only if a material with very low hardness is used would this failure be a concern but, then, this low hardness would probably lead to pull-through failure at the edges of the fastener head in Figure 11.7 before failure of the bottom of the flange in compression.

The bending failure of the flange caused by the eccentricity of the applied load F is, usually, the critical failure mode. For this reason it deserves a more detailed examination. The horizontal flange is treated as a cantilever beam of length $e - t/2 - D_o/2$ where e, t and D_o are defined in Figures 11.4 and 11.7. A shear load F is applied at the free end of the beam.

If the eccentricity e is small, and the beam undergoes small deflections, the bending moment at the beam root, which is the outer edge of the fastener head in the horizontal flange, is given by classical beam theory:

$$M = F \left(e - \frac{t + D_o}{2} \right) \tag{11.13}$$

If the per unit width moment M/w given by Equation (11.13) exceeds the bending moment M_x that would cause first-ply-failure of the flange layup, the clip fails in bending at the edge of the fastener. Note that w is the clip width perpendicular to the plane of Figures 11.4 and 11.6.

If the eccentricity is large the beam undergoes large deflections. This means that, if it were unconstrained by the vertical web attachment, the tip of the beam would move up and to the left in Figure 11.8. Its projected horizontal length would be less than the beam length by an amount δ_x shown at the bottom of Figure 11.8. In reality, however, the attachment of the vertical web reduces δ_x significantly. The radius of the clip corner 'opens up' and the beam elongates along with bending. A complete solution would have to account for these effects and would need to model where the vertical web is attached to the adjacent component and how the radius region of the clip 'opens up' under shear and axial loads. This can be very complicated. Instead a simpler model is sought.

If the beam is modelled using linear beam theory without some consideration of the constraint provided by the attachment of the vertical web, the results are very conservative. For large values of the eccentricity e (greater than 2.5 cm for typical clip layups) the maximum bending moment in the flange, predicted by simple beam theory, would lead to failure at one-third the experimentally measured values. If the beam is modelled using large deflections but still without some consideration of the constraint provided by the attachment of the vertical web, the results are conservative by approximately a factor of 2 for large values of the eccentricity. A reliable solution, with reasonable accuracy for typical clip configurations, can be obtained by combining the large deflection solution of a cantilever beam under a shear load with the effect of an axial tensile load as shown in Figure 11.9.

The solution starts with the large deflection solution of a cantilever beam under shear load F at the tip. The solution is in terms of elliptic integrals and can be summarized as follows [6, 7].

The angle φ_o between the tangent at the beam tip and the horizontal axis, when only F is acting, is determined by solving iteratively the equation

$$K \left(\sqrt{\frac{1 + \sin \varphi_o}{2}} \right) - F \left(\sin^{-1} \left(\frac{1}{\sqrt{1 + \sin \varphi_o}} \right), \sqrt{\frac{1 + \sin \varphi_o}{2}} \right) = \sqrt{\frac{F L^2}{w D_{22}}} \tag{11.14}$$

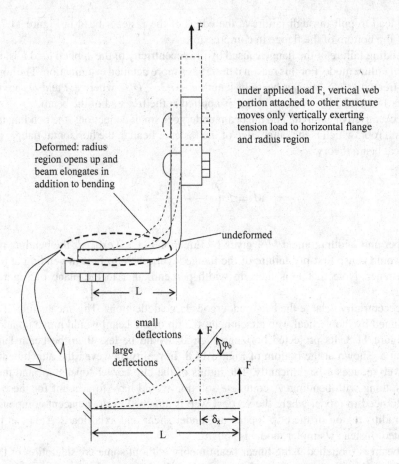

Figure 11.8   Clip flange deformations modelled as a cantilever beam under tip load

where w is the width of the clip and D_{22} is the 22 entry of the bending stiffness matrix with 2 coinciding with the horizontal direction in Figure 11.4. Also, $K(k)$ is the complete elliptic integral of the first kind and $F(\theta,k)$ is the incomplete elliptic integral of the first kind with θ in radians.

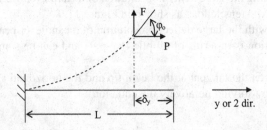

Figure 11.9   Cantilever beam under shear and axial loads

The projected length of the beam $L - \delta_y$ is then given by

$$L - \delta_y = \sqrt{\frac{2wD_{22}}{F}} \sqrt{\sin \varphi_o} \tag{11.15}$$

Then, the bending moment at the root of the beam is not FL which is the value of the linear solution given by Equation (11.13), but

$$M = F(L - \delta_y) = F\sqrt{\frac{2wD_{22}}{F}} \sqrt{\sin \varphi_o} \tag{11.16}$$

which, for sufficiently large values of e, can be significantly smaller than the linear solution from Equation (11.13) and thus can give higher failure loads for the clip than the linear solution.

At this point, the effect of a horizontal load P is approximated. Referring to Figures 11.8 and 11.9, it is assumed that the effect of the radius region is to exert a follower load on the tip of the flange, that is, a load tangent to the flange tip. This load has a vertical component F and a horizontal component P. Knowing F and φ_o, the value of P can be determined:

$$P = \frac{F}{\tan \varphi_o} \tag{11.17}$$

In a cantilever beam under shear and axial load, the effect of the axial load, if it is tensile as it is in our case, is to decrease the root moment [8].

$$M_{actual} = qM \tag{11.18}$$

where q is a factor depending on the value of $\sqrt{PL^2/(wD_{22})}$ and is tabulated. For most cases of interest, q is found to be in the range $0.55 < q < 0.7$.

The analysis procedure is then as follows:

1. For a given clip layup and geometry, determine the bending moment per unit width M_{fail} that causes first-ply-failure of the flange.
2. Select inner corner radius.
3. Apply a load F and iterate on Equation (11.14) to determine φ_o.
4. Use Equations (11.15) and (11.16) to obtain the root moment M without correction for axial load P.
5. Use Equation (11.18) to obtain the corrected root moment M_{actual}.
6. If $M_{actual}/w < M_{fail}$ from step 1, go to step 3 and increase load F. If $M_{actual}/w > M_{fail}$, go to step 3 and decrease load F. Repeat until the two moments are within a predetermined tolerance. The current value of F is the load that fails the flange in bending.
7. Apply Equation (11.11) to check if the corner delaminates. If it does (left-hand side of Equation 11.11 is greater than 1) go to step 2 and repeat step 7 until a value of R_i is found that prevents the delamination of the corner.

It is important to note that if the corner delaminates for a certain R_i value, it may not be possible to find another greater R_i value that moves the delamination load above the load causing bending failure of the clip flange. This happens if the load causing bending failure is quite high. Then, for the given layup, delamination will always happen before bending failure. In such cases, changing the layup and thickness of the clip may be necessary.

The above procedure covers bending and onset of delamination failure of the clip flange when the eccentricity e is large and the flange undergoes large deflections. If e is small, that is, $L - \delta_y \approx L$, the linear solution for beam bending with Equation (11.13) should be used. Comparison of the approach to test results for aluminium clips shows the method to be within 5% for low eccentricities and within 30% (conservative) for high eccentricities.

In addition to flange bending and onset of delamination of the corner, the flange should be checked for pull-through failure. Finally, the fastener(s) used in the clip should also be checked for failure. For example, there should be no tension failure of the fastener under load F_1 given by Equations (11.1) and (11.2).

To demonstrate the approach and give a feel for typical results, two different composite layups are selected for comparison to an aluminium clip of the same weight.

Two composite materials are considered with properties given in Table 11.1.

It is assumed that the fastener in the horizontal flange has a diameter of 6.35 mm which, for standard fasteners, implies an outer diameter for the fastener head D_o of 12.95 mm.

The two composite layups were [(±45)/(0/90)$_2$/45/−45]s and [(±45)$_3$/45/−45]s with the 90 direction coinciding with the circumferential direction of the clip corner in Figure 11.8. Recall (see Chapter 3) that (±45) and (0/90) are fabric plies. Both layups have the same thickness, 1.756 mm. Considering the previous discussion and the importance of the bending stiffness of the flange one would try to have as many 90° uni-directional tape plies as possible, especially on the outside to maximize the bending strength and stiffness. However, it is rather difficult to form a 90° ply around a 90° bend especially if the corner radius is relatively small.

Table 11.1 Composite materials used in design example

Property	Uni-directional tape	Plain weave fabric
E_x (GPa)	138	68.9
E_y (GPa)	11.4	68.9
ν_{xy}	0.29	0.05
G_{xy} (GPa)	4.83	4.83
t_{ply} (mm)	0.1524	0.1905
X^t (MPa)	2068	1034
X^c (MPa)	1379	1034
Y^t (MPa)	82.72	1034
Y^c (MPa)	330.9	1034
S (MPa)	117.2	117.2
ν_{xz}	0.29	0.29
ν_{yz}	0.29	0.29
G_{xz} (GPa)	4.83	4.83
G_{yz} (GPa)	4.83	4.83
Z_z (MPa)	82.73	82.73
$Z_{xz} = Z_{yz}$ (MPa)	117.2	117.2

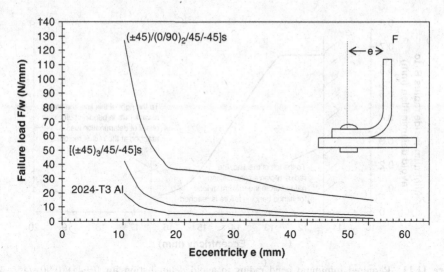

Figure 11.10   Tension clip failure loads for different load eccentricities

This may cause bridging and delaminations (see Chapter 12, guideline 20). For this reason, (0/90) fabric plies are used in the first laminate. These increase the bending strength over an all 0° or all 45° layup. 45° plies are added because they conform easily to 90° corners and provide strength and stiffness in other directions to protect against secondary load cases. The second layup is made exclusively of 45° plies. This is easy to fabricate and has very good shear stiffness and strength. It would make a very good shear clip (see next section). It is included here for comparison purposes.

In addition to the two composite layups, a 2024-T3 aluminium tension clip with exactly the same weight is used for comparison. The thickness of the aluminium clip is 1.034 mm. The tension yield strength of 2024-T3 aluminium is 344.7 MPa. The two composite and one aluminium tension clips are compared in Figure 11.10 where the failure load in N/mm is plotted as a function of eccentricity e.

Note that at about $e = 18$ mm the difference between the large and small deflection solutions becomes significant and, for higher e values, the large deflection solution outlined above is used. Switching from one solution to the next was not done in a smooth continuous fashion in the case of $[(\pm 45)/(0/90)_2/45/-45]$s and this accounts for the apparent sharp kink of the graph at $e \approx 17$ mm.

As is seen from Figure 11.10, the composite clips are better than the aluminium clip with the $[(\pm 45)/(0/90)_2/45/-45]$s significantly outperforming the other two. These results suggest that the use of composite tension clips has a lot of potential for saving weight over standard aluminium clips. However, it is important to note that (a) the aluminium clip is designed for yielding and (b) the alloy used is not the strongest. If ultimate strength was used with 7075-T6 instead of 2024-T3, the Al curve in Figure 11.10 would move up by a factor of 1.6 bringing it only slightly below the $[(\pm 45)_3/45/-45]$s curve. Therefore, composite tension clips are very promising if delamination at the corner is avoided and their cost is not prohibitive.

The minimum inside radius required to avoid onset of delamination at the radius region before bending failure of the flange was also determined for the two composite layups. It

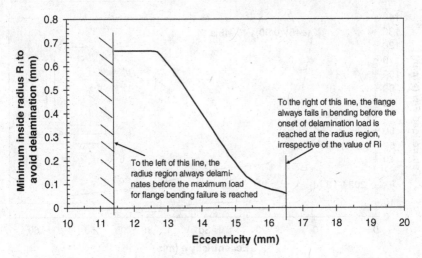

Figure 11.11 Required minimum bend radius to avoid delamination for $[(\pm45)/(0/90)_2/45/-45]$s tension clip

was found that for the $[(\pm45)_3/45/-45]$s any radius would guarantee no delamination for $e > 12.7$ mm. For $10 < e < 12.7$ mm, a minimum R_i of 0.016 mm is needed. The main reason for this insensitivity to the value of e is the fact that bending failure of the flange occurs at low loads that are below the load that would cause onset of delamination in the radius region.

The situation is significantly more interesting for the $[(\pm45)/(0/90)_2/45/-45]$s layup. The minimum R_i required for this case is shown in Figure 11.11.

There are two cut-off values in Figure 11.11. To the left of, approximately, $e = 11.5$ mm, no radius will avoid onset of delamination. The applied load that causes bending failure of the flange is significantly higher than the load that causes onset of delamination at the corner. At the other extreme, to the right of $e = 16.5$ mm, any radius will work. The failure load of the flange is lower than the load for delamination onset. In-between these two extremes, the minimum required R_i value is given by the curve in Figure 11.11.

11.2.1.2 Shear Clips

Depending on the orientation of the parts connected by the shear clip, two major cases are distinguished. These are shown in Figure 11.12.

In both cases, in addition to the main shear loading caused by the force F, there are bending or twisting moments induced by the eccentricities e_1 and e_2. The lines of action of F are through the centre of gravity of the connection (e.g. fastener pattern) of each side of the clip (web and flange) with the adjacent structure.

The two moments Fe_1 and Fe_2 can drastically reduce the load-carrying ability of the clip. Even if the in-plane shear loads F/w on each side of the clip do not cause failure, it is more than likely that the twisting moments Fe_1 and Fe_2 will. However, this does not account for the stiffness of the attaching parts A and B in Figure 11.12. The relative stiffnesses of the parts

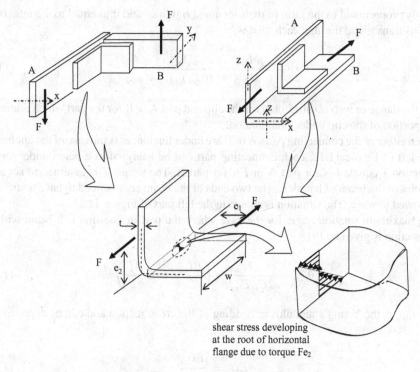

shear stress developing
at the root of horizontal
flange due to torque Fe_2

Figure 11.12   Two shear clip applications

attached and the clip will determine what fraction of these moments is actually transmitted through the clip.

Focusing on the left case in Figure 11.12, the force F causes torsion of part A about the y axis and bending about the x axis. Also, it causes torsion of part B about the x axis and bending about the y axis. The situation in the right-hand side is slightly different. The force F causes bending of A or B about the z and x axes. If the two parts A and B are sufficiently stiff to resist these applied moments and torques, the only load going through the shear clip is shear in the plane of the web and flange, for which such a configuration is ideally suited.

Parts A and B have very high bending stiffness about the axis of their highest moment of inertia, for example, axis x for part A and y for part B for the left application in Figure 11.12. Therefore, it is more important to check if their bending stiffness is high enough about the axis of lowest moment of inertia and thus sufficient to render the amount of torsion transmitted through the clip negligible. A similar situation occurs for the application on the right of Figure 11.12.

As a result, the weakest bending and torsional stiffnesses of parts A and B must be compared to the bending stiffness of the clip flange and web to determine if and how much of a bending load is transmitted through the clip.

For a beam under torsion (parts A or B on the left of Figure 11.12), the maximum rotation is compared to the maximum rotation of the web or flange of the clip. The ratio of the two is inversely proportional to the ratio of their torsional rigidities and thus equal to the ratio of the moments transmitted through each, that is,

$$\frac{M_{\text{clip}}}{M_{\text{A,B}}} = \frac{(\theta_{\max})_{\text{A,B}}}{(\theta_{\max})_{\text{clip}}} \tag{11.19}$$

where the flange or web is substituted for the clip and part A or B for the part which is attached to the portion of the clip under examination.

When either of the connecting parts A or B are under torsion, as is the case of the application on the left of Figure 11.12, each connecting part can be analysed as a beam under torsion. The torsion T equals Fe_2 for part A or Fe_1 for part B. The torque T is assumed to act at the mid-point of the beam of length L. The two ends of the beam are assumed to have no twist but are allowed to warp. The situation is shown in the left part of Figure 11.13.

The maximum rotation angle for the case where the part in question is a beam with 'C' cross-section is given by [9]

$$\theta_{\max} = \frac{Fe_2}{2C_w E_{\text{bcs}} \beta^3} \left(\frac{\beta L}{2} - \tanh \frac{\beta L}{2} \right) \tag{11.20}$$

where E_{bcs} is the Young's modulus in bending of the cross-section and can be approximated by

$$E_{\text{bcs}} = \frac{(EI)_{\text{cs}}}{I_{\text{cs}}} \tag{11.21}$$

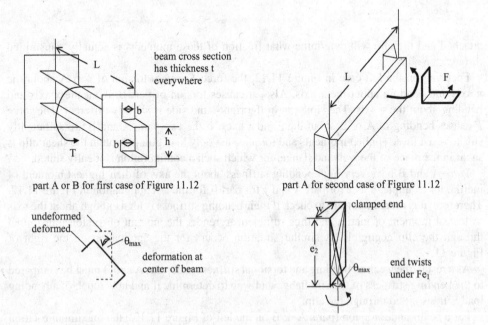

part A or B for first case of Figure 11.12   part A for second case of Figure 11.12

Figure 11.13 Modelling part A or B in situations where a shear clip is used

with $(EI)_{cs}$ the (minimum) bending stiffness of the entire cross-section (see Chapter 8.2) and I_{cs} the (minimum) moment of inertia of the cross-section.

Note that the length L of the beam between supports appears in Equation (11.20). For example, for the middle shear clip in Figure 11.1, L is the distance between the two attachments on either side of that clip as shown in the figure. This dependence on beam length should be expected because as the unsupported beam length increases the maximum rotation θ_{max} increases.

The warping constant C_w in Equation (11.20) is given ('C' cross-section with properties shown in Figure 11.13) by

$$C_w = \frac{h^2 b^3 t}{12} \frac{2h + 3b}{h + 6b} \tag{11.22}$$

The parameter β combines the torsional and bending rigidities. For a cross-section made of composite materials, an approximation is as follows (valid for open cross-sections):

$$\beta = \sqrt{\frac{\sum \dfrac{4b_i}{(d_{66})_i}}{C_w E_{bcs}}} \tag{11.23}$$

where b_i and $(d_{66})_i$ are the width and 66 entry of the inverse of the D matrix of the ith member of the cross-section. The summation in the numerator is taken over all members of the cross-section.

When part A on the left of Figure 11.12 is under torsion, the flange of the clip that is attached to part A is also under a torque Fe_2. Treating the flange as a beam of length e_1 fixed at one end, the maximum rotation the flange would undergo is given by

$$\theta_{max} = \frac{T e_1}{GJ} = \frac{Fe_2 e_1 (d_{66})_{clipflange}}{4w} \tag{11.24}$$

where GJ is the torsional rigidity of the flange under question.

Using Equations (11.20) and (11.24) to substitute in Equation (11.19),

$$\frac{M_{clip}}{M_A} = \frac{2w \left(\dfrac{\beta L}{2} - \tanh \dfrac{\beta L}{2} \right)}{e_1 (d_{66})_{clipflange} \beta \sum \dfrac{4b_i}{(d_{66})_i}} \tag{11.25}$$

with an analogous expression for the web of the clip and part B where e_1 is replaced by e_2 and quantities in the right-hand side of Equation (11.25) refer to part B instead of part A.

Equation (11.25) assumes that the clip is at the mid-point of a beam (part A or B) of length L. If the clip is closer to one of the two ends θ_{max} is smaller than what Equation (11.20) predicts and, thus, the distribution of moments between the clip and its support is conservative where the twisting moment that is assumed to be exerted on the clip is higher than its actual value.

The situation on the right-hand side of Figure 11.12, instead of torsion, involves bending of the clip flange or web and parts A or B. Each of the two parts, A and B, is treated as a simply supported beam of length L with a concentrated moment applied at some point along its length. As in the previous case with torque loads, the beam length is the distance between the two supports on either side of the clip location. The bending moments are Fe_1 for part A and Fe_2 for part B.

In this case the maximum deflection of each of the two beams A or B under a concentrated bending moment is compared to the maximum deflection of the corresponding portion of the clip (web or flange) under a twisting moment. Using part A on the right of Figure 11.12 as an example, and treating it as a simply supported beam of length L under a concentrated moment Fe_1, the largest deflection occurs at a distance $0.565L$ from one end when this moment acts at the other end. This case conservatively covers all possible locations of the clip between the two ends. The largest deflection of the beam for this case is [10]

$$u_{max} = 0.0642 \frac{Fe_1 L^2}{(EI)_{cs}} \qquad (11.26)$$

with $(EI)_{cs}$ the bending stiffness of part A when it bends about the z axis in Figure 11.12.

The corresponding maximum deflection in the (vertical) web of the clip, if the entire twisting moment Fe_1 were acting, is obtained by considering the situation in the bottom right of Figure 11.13:

$$u_{clip} = \frac{w}{2} \tan \theta_{max} = \frac{w}{2} \tan \left(\frac{Fe_2 e_1 (d_{66})_{clipweb}}{4w} \right) \qquad (11.27)$$

where Equation (11.24) was used. Assuming small angles, this equation can be simplified:

$$u_{clip} = \frac{Fe_2 e_1 (d_{66})_{clipweb}}{8} \qquad (11.28)$$

Combining Equations (11.26) and (11.27) gives the ratio of bending moment transmitted to the clip (web) divided by the bending moment transmitted to part A:

$$\frac{M_{clip}}{M_A} = \frac{0.0642 \dfrac{Fe_1 L^2}{(EI)_{cs}}}{\dfrac{Fe_2 e_1 (d_{66})_{clipweb}}{8}} = \frac{(0.51) L^2}{e_2 (d_{66})_{clipweb} (EI)_{cs}} \qquad (11.29)$$

A completely analogous relation can be obtained for the clip flange and part B. Depending on which of the two applications in Figure 11.12 is under consideration, Equation (11.21) or (11.29) is used to determine the amount of twisting moment in the respective flange or web of the clip as a ratio to the corresponding twisting or bending moment in the part providing backup support.

An example of the application on the right of Figure 11.12 is given here. The same two shear clips are considered as in the previous section: $[(\pm 45)/(0/90)_2/45/-45]s$ and $[(\pm 45)_3/45/-45]s$. The compliance d_{66} for each clip is 0.1077 and 0.0677 $(Nm)^{-1}$, respectively. The ratio of the moments from Equation (11.29) is shown in Figure 11.14.

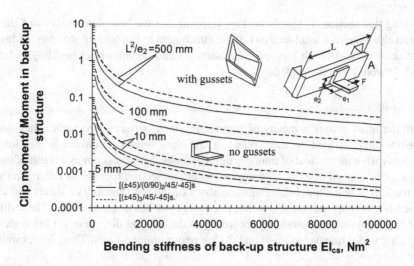

Figure 11.14 Shear clip bending loads as a function of stiffness of the backup structure

For each value of the parameter L^2/e_2, two curves are shown, one for each clip layup. As the ratio L^2/e_2 increases, the moment ratio increases. For high values of L^2/e_2 the moment ratio may even exceed 1 suggesting that the moment in the clip is bigger than the moment in the backup structure. In general, cases where the moment ratio is greater than 0.1 should be avoided unless detailed analysis of the clip shows it does not fail under the extra bending load. In addition to keeping L^2/e_2 as small as possible, another way to delay clip failure under bending loads in the web or flange is to add gussets. These increase the bending stiffness and reduce the bending stresses in the web or flange. In all cases in Figure 11.14, the curve for the $[(\pm45)/(0/90)_2/45/-45]$s layup is below the curve for the $[(\pm45)_3/45/-45]$s layup. This means that, for a given bending stiffness of the backup structure, the $[(\pm45)/(0/90)_2/45/-45]$s clip is softer and will absorb a lower fraction of the total moment Fe_1. At the same time, the $[(\pm45)/(0/90)_2/45/-45]$s is weaker in shear than the $[(\pm45)_3/45/-45]$s clip. As a result, it may not be clear which of the two will perform better as a shear clip without detailed calculations.

To examine this tradeoff between lower stiffness for lower twisting moment but also lower shear strength, a numerical example is done as follows. Suppose that $e_1 = e_2 = 20$ mm, $w = 30$ mm and $L = 100$ mm. This gives $L^2/e_2 = 500$ mm. Also assume that $EI_{cs} = 40,000$ N m^2. From Figure 11.14, the moment ratio for the $[(\pm45)_3/45/-45]$s clip is 0.08 and for the $[(\pm45)/(0/90)_2/45/-45]$s clip it is 0.04. Then, the web for each of the two clips must be analysed with a shear load $N_{xy} = F/w$ and a twisting moment $M_{xy} = r_m Fe_1/w$ with r_m equal to 0.08 or 0.04 depending on the case.

The Tsai–Wu failure criterion (see Chapter 4.4) is now applied with $M_{xy}/N_{xy} = r_m e_1$ and the force F that causes failure for each clip under combined N_{xy} and M_{xy} is determined. It is found that the $[(\pm45)/(0/90)_2/45/-45]$s fails when $F = 4465$ N and the $[(\pm45)_3/45/-45]$s fails when $F = 3972$ N which is 11% lower. In this case, the reduction in twisting moment more than offset the reduction in shear strength for the $[(\pm45)/(0/90)_2/45/-45]$s clip.

In general, only if F is large and e_1 relatively small, most of the load is shear and very little is torsion so the clip with the higher shear strength would be expected to perform better.

In closing this section, it should be kept in mind that the complete analysis of the shear clip would also require a local analysis of the attachment to the backup structure. If fasteners are used, a bearing strength (see previous section) and an inter-rivet buckling check (see Section 8.7) would also be needed.

11.2.2 Lugs

Lugs are the most common means of transferring load between two parts. They are very effective in carrying load in their plane. They were briefly discussed in Section 5.1.2 (Figure 5.6) within the context of multiple failure modes that may occur as a result of one load condition, axial tension of a lug. The discussion in this section focuses on failure analysis of the lug itself. It is understood that failure analysis of the pin (bending or shear) will also be performed but, as pins are usually metal, pin analysis is not discussed here. Three different cases of lug analysis are considered according to the loading direction: (a) lug under axial loads, (b) lug under transverse loads and (c) lug under oblique loads. These are examined in some detail below.

11.2.2.1   Lug under Axial Loads

The main failure modes for a lug under axial load are net section failure, shear-out failure and bearing failure. Combinations of these failure modes, usually including delaminations, are also possible. These are shown in Figure 11.15.

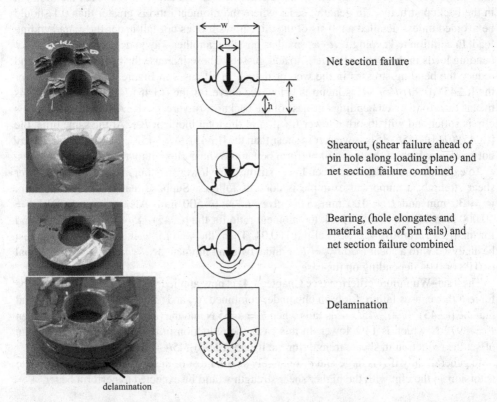

Net section failure

Shearout, (shear failure ahead of pin hole along loading plane) and net section failure combined

Bearing, (hole elongates and material ahead of pin fails) and net section failure combined

Delamination

delamination

Figure 11.15   Failure modes for a composite lug under axial load (see Plate 26 for the colour figure)

The net section analysis follows the procedure described in [11]. Superposing different problems with known solutions, as in [11], the normal stress σ in the net section region next to the lug hole, parallel to the applied load, can be determined:

$$
\sigma = \frac{1}{2}
\begin{bmatrix}
\sigma_a \left[1 + \dfrac{1}{2\left(\dfrac{x}{a}\right)^2} + \dfrac{3}{2\left(\dfrac{x}{a}\right)^3} - \dfrac{(K_{\text{TOH}}^\infty - 3)}{2} \left(\dfrac{5}{\left(\dfrac{x}{a}\right)^6} - \dfrac{7}{\left(\dfrac{x}{a}\right)^8} \right) \right] + \\[2em]
\dfrac{2F}{\pi^2 Dt}\left(3 + \dfrac{a^2}{x^2}\right) - \dfrac{F}{2\pi^2 Dt}\dfrac{x}{a}\left(3 - 2\dfrac{a^2}{x^2} - \dfrac{a^4}{x^4}\right)\ln\left(\dfrac{1 + 2\dfrac{a}{x} + \dfrac{a^2}{x^2}}{1 - 2\dfrac{a}{x} + \dfrac{a^2}{x^2}}\right)
\end{bmatrix}
\tag{11.30}
$$

where $a = D/2$, D is the diameter of the pin hole, t is the lug thickness, F is the applied load and x is the distance from the hole edge perpendicular to the load direction. In this expression, σ_a is the far-field, infinite plate-applied stress which has F as a reaction force in the lug hole. In addition, K_{TOH}^∞ is the stress concentration factor for an infinite orthotropic plate with an unloaded hole given by

$$
K_{\text{TOH}}^\infty = 1 + \sqrt{\frac{2}{A_{22}}\left(\sqrt{A_{11}A_{22}} - A_{12} + \frac{A_{11}A_{22} - A_{12}^2}{2A_{66}} \right)}
\tag{11.31}
$$

where A_{ij} are membrane stiffnesses for the laminate with the 1 direction aligned with the applied load.

The critical location is at the edge of the hole where the two net tension cracks initiate as shown at the top of Figure 11.15. Evaluating Equation (11.30) at that location, the critical stress at the edge of the pin hole is determined:

$$
\sigma_{\text{crit}} = \frac{1}{2}\left(K_{\text{TOH}}^\infty \sigma_a + \frac{8F}{\pi^2 Dt} \right)
\tag{11.32}
$$

This expression is valid for an infinite lug (edge distance e and lug width w in Figure 11.12 large). For a lug of finite width, a finite width correction factor [12] is applied:

$$
\frac{K_T}{K_{\text{TOH}}^\infty} = \frac{2 + \left(1 - \dfrac{D}{w}\right)^3}{3\left(1 - \dfrac{D}{w}\right)}
\tag{11.33}
$$

Combining Equations (11.32) and (11.33) and noting that $\sigma_a = F/(wt)$, the following expression is obtained:

$$
F = \frac{6F^{tu}wt}{K_{\text{TOH}}^\infty + \dfrac{8}{\pi^2}\dfrac{w}{D}} \frac{1 - \dfrac{D}{w}}{2 + \left(1 - \dfrac{D}{w}\right)^3} = \frac{6F^{tu}wt}{K_{\text{TOH}}^\infty + \dfrac{8}{\pi^2}\dfrac{w}{D}} \frac{\left(\dfrac{w}{D}\right)^3\left(\dfrac{w}{D} - 1\right)}{\dfrac{w}{D}\left(2\left(\dfrac{w}{D}\right)^3 + \left(\dfrac{w}{D} - 1\right)^3\right)}
\tag{11.34}
$$

where F^{tu} is the first-ply-failure strength of the lug laminate under tension.

After cancelling w and bringing D to the numerator of the second fraction, the final expression for the failure load in a lug failing in net tension is derived:

$$F = \frac{6F^{tu}(w-D)t}{K_{TOH}^{\infty} + \frac{8}{\pi^2}\frac{w}{D}} \frac{\left(\frac{w}{D}\right)^3}{\left(2\left(\frac{w}{D}\right)^3 + \left(\frac{w}{D}-1\right)^3\right)} \tag{11.35}$$

There are two limiting cases: As $D \to w$, F tends to zero as it should. As $D \to 0$, one cannot simply substitute in Equation (11.35) without recognizing that if there is no hole, K_{TOH}^{∞} tends to 1 and $8w/(\pi^2 D)$ also must be replaced by 1 since there is no hole to carry bearing stress $P/(Dt)$ any more. Then, F becomes $F^{tu}wt$ as it should.

One interesting implication of Equation (11.35) is that there is an optimum w/D ratio for which the load F to cause net section failure is maximized. For different values of K_{TOH}^{∞}, and therefore for different layups, the net section stress $F/((w-D)t)$ normalized by the tension strength F^{tu} is plotted in Figure 11.16. It can be seen that each curve has a maximum which means there is an optimum w/D ratio for each layup (value of K_{TOH}^{∞}) that maximizes the load at which the lug would fail in net tension. Since, however, the curves are all close to horizontal, the difference between using the optimum w/D value and other values in the vicinity is quite small. It should also be pointed out that optimizing for net section failure does not guarantee that the lug will not fail earlier in one of the two other failure modes (shear-out or bearing) examined immediately below.

The analysis for shear-out focuses on determining the average shear stress along the two planes of length ℓ shown in Figure 11.17.

From basic geometry,

$$\ell = \frac{1}{2}\sqrt{w^2 - D^2} = \frac{D}{2}\sqrt{\left(\frac{w}{D}\right)^2 - 1} \tag{11.36}$$

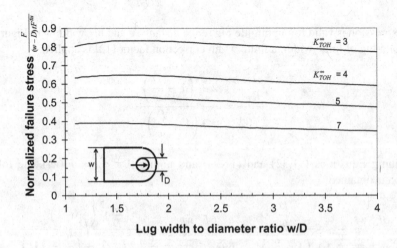

Figure 11.16   Failure load as a function of lug geometry (net tension failure)

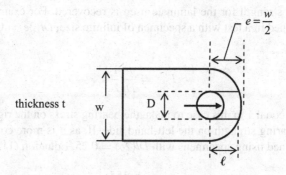

Figure 11.17 Geometry for shear-out lug analysis

Then, denoting the first-ply-failure stress of the lug laminate under pure shear by τ_{ult}, the value of F that would cause shear-out failure of the lug is computed from

$$\tau_{\text{ult}} = \frac{F}{2\ell t} = \frac{F}{2t\dfrac{D}{2}\sqrt{\left(\dfrac{w}{D}\right)^2 - 1}} = \frac{F}{Dt\sqrt{\left(\dfrac{w}{D}\right)^2 - 1}} \tag{11.37}$$

where F is divided by 2 to account for the two shear-out surfaces in Figure 11.17.

Therefore, the load F to cause shear-out failure is given by

$$F = \tau_{\text{ult}}Dt\sqrt{\left(\frac{w}{D}\right)^2 - 1} \tag{11.38}$$

Note that the equations shown here are for the special case of the lug of Figure 11.17 where $e = w/2$. If this is not the case, the value of ℓ corresponding to the case of interest must be computed.

The analysis for bearing failure uses the bearing stress from Equation (11.5) adjusted for finite width using Equation (11.33). In this case, however, the finite width correction factor involves the edge distance e instead of the lug width w (see top of Figure 11.15). Then, the bearing stress in the lug can be written as:

$$\sigma_{\text{brg}} = \frac{F}{Dt}\frac{2 + \left(1 - \dfrac{D}{2e}\right)^3}{3\left(1 - \dfrac{D}{2e}\right)}\lambda \tag{11.39}$$

where the first fraction in the right-hand side is the standard expression for bearing stress (see also Equation 11.5) and the second fraction is the finite width correction factor with $2e$ replacing w in Equation (11.33). This assumes the lug extends to the side opposite to the edge distance e (parallel to the direction of loading) for a length at least equal to e. If this is not true, that is, if the lug is shorter than $2e$, the equations presented here must be adjusted. The parameter λ is an 'equivalence' factor accounting for the geometry of the specimen used so that the bearing strength for the laminate used is recovered. For example, if the bearing strength were obtained in a test with a specimen of infinite size, $D/(2e) = 0$, Equation (11.39) would read

$$F^{bru} = \frac{F}{Dt}\frac{1}{\lambda}$$

so λ would have to equal 1 in this case to make the bearing stress on the right-hand side equal to the measured bearing strength on the left-hand side. If, as it is more common, the bearing strength were obtained using specimens with $D/(2e) = 0.25$, Equation (11.39) would read

$$F^{bru} = \frac{F}{Dt}\frac{155}{144}\lambda$$

and λ would have to equal $144/155 = 0.929$. The value of λ as a function of the value of D/e that is used to generate the bearing strength of the layup used is given in Figure 11.18.

Equation (11.39) can now be used to solve for the force F that would cause bearing failure of the lug by setting the bearing stress on the left-hand side equal to the bearing strength of the laminate used F^{bru}. Then

$$F = \frac{F^{bru}Dt}{\lambda}\frac{3\left(\frac{e}{D}\right)^2\left(\frac{e}{D}-\frac{1}{2}\right)}{2\left(\frac{e}{D}\right)^3 + \left(\frac{e}{D}-\frac{1}{2}\right)^3} \tag{11.40}$$

Then, the approach for analysing a lug under axial load consists of calculating the failure loads for net section, shear-out and bearing failures, respectively, using Equations (11.35), (11.38) and (11.40). The lowest is the lug failure load. A comparison of predicted failure loads to test results is shown in Figure 11.19. A wide range of graphite/epoxy lugs with w/D from 1.8 to 3, thickness from 6.8 to 12 mm, and layups 25/50/25, 50/33/17, and 63/25/12 (%0/%45/%90) using combinations of uni-directional tape and plain weave fabric were included in the tests.

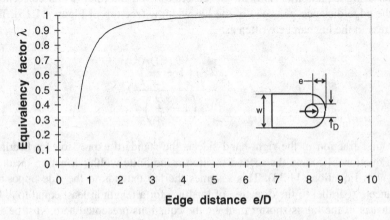

Figure 11.18 Factor λ used in Equations (11.39) and (11.40)

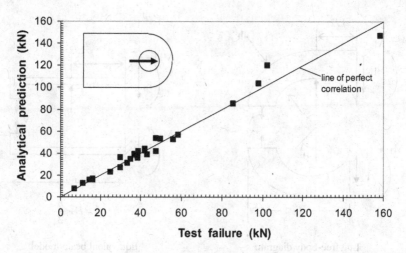

Figure 11.19   Analysis–test correlation for composite lugs under axial loads

The straight line shown in Figure 11.19 is the perfect correlation line. Symbols further away from the line imply bigger discrepancies between analytical predictions and test results. If a symbol is above the perfect correlation line, the analytical failure load is higher than the test result and if below, the analytical failure load is lower than the test result. The results in Figure 11.19 show very good agreement between analysis and test results with the biggest deviation being 17.3%.

For optimum design, the material selection, stacking sequence and geometry must be such that the three failure loads (net section, shear-out and bearing) are as close to each other as possible. Then, the lug is not overdesigned for any of the three and carries no unnecessary weight.

11.2.2.2   Lug under Transverse Loads

In a transverse loading situation, the load is still in the plane of the lug but at 90° to the axial load examined in the previous section. This is shown in Figure 11.20. The analysis of this

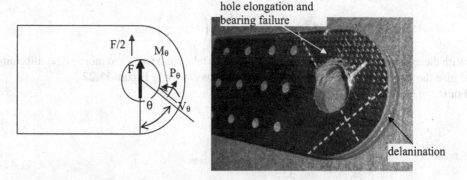

Figure 11.20   Lug under transverse load (see Plate 27 for the colour figure)

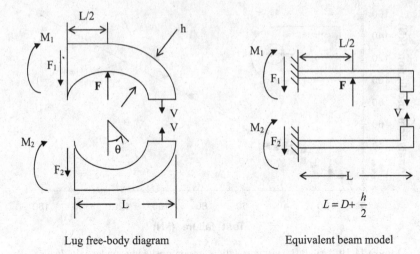

Lug free-body diagram   Equivalent beam model

Figure 11.21 Simplified model of lug under transverse loading

situation requires a number of assumptions and approximations. The force F is split unevenly between the upper and lower ligaments of the lug. The upper and lower parts of the lug are modelled as beams. This is shown in Figure 11.21.

The structure is statically indeterminate, however. Therefore, a displacement compatibility condition is imposed by requiring that, at $\theta = 90°$, the vertical deflection of the top part of the lug matches that of the bottom part of the lug. This leads to

$$V = \frac{5}{32}F$$

$$F_1 = \frac{27}{32}F$$

$$F_2 = \frac{5}{32}F \tag{11.41}$$

$$M_1 = \frac{11}{32}FL = \frac{11}{32}F\left(D + \frac{h}{2}\right)$$

$$M_2 = \frac{5}{32}FL = \frac{5}{32}F\left(D + \frac{h}{2}\right)$$

With these known, any section cut of the lug can be taken and force and moment equilibrium will give the internal forces P_θ, V_θ and moment M_θ as shown in Figure 11.22.

For $0 \le \theta \le 180$:

$$P_\theta = \frac{5}{32}F \sin\theta$$

$$V_\theta = \frac{5}{32}F \cos\theta \tag{11.42}$$

$$M_\theta = \frac{5}{32}F\left(\frac{D+h}{2}\right)(1 - \sin\theta)$$

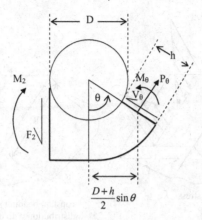

Figure 11.22 Forces and moments in the lower part of the lug

The stresses caused by the internal load from Equations (11.42) or the end loads from Equations (11.41) can now be determined. The critical location will depend on the lug layup since the ultimate normal or shear stress changes with the value of θ for a given lug layup. However, examining Equations (11.42) it can be shown that the two most critical locations are the top left edge where M_1 and F_1 act, and the $\theta = 0$ location.

One additional assumption is necessary to complete the analysis. If the bending moment M_θ of M_1 produces the linear normal stress distribution predicted by beam theory, a problem arises. The resulting peak stress is, for most typical lug configurations, very high. Using the tension or compression strength values obtained from standard coupon specimens tested under uniaxial loading as the strength values for the laminate is very conservative. It is known that the outer fibre tension or compression strength under bending is significantly higher than the corresponding strength obtained in a uniaxial test [13]. The main reason is that failure, especially in a composite, starts at a location of a small defect. The probability of a defect that could cause failure being located in a confined region of rapidly changing stress such as in a bending situation is much lower than in a specimen with constant stress throughout the specimen cross-section as in a tensile or compressive coupon. Therefore, specimens tested in pure bending would exhibit higher failure stresses than specimens tested in pure tension or compression. There are other less important reasons related to specimen geometry and the test method itself [13] which are not discussed here. There are, therefore, two options. Either use the linear bending stress distribution at any location of interest but with updated increased tension and compression strength values obtained from bending tests (e.g. four-point-bending) or represent the bending moment as consisting of constant tension and compression stresses each acting over half the beam with the strength values obtained in uniaxial tests. These options are shown in Figure 11.23.

The second option is selected. This means that instead of the standard peak bending stress obtained from beam theory,

$$\sigma_{max} = \frac{6M}{h^2 t}$$

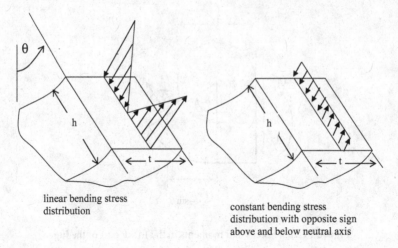

linear bending stress
distribution

constant bending stress
distribution with opposite sign
above and below neutral axis

Figure 11.23   Options for bending stress distribution in a lug under transverse load

an average constant stress over half the beam thickness is selected such that the total load over half the cross-section is the same in both cases. This will not cause the same moment at each θ location as the linear distribution. The bending moment is now smaller. This is an approximation that agrees well with test results (see below and also next section) but more tests are needed to validate this approach. Thus, the bending stress over half the cross-section to the right of Figure 11.23 is given by half the peak value from the linear bending stress distribution:

$$\sigma = \frac{3M}{h^2 t} \tag{11.43}$$

Taking now the internal loads from Equation (11.42), evaluating at the critical location $\theta = 0$ and calculating the maximum normal stress due to bending via Equation (11.42) and shear stress via the standard quadratic stress distribution gives the following critical stresses:

$$\tau_{max} = \frac{15}{64} \frac{F}{ht}$$
$$\sigma_{max} = \frac{15}{64} \frac{F}{ht} \left(\frac{D}{h} + 1 \right) \qquad \text{at } \theta = 0 \tag{11.44}$$

In addition, the top left edge of the lug where M_1 and F_1 are acting (Figure 11.21) must also be checked. The corresponding stresses are

$$\tau_{max} = \frac{81}{32} \frac{F}{wt}$$
$$\sigma_{max} = \frac{33}{8} \frac{F}{w^2 t} \left(D + \frac{h}{2} \right) \tag{11.45}$$

at edge of lug hole.

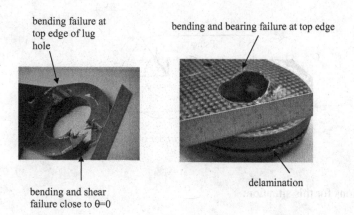

bending failure at
top edge of lug
hole

bending and bearing failure at top edge

delamination

bending and shear
failure close to θ=0

Figure 11.24   Failure modes of lugs tested under transverse loading (see Plate 28 for the colour figure)

The stresses from Equation (11.44) or (11.45) can be transformed to forces per unit width (N_x and N_{xy} if x is normal to the cross-section and y along h for Equation 11.44 and along w for Equation 11.45) by multiplying by t. Then, a first-ply-failure criterion can be applied (see Chapter 4). The lowest value of F obtained from Equation (11.44) or (11.45) is the lug failure load.

The approach was applied to lugs with quasi-isotropic layups consisting of uni-directional tape and fabric graphite/epoxy plies with $e/D = 1$ or 1.5 and $w/D = 2$ or 3. The predictions from the methodology just described were within 9% of the test results. Failed test specimens are shown in Figures 11.20 and 11.24.

As can be seen in Figure 11.24, additional failure modes such as bearing and delamination may also occur and, in some cases, may be the critical failure modes. For bearing, the procedure given in the previous section with w/D replacing e/D can be used. For delamination, three-dimensional stress solutions (see Section 9.2.2), especially near the edges of the lug are necessary.

A final note relating to the effect of layup is in order. As the layup changes, the relative importance of the shear and normal stresses at the two locations selected above changes. For example, an all 0 laminate has very low shear strength but very high bending strength and an all ±45 laminate has high shear strength and low bending strength. It is conceivable, therefore, that the two locations selected above will not cover all possible laminates. While they are valid for quasi-isotropic laminates and laminates that do not differ much from quasi-isotropic, for 'extreme' stacking sequences, Equations (11.42) should be applied at different θ locations and the local stresses should be obtained to see if the critical location(s) is(are) different than the two determined above.

11.2.2.3   Lug under Oblique (Combined) Loads

The most general case of loading a lug in its plane is that of an oblique load as shown in Figure 11.25. For metal lugs, the use of an interaction curve [14] has been shown to give very

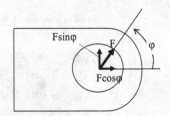

Figure 11.25 Lug under oblique load

good predictions for this situation:

$$\left(\frac{F\cos\varphi}{F_a}\right)^{1.6} + \left(\frac{F\sin\varphi}{F_{tr}}\right)^{1.6} = 1 \qquad (11.46)$$

For a metal lug, therefore, the applied load is resolved into an axial component $F\cos\varphi$ and a transverse component $F\sin\varphi$ as shown in Figure 11.25. Then, each component load is treated as an individual load case and the axial failure load F_a and transverse failure load F_{tr} are determined. Then, the component loads and the failure loads for the individual load cases are combined in the interaction curve of Equation (11.46). If the left-hand side is less than 1 there is no failure and if greater than 1, the lug fails.

The exact approach may also be used for composite lugs provided the axial and transverse failure loads are obtained using the methods in Sections 11.2.2.1 and 11.2.2.2. Limited test data for quasi-isotropic lugs made with graphite/epoxy fabric material show good agreement with predictions with the analysis being no more than 4% lower than test results. This is shown in Figure 11.26.

In closing this section, several comments specific to the design of lugs are in order:

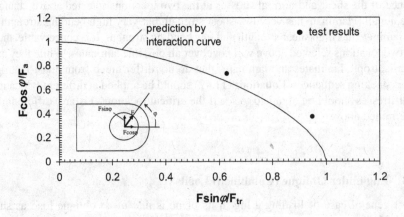

Figure 11.26 Analytical predictions versus test results for lugs under oblique loading

- Lugs are ideal for carrying load in their plane. However, in practice, no matter how careful one is to ensure that the primary load is in the plane of the lug, there are always smaller loads perpendicular to the lug plane. To cover for these cases, it is common to take a percentage of the maximum in-plane load, usually around 10%, and apply it at the lug pin, perpendicular to the lug plane. Then, a bending analysis of the lug is necessary to make sure this secondary load will not cause premature failure
- If one designs a quasi-isotropic composite lug using average strength values and minimizes the weight, the resulting design is, typically, 20% lighter than the equivalent aluminium design (using quasi-isotropic layup for the composite). If, instead one uses B-Basis values (see Section 5.1.3.3) the resulting design is about 15% heavier than the aluminium design. It is, therefore, very important to select the lug layup and geometry very carefully in order to generate a design that is lighter than a baseline aluminium design.

11.3   Other Fittings

As mentioned at the beginning of this chapter, most fittings are special-purpose parts that are specific to the geometry and loading situation of the parts they connect. Therefore, a description of all fittings and a creation of appropriate generic analysis methods is, if not impossible, definitely inefficient. Here, a few more fitting types are mentioned to give a better overview of this class of structural parts.

11.3.1   Bathtub Fittings

These are similar to clips but are used in more highly loaded situations. They are, essentially, clips with reinforcing and stiffening walls (gussets). These were briefly introduced during the discussion of tension clips with high bending moments applied to them (see also Figure 11.14). Typical bathtub fittings are shown in Figure 11.27. There are methods to analyse metal bathtub fittings but their reliable extension to composites has not been done yet (at least not in open literature). Finite-element methods may be the most accurate alternative in analysing such fittings.

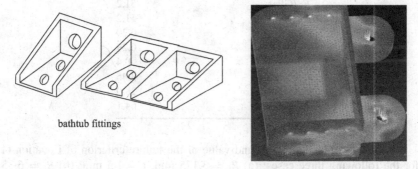

bathtub fittings

root fitting with stress contours
from finite element analysis

Figure 11.27   Other common fittings (see Plate 29 for the colour figure)

11.3.2   Root Fittings

These are fittings typically found at the roots of parts such as spars and serve to connect spars, skins and adjacent structure together. They typically integrate other fittings such as clips, lugs, or bathtub fittings. An example is shown in Figure 11.27. Other than the local analysis of smaller fittings already discussed in this chapter, the finite-element method is the most reliable approach for their analysis.

Exercises

11.1   Consider a composite corner as shown in the figure with applied moment $M_o = -53.7$ N mm/mm and shear $S_o = -7.8$ N/mm.

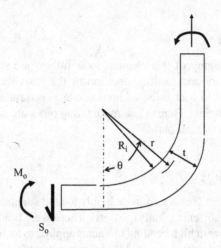

The layup used has the following properties:

E_r (GPa)	11.4
E_x (GPa)	51.2
E_θ (GPa)	51.2
ν_{xr}	0.29
$\nu_{x\theta}$	0.29
$\nu_{\theta r}$	0.4
$G_{r\theta}$ (GPa)	3.45
Z_r (MPa)	28.9
$Z_{r\theta}$ (MPa)	64.1

Determine the location (r,θ) and value of the failure criterion of Equation (11.11) for the following three cases: (a) $R_i = 3.175$ mm, $t = 1.5$ mm; (b) $R_i = 6.35$ mm, $t = 1.5$ mm; and (c) $R_i = 9.525$ mm, $t = 1.5$ mm.

11.2   Repeat exercise 11.1 but now the thickness for the three cases is 2.5 mm. Compare the results with exercise 11.1.

References

[1] Bruhn, E.F., *Analysis and Design of Flight Vehicle Structures*, Tri-state Offset Company, Cincinnati, OH, 1973, Sections D3.2 and D3.3.

[2] Lehknitskii, S.G., *Anisotropic Plates*, Translated by S.W. Tsai and T. Cheron, Gordon and Breach Science Publishers, New York, 1968, Sections 23 and 24.

[3] Weber, J.P.H., Through-the-Thickness Stresses and Failure in the Corner Radius of a Laminated Composite Section, *ESDU data item 94019* (1994).

[4] Brewer, J.C. and Lagacé, P.A., Quadratic Stress Criterion for Initiation of Delamination, *Journal of Composite Materials*, **22**, 1141–1155 (1988).

[5] Chang, F.K. and Springer, G.S., The Strengths of Fiber Reinforced Composite Bends, *Journal of Composite Materials*, **20**, 30–45 (1986).

[6] Bisshopp, K.E. and Drucker, D.C., Large Deflections of Cantilever Beams, *Quarterly of Applied Mathematics*, 272–275 (1945).

[7] Beléndez, T., Neipp, C. and Beléndez, A., Large and Small Deflections of a Cantilever Beam, *European Journal of Physics*, **23**, 371–379 (2002).

[8] Young, W.C. and Budynas, R.G., *Roark's Formulas for Stress and Strain*, 7th edn, McGraw-Hill, New York, Chicago, San Francisco, 2002, Section 8.17, Table 8.7a, Case 1a.

[9] Young, W.C. and Budynas, R.G., *Roark's Formulas for Stress and Strain*, 7th edn, McGraw-Hill, New York, Chicago, San Francisco, 2002, Section 10.7, Table 10.3, Case 1e.

[10] Young, W.C. and Budynas, R.G., *Roark's Formulas for Stress and Strain*, 7th edn, McGraw-Hill, New York, Chicago, San Francisco, 2002, Section 8.17, Table 8.1, Case 3e.

[11] Kassapoglou, C. and Townsend, W.A., Failure Prediction of Composite Lugs under Axial Loads, *AIAA Journal*, **41**, 2239–2243 (2003).

[12] Tan, S.C., "Finite Width Correction Factors for Anisotropic Plate Containing a Central Opening", *J. Composite Materials*, **22**, 1080–1097 (1988).

[13] Whitney, J.M. and Knight, M., The Relationship between Tensile Strength and Flexure Strength in Fiber-Reinforced Composites, *Experimental Mechanics*, **21**, 211–216 (1980).

[14] Bruhn, E.F., Analysis and Design of Flight Vehicle Structures, *Tri-states Offset Company*, Section D1.13 (1973).

12

Good Design Practices and Design 'Rules of Thumb'

Throughout the previous chapters, several guidelines that result in robust designs have been presented and, in some cases, analytical models that support them were given. In this chapter, all the rules already mentioned in this book are collected and some new ones added to provide a framework within which most composite designs can perform successfully.

Design guidelines are a result of analysis and trending, test results and experience. As such, they typically have a range of applicability (especially in terms of the stacking sequences to which they apply) outside of which they may or may not be as successful. There is no reason why any and all of the guidelines should be closely followed. Deviations and departures from them are often necessary. As long as the reasons for deviation are understood and test results and accurate analysis are available to support that deviation, there is no reason to limit the designs by following these guidelines. In fact, there is a motivation to open up or reformulate some of these guidelines in order to generate more efficient and/or more robust designs in the future [1].

The most important guidelines with a brief discussion are listed below. Other guidelines and/or variations of the ones presented below can be found in the literature, for example, in reference [2].

12.1 Layup/Stacking Sequence-Related

1. The *layup* (stacking sequence) of a laminate *should be symmetric*. This eliminates unwanted (and difficult to analyze) membrane/bending coupling (B matrix is zero).
2. The *layup should be balanced* (for every $+\theta$ ply there should be a $-\theta$ ply of the same material and thickness somewhere in the laminate). This eliminates stretching/shearing coupling ($A_{16} = A_{26} = 0$).
3. *Bending/twisting coupling should be avoided.* One way to achieve this is to use antisymmetric layups, but this violates guideline number 1. Another is to use fabric materials and unidirectional materials exclusively in the 0 and 90° directions. When these options are not possible, layups where D_{16} and D_{26} are small compared with the remaining terms

Design and Analysis of Composite Structures: With Applications to Aerospace Structures, Second Edition. Christos Kassapoglou.
© 2013 John Wiley & Sons, Ltd. Published 2013 by John Wiley & Sons, Ltd.

of the D matrix should be preferred. To that end, $+\theta$ and $-\theta$ plies should be grouped together. Also, special classes of laminates with negligible D_{16} and D_{26} are possible [3].

4. *The 10% rule.* At least 10% of the fibres in every layup should be lined up with each of the four principal directions: $0°, 45°, -45°$, and $90°$. This protects against secondary load cases, which have small load magnitudes and thus are not included in the design effort, but could lead to premature failure if there are no fibres in one of the four principal directions. In some cases, instead of 10% other values (12, 15%) are also used.

5. *Minimization of the number of unidirectional plies with same orientation next to each other.* If there is a number of unidirectional (UD) plies of the same orientation next to each other, then a matrix crack forming in them can grow easily in the matrix and extend from one end of the identical ply stack to the other without being arrested. Such cracks can be caused by thermal stresses during cure or due to transverse loading during service (transverse to the orientation of the fibres in the ply stack in question). It is recommended to avoid ply stacks of the same ply orientation that exceed 0.6–0.8 mm (corresponding to 4–5 plies for typical UD materials). Interrupting the ply stack with plies of different orientation (preferably with at least 45° difference from the plies in the ply stack) provides a means to arrest microcracks. The probability of microcracks coalescing and/or creating delaminations is minimized.

12.2 Loading and Performance-Related

6. To improve the *bending stiffness of a one-dimensional composite structure* place 0° plies as far away from the neutral axes as possible (this maximizes D_{11}).

7. *Panel buckling and crippling improvement.* Place 45/−45 plies as far away from the neutral axis as possible (this maximizes D_{66}).

8. *Fastener rule 1.* Maintain skin thickness/fastener diameter ratio $<1/3$ to minimize fastener bending (Figure 12.1a).

9. *Fastener rule 2 (countersunk fasteners).* Maintain skin thickness/to countersunk depth $> 3/2$ to avoid pulling the fastener through the skin under out-of-plane loads (Figure 12.1b).

10. *45° fabric plies on the outside.* To improve damage resistance, that is to limit the amount of damage caused by low speed impact, fabric plies should be placed on the outside of a stacking sequence. They limit the amount of fibre splitting and help contain splits created in the first (impacted) or last ply.

11. *Skin layup should be dominated by 45/−45 plies.* Using 45° and −45° plies improves the shear stiffness and strength of the layup. This is also a good rule to follow for beam or stiffener webs under shear loads in the plane of the web.

12. *Fastener rule 3.* For improved load transfer around fasteners in bolted joints, at least 40% of the fibres should be in the $+45°$ and $-45°$ directions relative to the applied axial load.

13. *Fastener rule 4.* To avoid interaction and increased stress concentrations *fastener spacing* should be at least $4–5D$ where D is the fastener diameter (Figure 12.1c). This only ensures that the full by-pass load is developed between fasteners, and the load distribution around one fastener does not affect that around its neighbours. This decreases the stress concentration effect. It does not account for other considerations such as inter-fastener buckling (see Section 8.7) or potential improvements in bolted joint performance with lower spacings alluded to in Section 8.7. Specific requirements of each design might supersede this guideline.

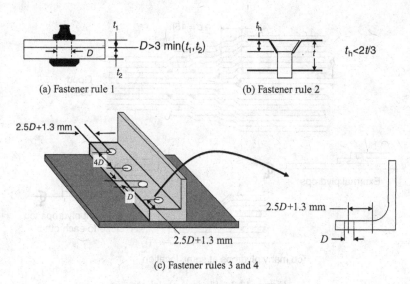

(a) Fastener rule 1   (b) Fastener rule 2

(c) Fastener rules 3 and 4

Figure 12.1 Fastener rules of thumb

14. *Fastener rule 5 (edge distance).* To minimize edge effects (so that the load distribution around the fastener approaches that of a fastener in an infinite plate) the edge distance between a fastener and the edge of a part should be no less than $2.5D + 1.3$ mm where D is the fastener diameter (see Figure 12.1c). This includes the distance of a fastener from the tangency point of the radius region of a web transitioning to a flange (see Figure 12.1c).

15. Plydrop guidelines. See Figure 12.2.
    (a) Avoid external plydrops. The tendency to delaminate at the edge of the terminating ply is high. Plydrops should be as close to the mid-plane of the laminate as possible.
    (b) For more than one plydrop, try to drop plies symmetrically with respect to the mid-plane of the laminate.
    (c) Avoid dropping more than 0.5 mm worth of plies at the same location to minimize the interlaminar stresses created (see typical results in Section 9.2.2).
    (d) The distance between successive plydrops should be at least 10 or 15 times the dropped height to avoid constructive interference (enhancement of stresses) between the stresses at the different plydrop sites (see Section 9.2.2).

16. *Anti-peel fasteners.* For highly post-buckled stiffened panels with co-cured stiffeners, using two fasteners at each stiffener end postpones or eliminates the skin–stiffener separation failure mode (see Section 9.2.2 and Figure 9.2.1).

12.3   Guidelines Related to Environmental Sensitivity and Manufacturing Constraints

17. *Minimum gage.* For lightly loaded structure, the thickness should be no lower than 0.5–0.6 mm to keep moisture from seeping into the structure. For lower thicknesses, additional coating protecting against moisture should be used.

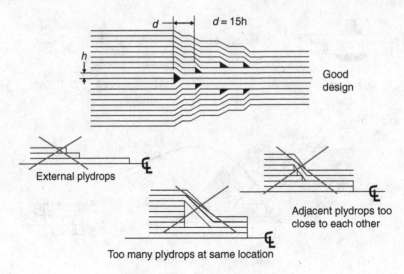

Figure 12.2 Plydrop guidelines

18. Minimum flange width
    (a) *Fastener rule 6.* If fastened, the minimum flange width is the sum of edge distances from guideline 14: $5D + 2.6$ mm from the flange edge to the tangency point of the web-to-flange transition.
    (b) If *co-cured or bonded* the minimum flange width for lightly loaded structure is 12.7 mm and for highly loaded structure is 19 mm. These values are the minimum required for the load shearing through the flange to reach at least 95% of its far field value.
19. *Minimum web height.* To avoid damage during handling and to make fabrication easier, the minimum web height should be 18 mm. This is particularly important for stiffeners with flanges at both ends of the web (I, J, C, Z) where access to the web is limited.
20. *Bridging avoidance.* Avoid 90° plies going around corners (see Figure 12.3), in particular when convex tools are used during layup. It is very hard to make the stiff 90° fibres

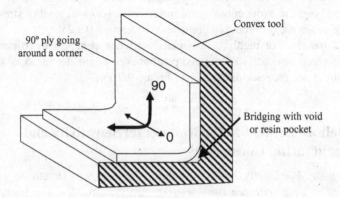

Figure 12.3 Bridging at a corner

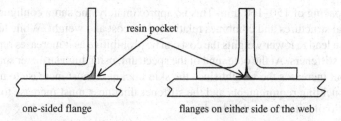

one-sided flange   flanges on either side of the web

Figure 12.4   Exposed resin pockets may lead to delaminations easily

conform to the shape of the tool and, usually, bridging occurs where a void and/or resin pocket is created.

12.4   Configuration and Layout-Related

21. *Preferred stiffener shapes.* Unless the structure is lightly loaded, stiffeners with a one-sided flange (L, C, Z) on the skin side should be avoided and stiffeners with flanges on either side of the web should be preferred (T, I, J, Hat). This protects the resin pocket present at the web flange corner (see Figure 12.4) from moisture and contamination and minimizes the possibility that matrix cracks may develop there and coalesce into delaminations under fatigue loading.

22. *Stiffener and frame spacing.* While the optimum spacing of frames and stiffeners will be dictated by the design loads and cost and weight considerations, a configuration that has been found to be robust and reasonably efficient is frame spacing of 500–510 mm and

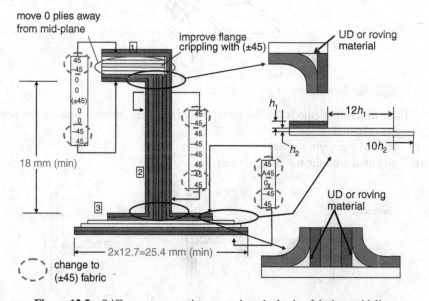

Figure 12.5   Stiffener cross-section created on the basis of design guidelines

stiffener spacing of 150–160 mm. This is, approximately, the same configuration used in many metal structures and combines relatively low cost and weight. While lower stiffener spacing can lead to lower weights the cost can be prohibitive as it increases rapidly with the number of stiffeners. At the other end of the spectrum, using high stiffener spacing reduces the cost, but increases the weight since the skin thickness must increase to meet buckling and post-buckling requirements and the stiffener thickness must increase to maintain the desired ratio of stiffener to skin loads.

At this point, with all the design guidelines in place, the J stiffener cross-section that has been used all along as an example last discussed in Section 9.2.2 (Figure 9.20) can be revisited and the preliminary configuration can be finalized. This is shown in Figure 12.5. This is preliminary in the sense that no specific load has been used to design it. The final dimensions and layup would depend on the applied loads. What is shown in Figure 12.5 is just a good starting point applicable to lightly loaded stiffeners. Note that what is shown in Figure 12.5 does not satisfy the 10% rule.

Exercises

12.1   Consider three composite parts intersecting at right angles as shown in the figure below. They are under tension and shear loads as shown.

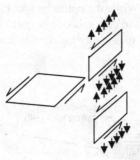

Three different methods for assembling them together are proposed, shown below:
(a)   bolted connections,
(b)   bonded connections, and
(c)   co-cured with the use of a 3-D preform.

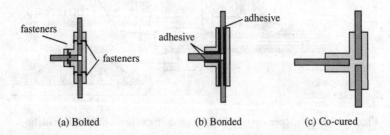

(a) Bolted   (b) Bonded   (c) Co-cured

Discuss the merits and disadvantages of each of the three approaches from a weight and cost perspective. Combine the material from this chapter with that from Chapter 2. Include in your discussion (but do not limit it to) assembly cost associated with fastener installation, weight impact of use of fasteners, bearing load requirements, bondline thickness control, inspection issues of adhesive, use of RTM with 3-D preforms and associated tooling cost, effect of process on final strength and stiffness, etc.

12.2 Referring to Figure 12.4, determine which of the design guidelines presented in this chapter are not satisfied and discuss the implications. (For example, the web layup does not satisfy the 10% rule.)

References

[1] Abdalla, M.M., Kassapoglou, C. and Gurdal, Z., Formulation of Composite Laminate Robustness Constraint in Lamination Parameters Space, *Proc. 50th AIAA/ASME/ASCE/AHS/ASC Structures, Structural Dynamics and Materials Conf.*, Palm Springs CA, 4–7 May 2009.

[2] Beckwith, S.W., Designing with Composites: Suggested 'Best Practices' Rules, *SAMPE Journal*, **45**, 36–37 (2009).

[3] Caprino, F. and Crivelli Visconti, I., A Note on Specially Orthotropic Laminates, *Journal of Composite Materials*, **16**, 395–399 (1982).

13

Application – Design of a Composite Panel

Perhaps the best way to see how the concepts and methods presented in previous chapters can be used is through a typical application. This provides insight into how the geometry and layup for different components can be selected avoiding failure in any of the expected failure modes while maintaining competitive weight and cost. A formal optimization for weight and/or cost is not done here. Instead, cost and weight trades among different designs will be performed to highlight the advantages and disadvantages of different concepts and to show how different decisions on design concept, geometry and layup affect the design outcome.

The application selected is a fuselage panel as shown in Figure 13.1.

Two separate loading conditions are considered:

(a) Uniform pressure of 12,500 Pa
(b) Applied in-plane loads $N_x = -350$ N/mm, $N_{xy} = 175$ N/mm

In addition, the out-of-plane deflection for the pressure load case must not exceed 2.0 mm. The panel dimensions are 1.5 m × 0.75 m. Three different design concepts, monolithic, stiffened and sandwich panel will be considered. The panel is attached to adjacent structure with fasteners. In all cases, for simplicity, the panel is assumed to be flat.

Two composite materials, one unidirectional tape and one plain-weave fabric, and three core materials are available. The material properties are given in Table 13.1. An adhesive is also available for bonding sandwich facesheets to the core with mass per unit area equal to 0.147 kg/m^2.

13.1 Monolithic Laminate

The simplest design is that of a monolithic laminate. It is expected to be the heaviest because of the large number of plies that will be needed to meet the 2.0 mm maximum deflection requirement. At the same time, even though the number of plies will be high, the cost is

Design and Analysis of Composite Structures: With Applications to Aerospace Structures, Second Edition. Christos Kassapoglou.
© 2013 John Wiley & Sons, Ltd. Published 2013 by John Wiley & Sons, Ltd.

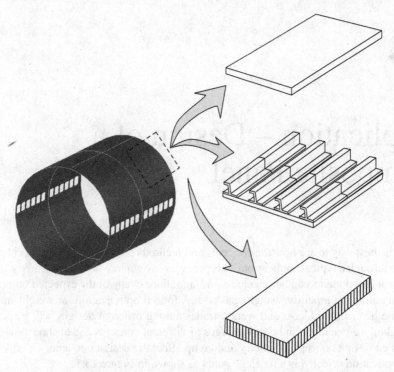

Figure 13.1 Portion of fuselage to be designed

Table 13.1 Properties of candidate materials

Property	Unidirectional tape	Plain-weave fabric
E_x (GPa)	137.9	70.3
E_y (GPa)	11.7	70.3
ν_{xy}	0.29	0.05
G_{xy} (GPa)	4.82	4.89
t_{ply} (mm)	0.1524	0.1905
Tension strength along fibres (tape) or warp dir. (fabric) X^t (MPa)	2068	1378
Compression strength transverse to fibres (tape) or fill dir. (fabric) X^c (MPa)	1723	1206
Tension strength transverse to fibres (tape) or fill dir. (fabric) Y^t (MPa)	96.5	1378
Compression strength transverse to fibres (tape) or fill dir. (fabric) Y^c (MPa)	338	1206
Shear strength S (MPa)	124	121
Bearing strength $\sigma_{br}{}^u$ (MPa)	772	827
Density (kg/m^3)	1609	1609

(Continued)

Table 13.1   (*Continued*)

Property	Core A	Core B	Core C
E_c (MPa)	131	169	214
G_{xz} (MPa)	41.4	62.0	82.7
G_{yz} (MPa)	20.7	31.0	42.0
Flatwise tension strength (kPa)	965.2	1310	1723
Flatwise compression strength (kPa)	1379	2068	2895
Transverse shear strength xz (kPa)	2413	2964	3447
Transverse shear strength yz (kPa)	1723	3447	4343
Density (kg/m³)	48.2	64.2	80.3

expected to be low because of the simplicity of the design. Laying down flat plies by hand or via robotic equipment can take very little time.

The situation is shown in Figure 13.2. The pressure case is used first to design the panel and the combined compression and shear will be checked after the design for the pressure case has been finalized. For the pressure case, the approach given in Section 5.3.2 for a point load is used here with the only change of replacing the point load by a distributed load.

The governing equation, see Equation (5.16a) has the form:

$$D_{11}\frac{\partial^4 w}{\partial x^4} + 2(D_{12} + 2D_{66})\frac{\partial^4 w}{\partial x^2 \partial y^2} + D_{22}\frac{\partial^4 w}{\partial y^4} = p_o \qquad (13.1)$$

with p_o the applied pressure load of 13,800 Pa.

In deriving Equation (13.1), it was assumed that $D_{16} = D_{26} = 0$.

A solution to Equation (13.1) can be found in terms of Fourier series. The out-of-plane deflection w can be written as:

$$w = \sum\sum A_{mn} \sin\frac{m\pi x}{a} \sin\frac{n\pi y}{b} \qquad (13.2)$$

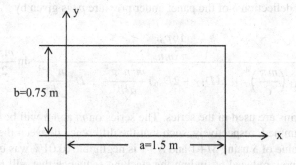

Figure 13.2   Panel dimensions and coordinate system

where A_{mn} are unknown coefficients and a and b are the length and width of the panel. It should be noted that Equation (13.2) satisfies the simply supported boundary condition of zero w at all edges of the panel.

In order to use Equation (13.2), the applied load p_o must be expressed as a Fourier series:

$$p_o = \sum\sum B_{mn} \sin\frac{m\pi x}{a} \sin\frac{n\pi y}{b} \tag{13.3}$$

The unknown constants B_{mn} are obtained as Fourier series coefficients. Both sides of Equation (13.3) are multiplied by $\sin q\pi x/a \sin r\pi y/b$ and integrated over the panel domain (q and r are integers):

$$\int_0^a\int_0^a p_o \sin\frac{q\pi x}{a} \sin\frac{r\pi y}{b} dy dx = \int_0^a\int_0^a B_{mn} \sin\frac{m\pi x}{a} \sin\frac{n\pi y}{b} \sin\frac{q\pi x}{a} \sin\frac{r\pi y}{b} dy dx \tag{13.4}$$

The integrals on the right-hand side are non-zero only when $m = q$ and $n = r$. Then, performing the integrations and solving for B_{mn}:

$$B_{mn} = \frac{16 p_o}{\pi^2 mn} \tag{13.5}$$

with $m,n = 1,3,5,\ldots$ Using Equations (13.2), (13.3) and (13.5) to substitute in Equation (13.1):

$$\sum\sum \left[D_{11}\left(\frac{m\pi}{a}\right)^4 + 2\left(D_{12} + 2D_{66}\right)\frac{m^2 n^2 \pi^4}{a^2 b^2} + D_{22}\frac{n^4 \pi^4}{b^4} \right] A_{mn} \sin\frac{m\pi x}{a} \sin\frac{n\pi y}{b}$$

$$= \sum\sum \frac{16 p_o}{\pi^2 mn} \sin\frac{m\pi x}{a} \sin\frac{nxy}{b} \tag{13.6}$$

Matching coefficients of $\sin m\pi x/a \sin n\pi y/b$ leads to the expression:

$$A_{mn} = \frac{\dfrac{16 p_o}{\pi^2 mn}}{D_{11}\left(\dfrac{m\pi}{a}\right)^4 + 2\left(D_{12} + 2D_{66}\right)\dfrac{m^2 n^2 \pi^4}{a^2 b^2} + D_{22}\dfrac{n^4 \pi^4}{b^4}} \tag{13.7}$$

Then, the centre deflection δ of the panel under pressure p_o is given by

$$\delta = \sum\sum \frac{\dfrac{16 p_o}{\pi^2 mn}}{D_{11}\left(\dfrac{m\pi}{a}\right)^4 + 2\left(D_{12} + 2D_{66}\right)\dfrac{m^2 n^2 \pi^4}{a^2 b^2} + D_{22}\dfrac{n^4 \pi^4}{b^4}} \sin\frac{m\pi}{2} \sin\frac{n\pi}{2} \tag{13.8}$$

where only odd terms are used in the series. The series on m and n will be truncated after M and N terms are summed, respectively, such that the difference between the value of δ using M and N and the value of δ using $M+1$ and $N+1$ is negligible (0.01% was used here).

The problem is now reduced to finding the stacking sequence that will result in the value of δ being less than or equal to 2 mm. One way to do this without resorting to optimization software is to maximize the denominator of the right-hand side of Equation (13.8). For a given stacking sequence, if $m = n$, the coefficient of D_{22} is greater than the coefficients of all other D_{ij} terms in the denominator because $b < a$. The next highest coefficient, for sufficiently large values of $m = n$, is that of D_{66}. Therefore, the selected stacking sequence must have as high as possible a value of D_{22} and, as a second order effect, a high value of D_{66}. This means that the laminate must have a lot of 90° plies away from the neutral axis (for maximum D_{22}) and a lot of 45/−45° plies away from the neutral axes (for maximum D_{66}).

The laminate is constrained to consist of only 0, +45, −45 and 90° plies. With reference to Chapter 12, the following guidelines/rules of thumb are also enforced:

- The laminate is symmetric and balanced (Section 12.1, Guidelines 1 and 2).
- +45 and −45 unidirectional tape plies are next to each other to minimize D_{16} and D_{26} (Section 12.1, Guideline 3).
- The outermost plies are (±45) plain-weave fabric plies for improved damage resistance (Section 12.2, Guideline 10).
- At least 10% of the fibres are in the 0, +45, −45 and 90 directions to protect against secondary load cases (Section 12.1, Guideline 4).
- No more than four unidirectional tape plies of the same orientation are stacked next to each other to minimize the effect of microcracks (Section 12.1, Guideline 5).
- At least 40% of the plies are in the +45/−45 directions near the edges of the panel for improved bearing strength (Section 12.2, Guideline 12).

An attempt to combine all the rules and requirements just mentioned is the following laminate (0 direction is along the x axis in Figure 13.2):

$$[(\pm 45)_2/90_4/(0/90)/90_4/0/45/-45/90_4/(0/90)/90_4/0/90_4/45/-45/90_4/0/90_4/45/$$
$$-45/90_4/0]s$$

This is a 92-ply laminate with thickness $h = 14.3$ mm. It has a maximum deflection under pressure, as predicted by Equation (13.8), of 1.991 mm. Note that plies in parentheses, (±45) and (0/90), are single plies of the fabric material.

It should be checked that there is no failure anywhere in the panel under either of the loading conditions. For the pressure load case, with w determined from Equations (13.3) and (13.7), the bending moments can be obtained using Equation (3.46) which are expanded below:

$$M_x = -D_{11}\frac{\partial^2 w}{\partial x^2} - D_{12}\frac{\partial^2 w}{\partial y^2} - 2D_{16}\frac{\partial^2 w}{\partial x \partial y}$$

$$M_y = -D_{12}\frac{\partial^2 w}{\partial x^2} - D_{22}\frac{\partial^2 w}{\partial y^2} - 2D_{26}\frac{\partial^2 w}{\partial x \partial y} \qquad (13.9)$$

$$M_{xy} = -D_{16}\frac{\partial^2 w}{\partial x^2} - D_{26}\frac{\partial^2 w}{\partial y^2} - 2D_{66}\frac{\partial^2 w}{\partial x \partial y}$$

Using Equation (13.3) to substitute in Equation (13.9):

$$M_x = \sum\sum\left(D_{11}\left(\frac{m\pi}{a}\right)^2 + D_{12}\left(\frac{n\pi}{b}\right)^2\right)A_{mn}\sin\frac{m\pi x}{a}\sin\frac{n\pi y}{b}$$

$$-\sum\sum 2D_{16}\frac{mn\pi^2}{ab}A_{mn}\cos\frac{m\pi x}{a}\cos\frac{n\pi y}{b}$$

$$M_y = \sum\sum\left(D_{12}\left(\frac{m\pi}{a}\right)^2 + D_{22}\left(\frac{n\pi}{b}\right)^2\right)A_{mn}\sin\frac{m\pi x}{a}\sin\frac{n\pi y}{b}$$

$$-\sum\sum 2D_{26}\frac{mn\pi^2}{ab}A_{mn}\cos\frac{m\pi x}{a}\cos\frac{n\pi y}{b} \qquad (13.10)$$

$$M_{xy} = \sum\sum\left(D_{16}\left(\frac{m\pi}{a}\right)^2 + D_{26}\left(\frac{n\pi}{b}\right)^2\right)A_{mn}\sin\frac{m\pi x}{a}\sin\frac{n\pi y}{b}$$

$$-\sum\sum 2D_{66}\frac{mn\pi^2}{ab}A_{mn}\cos\frac{m\pi x}{a}\cos\frac{n\pi y}{b}$$

These equations should be evaluated at every point in the panel and then a first-ply-failure criterion (see Chapter 4) applied to determine if the panel fails. However, the critical locations can be determined here by inspection of Equations (13.10). The D_{ij} terms for this laminate are:

$$D_{11} = 7545\text{ Nm}$$

$$D_{12} = 2534\text{ Mm}$$

$$D_{16} = 25.3\text{ Nm}$$

$$D_{22} = 25432\text{ Nm}$$

$$D_{26} = 25.3\text{ Nm}$$

$$D_{66} = 2876\text{ Nm}$$

It can be seen that D_{16} and D_{26} are more than two orders of magnitude smaller than the other terms in the D matrix. This is a result of the fact that few 45 and -45 plies were used and they were placed next to each other in the stacking sequence. Therefore, the terms involving D_{16} and D_{26} in Equation (13.10) can be neglected to obtain:

$$M_x = \sum\sum\left(D_{11}\left(\frac{m\pi}{a}\right)^2 + D_{12}\left(\frac{n\pi}{b}\right)^2\right)A_{mn}\sin\frac{m\pi x}{a}\sin\frac{n\pi y}{b}$$

$$M_y = \sum\sum\left(D_{12}\left(\frac{m\pi}{a}\right)^2 + D_{22}\left(\frac{n\pi}{b}\right)^2\right)A_{mn}\sin\frac{m\pi x}{a}\sin\frac{n\pi y}{b} \qquad (13.10a)$$

$$M_{xy} = -\sum\sum 2D_{66}\frac{mn\pi^2}{ab}A_{mn}\cos\frac{m\pi x}{a}\cos\frac{n\pi y}{b}$$

It can now be seen that the magnitudes of M_x and M_y are maximized when M_{xy} is zero, at $x = a/2$ and $y = b/2$ (panel centre) and the magnitude of M_{xy} is maximized at $x = 0$, $y = 0$

(or $x = a$, $y = b$), at the panel edges where M_x and M_y are zero. Therefore, as a quick check, it suffices to check these locations separately. For a more accurate evaluation one should apply the failure criterion at every point in the panel to determine whether a combination of the bending moments, away of the respective maximum locations, is more critical.

Here, the Tsai-Wu failure criterion was applied. Note that the strength values in Table 13.1 were knocked down by multiplying each by $0.8 \times 0.8 \times 0.65$ to account for material scatter, environmental effects and impact damage (see Section 5.1.6). Summing up the series involved up to $m = n = 35$ shows less than 0.1% change so there is no need to go to higher number of terms. The resulting loads are:

$$M_x = 111.9 \text{ Nm/m}$$
$$M_y = 872.5 \text{ Nm/m}$$

at the centre of the panel and:

$$M_x = M_y = 0$$
$$M_{xy} = -163.6 \text{ Nm/m}$$

at the corners.

There is no first ply failure for either load case. For the first load case, the critical ply is number 81 with a reserve factor $R = 18.7$, that is the applied load must be increased proportionately by a factor of 18.7 for failure to occur. For the second load case, the critical ply is number 13 with, again, a factor $R = 18.7$. These translate to a margin of safety (see Section 5.1.7) of

$$MS_{TW} = \frac{\text{Allowable}}{\text{Applied}} - 1 = 18.7 - 1 = +17.7$$

In addition to the bending moments, there are transverse shear forces that develop at the panel edges. These can be determined by using Equations (5.4d) and (5.4e). Combining with Equation (13.10a):

$$Q_x = \sum\sum \left(D_{11} \left(\frac{m\pi}{a}\right)^3 + D_{12} \left(\frac{n\pi}{b}\right)^2 \left(\frac{m\pi}{a}\right) + 2D_{66} \frac{mn^2\pi^3}{ab^2} \right) A_{mn} \cos\frac{m\pi x}{a} \sin\frac{n\pi y}{b}$$

$$Q_y = \sum\sum \left(D_{12} \left(\frac{m\pi}{a}\right)^2 \left(\frac{n\pi}{b}\right) + D_{22} \left(\frac{n\pi}{b}\right)^3 + 2D_{66} \frac{m^2 n\pi^3}{a^2 b} \right) A_{mn} \sin\frac{m\pi x}{a} \cos\frac{n\pi y}{b}$$

$$\text{(13.11)}$$

Then Q_x is maximum at the intersection of the line $y = b/2$ (centreline of the panel) with the two edges $x = 0$ and $x = a$. The maximum value of Q_x there is $Q_x = 2764$ N/m. Similarly, the maximum value of Q_y occurs at the intersection of the other centreline of the panel, $x = a/2$ with the two edges $y = 0$ and $y = b$. Its maximum value is $Q_y = 7450$ N/m.

These maximum shear forces cause a transverse shear stress which, assuming a quadratic distribution through the thickness, has a maximum value equal to 1.5 times the average value or:

$$\tau_{xz\,max} = \frac{Q_{x\,max}}{h} = \frac{(1.5)2764}{0.0143} = 0.290 \text{ MPa}$$

$$\tau_{yz\,max} = \frac{Q_{y\,max}}{h} = \frac{(1.5)7450}{0.0143} = 0.781 \text{ MPa}$$

(13.11a)

These shear stress values should be compared to the transverse shear allowable of the material. In this case, as an approximation, the in-plane shear strength for the material(s) is used. And, for conservatism, the lowest between tape and fabric, 121 MPa in Table 13.1 is used. Obviously, there is plenty of positive margin (no failure) because the transverse shear stresses just computed are very low.

For the compression and shear load case, the critical plies are ply numbers 14 and 30 (and their symmetric counterparts in the lower half of the laminate). The R value is 5.17 indicating that the applied loads must be increased by a factor of 5.17 before failure occurs. Thus, the corresponding safety margin is:

$$MS_{c+s} = 5.17 - 1 = +4.17$$

The panel should also be checked for buckling under the in-plane compression and shear loads. Because of the high thickness of the panel, 14.3 mm, transverse shear effects may be significant and the buckling equations presented in Sections 10.2.1 and 10.2.2 should be used. Instead of the core thickness and shear stiffness, the thickness of the composite should be used in the correction factor.

Under pure compression, neglecting transverse shear effects, the buckling load for the panel is given by Equation (6.7):

$N_{xEcrit} = 786.9$ N/mm occurring for $m = 3$ (3 half-waves along the length of the panel

In order to use Equation (10.3), t_c is set equal to the laminate thickness, 14.3 mm and G_c is set equal to the transverse shear stiffness of the laminate. In general, this should be calculated by computing the compliances S_{ij} of the laminate and inverting the compliance tensor. An approximation can be obtained by setting G_c equal to the lowest transverse shear stiffness of the material. This should be measured experimentally but can be approximated as 70% of the in-plane shear stiffness G_{xy}. So $G_c = 0.7(4.82) = 3.37$ GPa.

Then, using Equation (10.2), the critical buckling load under compression is:

$$N_{xcrit} = \frac{786.9}{1 + \dfrac{(5/6)786.9(1000)}{0.0143(3.37 \times 10^9)}} = 776 \text{ N/mm}$$

Interestingly, the transverse shear correction amounts to only 1.4% reduction in the compression buckling load. It should be noted that, for this case, where the 'core' can carry bending loads, the correction factor $k = 5/6$ was used instead of 1 that is used for typical core materials.

Under pure shear, neglecting transverse shear effects, the buckling load can be estimated by using Equation (6.27) with the correction factor mentioned in Section 6.5:

$$N_{xyEcrit} = (0.7)\frac{9\pi^4 b}{32a^3}\left(D_{11} + 2(D_{12} + 2D_{66})\frac{a^2}{b^2} + D_{22}\frac{a^4}{b^4}\right) = 2049 \text{ N/mm}$$

Note that for a more accurate result, Equation (6.19) should be used. This buckling load is corrected for transverse shear effects using Equation (10.4) with $G_{xz} = G_{yz} = 3.37$ GPa. Then:

$$N_{xycrit} = \frac{2049}{1 + \dfrac{2049(1000)}{0.0143(3.37 \times 10^9)}} = 1965 \text{ N/mm}$$

Now, the interaction curve Equation (6.28) is used:

$$\frac{N_x}{N_{xcrit}} + \left(\frac{N_{xy}}{N_{xycrit}}\right)^2 = \frac{350}{776} + \left(\frac{175}{1965}\right)^2 = 0.459$$

which is less than 1 and, therefore, there is no buckling failure.

By multiplying the applied loads by a factor R in the above equation and setting it equal to 1:

$$\frac{350R}{776} + \left(\frac{175R}{1965}\right)^2 = 1$$

and solving for R, the factor by which the applied loads must be increased for buckling to occur, it is found that $R = 2.137$ and therefore,

$$MS_{buck} = 2.137 - 1 = +1.137$$

At this point, the only consideration left to complete the design of the monolithic panel is the attachment to adjacent structure. Fasteners are selected for the attachment. The fastener diameter and spacing must be selected such that (a) there is no pull-through failure under pressure loads (see Section 11.2.1.1 and Figure 11.7), (b) there is no bearing failure under the in-plane loads applied (see Section 11.2.1.1) and (c) there is no inter-rivet buckling failure under the compressive loads between fasteners (see Section 8.7).

The pull-through analysis is very conservative without experimentally measured allowables specific to the fastener type and material used. So only bearing and inter-rivet buckling are discussed here briefly. Both are a result of the second load case with combined shear and compression.

The most highly loaded fasteners would be the corner fasteners. Considering the fastener closest to the origin in Figure 13.2, the situation is as shown in Figure 13.3.

The in-plane loads on the corner fastener are a compression load F_c due to the applied N_x and two shear loads F_{sx} and F_{sy} from the shear load N_{xy} on both sides of the panel. The load

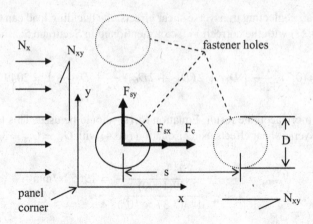

Figure 13.3 Loads on most highly loaded fastener

F_c can be obtained as the total compressive load on the panel edge divided by the number of fasteners along that edge:

$$F_c = \frac{N_x b}{\dfrac{b}{s}} = s N_x \tag{13.12}$$

where s is the fastener spacing and b/s the (approximate) number of fasteners alongside b. Similarly, the two loads due to N_{xy} are:

$$F_{sx} = \frac{N_{xy} a}{\dfrac{a}{s}} = s N_{xy}$$

$$F_{sy} = \frac{N_{xy} b}{\dfrac{b}{s}} = s N_{xy} \tag{13.13}$$

where the same fastener spacing s is used along both panel edges for convenience.

Then, the total load on the corner fastener is

$$P = \sqrt{(F_x + F_{sx})^2 + F_{sy}^2} = \sqrt{(s N_x + s N_{xy})^2 + s^2 N_{xy}^2} = s \sqrt{(N_x + N_{xy})^2 + N_{xy}^2} \tag{13.14}$$

This load must not cause bearing failure. For the panel to just fail in bearing, using Equation (11.5):

$$\frac{s \sqrt{(N_x + N_{xy})^2 + N_{xy}^2}}{Dt} = F^{bru} \tag{13.15}$$

where F^{bru} is the bearing strength of the laminate selected. This would be determined experimentally. A good approximation (valid for laminates with stiffness within $\pm 30\%$ that of a quasi-isotropic layup) is the value of 772 MPa given for the tape material in Table 13.1.

For inter-rivet buckling, Equation (8.56) gives:

$$\frac{F_c + F_{sx}}{wt} = \frac{c\pi^2 (D_{11})_e}{ts^2} \tag{13.16}$$

where w is the width of the 'flange' at the edge of the panel that carries the fastener loads.

The local thickness, t is expected to be different (lower) than the panel thickness h. Note that D_{11} refers to the laminate at the edge of the panel with the one direction coinciding with the axial load N_x. Following the Guideline 14 in Section 12.1 (see also Figure 12.1, w should equal at least $5D+2(1.3)$ mm. Here, to simplify the calculations $w = 5D$ is used (the extra 2.6 mm can be added afterwards). Then, using Equations (13.12), (13.13) and setting $c = 3$ for protruding head fasteners, Equation (13.16) becomes:

$$\frac{s (N_x + N_{xy})}{5D} = \frac{3 (D_{11})_e}{s^2} \tag{13.17}$$

One could use the same laminate for the edge as for the entire panel but this would be over-conservative. Instead, a laminate with about half the thickness of the basic laminate is considered. This is constructed by removing most of the 90° plies from the basic laminate, which have low strength in the x direction anyway. So the laminate considered is:

$$[(\pm 45)_2/(0/90)/0_4/45/-45/90/(0/90)/90/0/90/45/-45/90/0_2/90/45/-45/90/0]s$$

This laminate has 48 plies with thickness $t = 7.62$ mm and $D_{11} = 2767$ Nm. It may not be the optimum laminate from a bearing and inter-rivet buckling perspective since neither the compressions strength, which relates to the bearing strength, nor D_{11} are maximized. However, almost all its plies are plies in the layup used in the rest of the panel. This makes 'blending' from the edge of the panel to the centre easier, reduces manufacturing cost and minimizes problems of interlaminar stresses at plydrops between the thicker centre of the panel and the thinner edge.

Equations (13.15) and (13.17) can be solved simultaneously to determine the fastener spacing and diameter for a given layup selection at the panel edge. This, however, gives unrealistic geometries with very large D and very small s. The reason is because one of the two failure modes, inter-rivet buckling, will not be critical for any realistic geometry that meets the bearing requirement. Selecting s to equal $4.5D$ (see Guideline 13 in Section 12.1 and Figure 12.1) and selecting $D = 5$ mm, the left-hand side of Equation (13.15) gives the applied bearing stress:

$$\sigma_{brg} = \frac{s\sqrt{(N_x + N_{xy})^2 + N_{xy}^2}}{Dt} = 326.8 \text{ MPa}$$

which corresponds to a margin of safety:

$$MS_{brg} = \frac{772}{326.8} - 1 = +1.362$$

Similarly, the left-hand side of Equation (13.17) gives an applied inter-rivet stress of

$$\sigma_{ir} = \frac{s\left(N_x + N_{xy}\right)}{5D} = 0.473 \text{ MPa}$$

The right-hand side of Equation (13.17) gives an allowable inter-rivet buckling stress of:

$$\sigma_{ir}^u = \frac{3\left(D_{11}\right)_e}{s^2} = 16.4 \text{ MPa}$$

Combining, the margin of safety for inter-rivet buckling is:

$$MS_{ir} = \frac{16.4}{0.473} - 1 = +33.6$$

It is important to note now that the edge of the panel has about half the thickness of the rest of the panel. This saves weight but requires an additional iteration to double-check failure modes that may be affected by the change in thickness. For example, for transverse shear failure, Equations (13.11a) are now modified to read:

$$\tau_{xz\,max} = \frac{Q_{x\,max}}{h} = \frac{(1.5)2764}{0.00762} = 0.544 \text{ MPa}$$

$$\tau_{yz\,max} = \frac{Q_{y\,max}}{h} = \frac{(1.5)7450}{0.00762} = 1.465 \text{ MPa}$$

and the corresponding safety margins are still very high.

The weight of the panel can now be calculated accounting for the difference in thickness at the edges and including the extra 2.6 mm edge distance there. Then:

$$W_{monlithic} = 1609\left[\begin{array}{l}(0.75 - 2(0.0276))\,(1.5 - 2(0.0276))\,0.0143 + \\ (2\,(0.75 \times 0.0276) + 2(1.5 - 2(0.0276))0.0276)\,0.00762\end{array}\right] = 24.58 \text{ kg}$$

13.2 Stiffened Panel Design

The design of a stiffened panel requires a good tradeoff of the load between the skin and stiffeners and the sequencing of failure modes. This involves some iterations where a starting design is selected on the basis of some of the requirements and it is then refined until a final design is reached where all requirements are satisfied. The stiffened panel is a good example of a structure where one can divert load from one of the members (e.g. the skin) to another (e.g. the stiffeners) to change the critical failure mode, remove weight from one member and add less weight to the other so that, ultimately, a lower weight design is obtained.

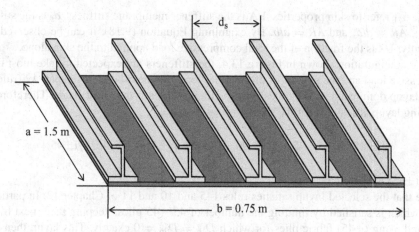

Figure 13.4 Stiffened panel with J stiffeners

For the stiffener design, a J-stiffener is selected. This is done for two reasons: First because, under certain assumptions and loading situations, the J-stiffener can be shown to be the most efficient stiffener design among the stiffener types examined in Chapter 8 [1]. Second, because it involves flanges on both ends of the web and brings out some of the aspects that simpler cross-sections ('T' and 'L') do not. The situation is shown in Figure 13.4. Note that the stiffeners are aligned with the long dimension of the panel (1.5 m) to carry the applied compression load efficiently.

To start the design, the second load case with combined compression and shear is used. As mentioned in Section 9.2, a design where the skin between stiffeners buckles first and global panel buckling follows at a higher load is preferred because it is easier to redistribute load and avoid total collapse of the structure when a failure is localized. In order to ensure this failure sequence, the bending stiffness of the stiffeners must be selected so that they provide simple support at the edges of the skin between them. For the purposes of the design it will also be assumed that frames or other reinforcing structures are placed at the two ends of the panel, also providing simple support to the skin.

It is also assumed that there is no buckling prior to reaching the ultimate load. Thus, the minimum bending stiffness EI of the stiffeners is obtained by using Equation (9.30) repeated here for convenience:

$$\text{EI} = D_{11}d_s \left[\sqrt{\frac{D_{22}}{D_{11}}} \left(2\lambda\overline{AR}^2 - \sqrt{\frac{D_{22}}{D_{11}}} (AR)^4 \right) + \frac{2(D_{12} + 2D_{66})}{D_{11}} \left(\lambda\overline{AR}^2 - (AR)^2 \right) - 1 \right]$$

(9.30)

where (see Section 9.2.1.1, Equation 9.29)

$$\lambda = \frac{\left(A_{11} + \dfrac{EA}{d_s} \right)}{A_{11}} \Rightarrow EA = A_{11}(\lambda - 1)d_s$$

(13.18)

and D_{ij}, A_{11} refer to skin properties, EA is the stiffener membrane stiffness, d_s is the stiffener spacing, $\overline{AR} = a/d_s$ and $AR = a/b$. By examining Equation (9.18), it can be observed that, physically, $1/\lambda$ is the fraction of the total compressive load applied to the skin alone.

In the configuration shown in Figure 13.4, the stiffeners are expected to take most of the compressive load and the skin most of the shear load. As mentioned in Chapter 12 (rule 11), a skin layup dominated by 45° plies is very efficient in carrying shear loads. Therefore, the following layup is selected for the skin:

$$[(\pm45)/0/45/-45/(\pm45)/90/(\pm45)/90/(\pm45)/-45/45/0/(\pm45)]$$

where (±45) is a fabric ply.

Note that the selected layup satisfies rules 1–5 and 10 and 11 of Chapter 12. In particular, rule 3 which is satisfied by limiting the number of 45/−45 plies, keeping them next to each other and using (±45) fabric plies for which $D_{16} = D_{26} = 0$ exactly. This layup then has a thickness of 2.1717 mm and:

$$A_{11} = 112776 \text{ N/mm}$$

$$D_{11} = 55868 \text{ N mm}$$

$$D_{12} = 21782 \text{ N mm}$$

$$D_{16} = 1743.2 \text{ N mm}$$

$$D_{22} = 31013 \text{ N mm}$$

$$D_{26} = 1743.2 \text{ N mm}$$

$$D_{66} = 22969 \text{ N mm}$$

with D_{16} and D_{26} less than 9% of the next higher value in the D matrix so that their effect can be neglected.

To generate the starting design, a number of stiffeners n_s is selected. This defines the stiffener spacing as

$$d_s = \frac{b}{n_s - 1} \tag{13.19}$$

where the 'minus one' in the denominator accounts for the fact that two of the stiffeners are right on the panel edges.

Then, different values of λ are selected and the corresponding EI and EA for each stiffener are determined from Equations (9.30) and (13.18). The resulting dependence of EI on λ and d_s is shown in Figure 13.5. Two trends are observed analogous to the results of Figure 9.8 in Section 9.2.1.3: First, as the stiffener spacing increases, the size of the skin between stiffeners increases and its buckling load decreases. Then, the required EI for the stiffeners decreases since the target bay buckling load of the skin goes down.

Second, as λ decreases, for a given stiffener spacing d_s and a given skin layup, which fixes A_{11}, the stiffener EA decreases. This means that the fraction of the load that is carried by the stiffeners decreases and more load is carried by the skin decreasing the skin bay buckling load

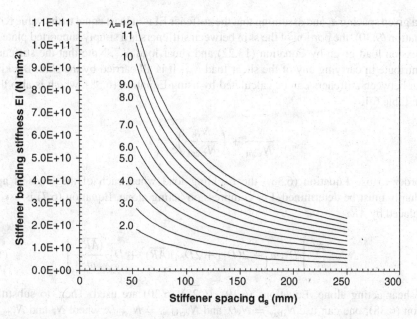

Figure 13.5 Minimum required stiffener bending stiffness as a function of stiffener spacing and the value of λ

for a given total applied compressive load. The effect is then the same as with the increase in stiffeners spacing: The bending stiffness of the stiffeners must decrease.

To proceed further, the compression load split between skin and stiffeners is needed. This can be determined by using Equations (8.12) and (9.18) renumbered here:

$$F_{\text{skin}} = \frac{A_{11}}{A_{11} + \dfrac{EA}{d_s}} F_{\text{TOT}} = \frac{1}{\lambda} F_{\text{TOT}} \tag{13.20}$$

$$F_{\text{stiffeners}} = \frac{(\lambda - 1)}{\lambda} F_{\text{TOT}} \tag{13.21}$$

where

$$F_{\text{TOT}} = bN_x$$

Combining Equations (13.20) and (13.21), the compressive load per unit width $N_{x\text{skin}}$ applied on the skin is found to be:

$$N_{x\text{skin}} = \frac{F_{\text{skin}}}{b} = \frac{A_{11}}{A_{11} + \dfrac{EA}{d_s}} \frac{bN_x}{b} = \frac{1}{\lambda} N_x \tag{13.22}$$

For a given spacing d_s and assuming that the stiffener EI is at least equal to the value required by Equation (9.30), the portion of the skin between stiffeners is a simply supported plate under compression load given by Equation (13.22) and shear load N_{xy}. Note that the stiffeners do not contribute in carrying any of the shear load N_{xy}. It is all carried by the skin. Buckling of the skin between stiffeners can be calculated by using Equation (6.38), which is also the last case of Table 6.1.

$$\frac{N_x}{N_{x\text{crit}}} + \left(\frac{N_{xy}}{N_{xy\text{crit}}}\right)^2 = 1 \qquad (6.38)$$

In order to use Equation (6.38), the buckling load when each of the loads is applied individually must be determined. For compression acting alone, Equation (6.7) is used with AR replaced by $\overline{AR}$:

$$N_o = \frac{\pi^2}{a^2}\left[D_{11}m^2 + 2(D_{12} + 2D_{66})(\overline{AR})^2 + D_{22}\frac{(\overline{AR})^4}{m^2}\right] \qquad (13.23)$$

For shear acting alone, Equations (6.19), (6.28)–(6.30) are used. Then, to substitute in Equation (6.38), one can use $N_{x\text{bay}} = N_x/\lambda$ and $N_{x\text{critbay}} = N_{x\text{crit}}/\lambda$ where N_x and $N_{x\text{crit}}$ refer to the entire panel and the subscript 'bay' refers to the portion of the skin between adjacent stiffeners. Thus:

$$\frac{N_{x\text{bay}}}{N_{x\text{critbay}}} = \frac{N_x}{N_{x\text{crit}}} \qquad (13.24)$$

Then, the left-hand side of Equation (6.38) can be evaluated for different values of the parameter λ and stiffener spacing d_s. Note that if the left-hand side is less than 1, the applied loads are less than the buckling loads and there is no buckling. This is shown in Figure 13.6 where a horizontal line is used to show buckling failure. To avoid bay buckling of the skin, the selected design must be below the horizontal line of Figure 13.6.

At this point, some considerations of the basic objectives weight and cost will help narrow the design space even further. To minimize the cost, the number of stiffeners must be minimized. This means that the largest possible stiffener spacing must be selected. It should be noted that this trend of lower cost for ever-increasing stiffener spacing may not hold for extreme values of d_s. In the limit, as $d_s \rightarrow b$ only two stiffeners are used, both at the panel edges. Since the skin layup is fixed at the current selected layup, the two stiffeners must be made very stiff to keep the skin from buckling early. This requires large thicknesses which, due to the increased number of plies, increase the cost significantly beyond the cost corresponding to smaller stiffener spacings (e.g. for the case of three instead of two stiffeners).

Keeping this constraint in mind, d_s values greater than 120 mm would give acceptable low cost values. At the same time, to minimize the weight, for a selected skin layup, the total cross-sectional area of the stiffeners must be minimized. This means, that EA for the stiffeners must be minimized. This is only approximate as E for the stiffeners is, in general, not constant but (may be) different for each member of the J cross-section. However, low EA values result in λ values that are close to 1 as is seen from Equation (13.18). This means that the compressive load in the skin is almost the same as that in the stiffeners. This is not efficient as it was already

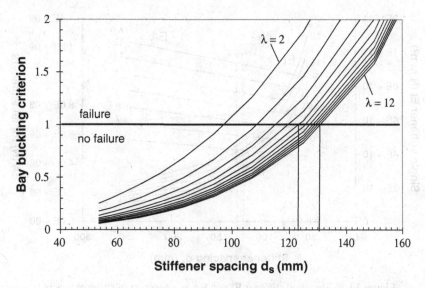

Figure 13.6 Bay buckling criterion as a function of stiffener spacing for various λ values

mentioned that most of the compressive load should be in the stiffeners. The skin is used primarily to carry the shear load. Therefore, relatively high λ values are preferred. Arbitrarily, the range $6 \leq \lambda \leq 12$ is selected. This is shown in Figure 13.6 as the region between the two vertical lines corresponding to $d_s > 120$ mm.

At this point, Equation (13.18) can be used to obtain EA for each stiffener and multiplying it by the number of stiffeners n_s determined by Equation (13.19) the total EA for all stiffeners can be obtained. This can then be plotted as a function of stiffener spacing along with the EI value (for each individual stiffener) obtained from Equation (9.30) in Figure 13.7. Again, the region of interest is delineated by the two vertical lines.

Figure 13.7 can now be used to define the starting values of EA and EI for the stiffeners. For seven stiffeners, $d_s = 125$ mm. To minimize the weight, the smallest EA in Figure 13.7 is selected which is the one corresponding to λ = 6. Reading off from Figure 13.7 or performing the calculations using Equations (9.30), (13.18) and (13.19), the following values are obtained for a single stiffener:

$$EA = 70485 \text{ kN}$$

$$EI = 2.348 \times 10^{10} \text{ N mm}^2$$

By trial and error, or other methods such as optimization using genetic algorithms, the starting geometry and layup shown in Figure 13.8 are selected.

This stiffener has EA = 70497 kN almost exactly matching the panel breaker requirement and EI = 7.96×10^{10} N mm^2 comfortably exceeding the minimum requirement. Then, using Equation (13.21) and the number of stiffeners, the force on each stiffener is determined to be 31,250 N. This is used to perform a crippling analysis with the aid of Equations (8.42) and (8.47). Note that Equations (8.43) and (8.48) are not used here because the material scatter is

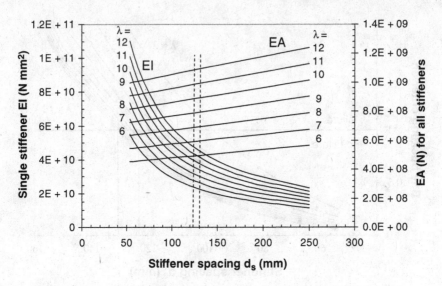

Figure 13.7   Required stiffener EI and EA as a function of stiffener spacing

already included in the knocked down properties used in the analysis. The crippling analysis is summarized in Table 13.2 below. All margins are positive and there is no crippling failure.

Each stiffener is checked for column buckling using Equation (8.21).:

$$F_{\text{crit}} = \frac{\pi^2 EI}{L^2} = \frac{\pi^2 79600}{1.5^2} = 349.2 \text{ kN}$$

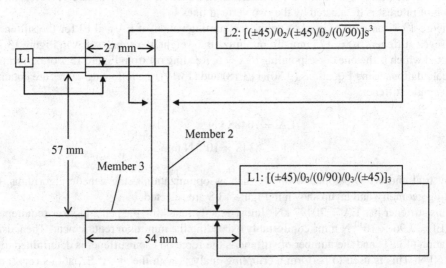

Figure 13.8   Stiffener geometry and layup

Table 13.2   Crippling analysis of stiffener members

Member	EA (N)	Force (N)	Stress (N/mm²)	b/t	OEF/NEF	$\sigma_{crip}/\sigma_{cu}$	σ_{crip} (N/mm²)	MS
1	12,147,501	5385	44.74	6.06	OEF	0.591	279.4	5.245
2	34,054,628	15,096	93.23	8.80	NEF	1.000	398.3	3.272
3	12,147,501	5385	44.74	6.06	OEF	0.591	279.4	5.245

The applied force per stiffener is given by summing the forces on all members (note that the force for member 3 in Table 13.2 must be multiplied by 2 to account for both horizontal flanges): 5385+15095+2(5385) = 31,250 N. Then, the column buckling margin is:

$$MS_{cbuck} = \frac{349,200}{31,250} - 1 = +10.17$$

If the stiffeners are not co-cured with the skin and fasteners are used, an inter-rivet buckling check is necessary. By equating the applied stress on the bottom horizontal flange of the stiffener to the inter-rivet buckling stress, see Equation (13.16):

$$44.74 \ (\text{N/mm}^2) = \frac{c\pi^2 (D_{11})_3}{t_3 s^2} \tag{13.25}$$

the maximum allowable fastener spacing s can be obtained. The subscript 3 in Equation (13.25) refers to member 3 of the stiffener cross-section. Solving Equation (13.25) with $c = 3$ (protruding head fasteners) and $(D_{11})_3 = 757525$ N mm and $t_3 = 4.458$ mm, gives

$$s = \sqrt{\frac{3\pi^2 757525}{44.74(4.458)}} = 335 \ \text{mm}$$

This, of course, is a very large spacing and any spacing smaller than this value will be acceptable. This implies that inter-rivet buckling is not a concern in this case.

The diameter of the fastener connecting the stiffener flange to the skin is selected so that there is no bearing failure. For a given value D of the fastener diameter, the fastener spacing s is selected by applying Equation (13.15):

$$\frac{F_{stiffener}}{\frac{a}{s} D t_3} = F^{bru} \tag{13.26}$$

with a/s the number of fasteners along a stiffener. Solving for s and substituting values with $D = 5$ mm:

$$s = \frac{F^{bru} a D t_3}{F_{stiffener}} = \frac{772(1500)5(4.458)}{31,250} = 825 \ \text{mm}$$

Again, the spacing is very large. Looking at the two values of s, 335 and 825 mm, one would have to pick the smallest of the two, 335 mm. However, this spacing, while compatible with bearing and inter-rivet buckling failure would definitely not be sufficient for joining the stiffeners to the skin. The simple support boundary condition required for skin buckling (between the stiffeners) would be violated. Also, under pressure loading, the skin would separate from the stiffeners in between fastener attachments. Therefore, the standard $4.5D$ spacing from Guideline 13, Section 12.1, is used here, giving a value of $s = 22.5$ mm.

The stiffeners and skin must also be checked for first-ply failure under the applied loads. For the compression and shear load case, the results from Table 13.2 can be used directly to calculate the first-ply failure for each member by dividing column 8 by column 7. Then, the margins can be obtained by dividing the first-ply-failure strength by the applied stress in column 4. The resulting margins are:

$$MS_{1,3} = \frac{427.7}{44.74} - 1 = +8.55$$

$$MS_2 = \frac{398.0}{93.23} - 1 = +3.269$$

For the strength check of the skin, the compression load on the skin is needed. The final selection of stiffener geometry and layups has not changed the value of λ because the final EA is identical to the target EA (70497 kN versus 70485 kN, respectively). Therefore, $\lambda = 6$. Then, using Equation (13.22):

$$N_{x\,skin} = \frac{350}{6} = 58.3 \text{ N/mm}$$

This combined with the full shear load $N_{xy} = 175$ N/mm are the applied loads for the skin strength analysis. Using the Tsai-Wu first-ply-failure criterion (see Chapter 4), the critical plies are numbers 4 (and 8) with $R = 1.51$. Thus, the following margin of safety is obtained:

$$MS_{c+s} = 1.510 - 1 = +0.510$$

The last check for this load case is buckling of the skin between the stiffeners and buckling of the panel as a whole. Both failure modes were already accounted for during the determination of the stiffener spacing and geometry when Equation (9.30) was used. However, the final geometry for the stiffeners matched EA exactly but had EI greater than the requirement, 7.96×10^{10} N mm^2 versus 2.348×10^{10} N mm^2 which increases the margin of safety for the overall panel buckling (the MS for the bay buckling stays equal to 0 since EA is matched exactly). For buckling of the overall panel, Equation (13.23) is used with the adjustments for D_{11} from Equation (9.17):

$$(D_{11})_{panel} = 55868 + \frac{7.96 \times 10^{10}}{125} = 636.9 \times 10^6 \text{ N mm} \qquad (13.27)$$

Using now Equation (13.23), with the updated D_{11} value, it is found that, for the panel as a whole (skin and stiffeners included), $N_{x\text{crit}} = 2798$ N/mm and $N_{xy\text{crit}} = 216.4$ N/mm. Then, using Equation (6.28),

$$\frac{N_x}{N_{x\text{crit}}} + \left(\frac{N_{xy}}{N_{xy\text{crit}}}\right)^2 = \frac{58.3}{2798} + \left(\frac{175}{216.4}\right)^2 = 0.675$$

which is lower than 1 and, therefore, there is no failure. To make the left-hand side of the above equation exactly equal to 1, the applied loads must be scaled by a factor $R = 1.234$. Then, the corresponding margin of safety for this case is:

$$MS_{p\text{buck}} = 1.234 - 1 = +0.234$$

At this point the analysis for the compression and shear case is complete. The pressure load case must also be checked. The procedure follows Equations (13.1)–(13.11) with the D_{11} given by Equation (13.27) and the remaining terms of the D matrix being the same as for the skin of the skin-stiffened panelled as given above. Then, the maximum deflection from Equation (13.8) is found to be 1.29 mm, below the maximum allowable of 2 mm.

It is important to note that the calculation of maximum deflection assumed a panel where the stiffeners are smeared into the skin. This gives a conservative approximation for the deflection of the stiffeners but does not accurately account for the skin deflection between the stiffeners. Under pressure, the stiffeners move a certain amount and the skin, having a much lower bending stiffness, will move even further causing the so-called 'pressure pillowing' phenomenon. To determine more accurately the final skin deflection, one could model each bay as being infinitely long along the stiffeners with the skin undergoing cylindrical bending. This solution is analogous to the deflection calculation in Section 13.1, but has no dependence on the x coordinate, which is omitted here for simplicity. It is expected that the extra deflection of the skin between stiffeners will be less than the margin between the 1.29 mm just calculated and the 2 mm allowable deflection so that the current design still meets the deflection requirement.

The moments M_y and M_{xy} turn out to be very small. Only the M_x moment is appreciable. Using the first of Equations (13.9), it is found to be:

$$M_x = 3507 \text{ Nm/m}$$

This moment is carried primarily by the stiffener at the centre of the panel. There is a small fraction carried by the skin, according to the ratio of the bending stiffness of the stiffeners EI and that of the skin, bD_{11} with D_{11} here referring only to the skin. Conservatively, it can be assumed that it is all carried by the middle stiffener (this is equivalent to assuming that the skin fails under its own bending moment and all the load must be carried by the stiffener). For the geometry and layups of Figure 13.8, it can be shown that the neutral axis is at 27.66 mm from the bottom. Then, the maximum compressive strain in the stiffener is at the top flange with:

$$\varepsilon_{c\,\text{max}} = \frac{b_3 M_x \bar{y}}{EI} = \frac{(0.027)3507(0.02766)}{79,600} = 32.9 \ \mu\text{strain}$$

where the moment M_x, which is per unit width, was conservatively assumed to be constant over the width b_3 of the upper flange of the stiffener.

Clearly, this strain is negligibly small and there is neither material failure nor crippling failure of the upper flange under this loading condition. The web of the stiffener would have to be analysed also for crippling under the bending load with the portion from the neutral axis to the top of the stiffener treated as no-edge-free flange (see Section 8.5.3). Since, however, no-edge-free crippling has higher crippling strength than the one-edge-free crippling which would be the case for the upper flange on which the strains were just calculated to be very small, there is no need to perform the calculations for the web in detail.

The final check for the pressure case is failure of the skin under transverse shear loading. Using Equations (13.11), the maximum shear forces are found to be:

$$Q_{x\text{max}} = 19485 \text{ N/m}$$

$$Q_{y\text{max}} = 415 \text{ N/m}$$

Assuming, conservatively that these are both carried by the skin alone, the corresponding maximum shear stresses are:

$$\tau_{xz\,\text{max}} = \frac{Q_{x\,\text{max}}}{t_{skin}} = \frac{19485}{0.00217} = 8.98 \text{ MPa}$$

$$\tau_{yz\,\text{max}} = \frac{Q_{y\,\text{max}}}{t_{skin}} = \frac{415}{0.00217} = 0.19 \text{ MPa}$$

As was the case for the monolithic panel, with a transverse shear strength (allowable) of 125 MPa, the margins of safety for both shear stresses are positive. The lowest of the two is:

$$MS_{trshear} = \frac{125}{8.98} - 1 = +12.9$$

The only aspect of the design left is the attachment. Here, the exact same approach as in the case of the monolithic panel can be followed. In fact, the same design for the edges of the panel as for the monolithic panel can be used, or, because the bearing and inter-rivet buckling margins for the monolithic panel were high, a re-design of the area can be done, using the bottom flange of each stiffener and the skin as the edge doubler for fastener installation. The analysis procedure is analogous to that of the monolithic panel.

As an alternative, not examined here in detail, bonding could be used as the method of attachment to the adjacent structure. Some considerations that should be taken into account in such a case are briefly presented here. Bonding is most effective when an adhesive is used and, in particular, when the adhesive is loaded in shear. Out-of-plane (peeling) loads on the adhesive should be avoided. Extra care should be taken to maintain a constant bondline thickness which means film adhesives are preferred over paste adhesives. The contact area is calculated by dividing the applied shear load that is transmitted by the bondline by the adhesive shear strength. This is an average stress calculation. The shear stress in the adhesive peaks at the edges (as much as 70% higher than the average for typical bondlines if the adherends are not tapered) and detailed calculations should be done to make sure the peak stresses do not cause failure. These peak stresses may also be significant under fatigue loading. It is recommended that the peak shear stresses in the adhesive be kept below an endurance limit or a value that gives sufficient number of cycles (usually at least four lifetimes) before the adhesive cracks. As mentioned in Sections 5.1.5 and 12.3 and exercise 12.2, the inspection of adhesively bonded joints is not reliable enough in verifying the bond strength and quality. Therefore, provisions for being able to take the applied (limit) loads with half of the bondline ineffective must be made. These, typically, include secondary load paths, overdesigned bondlines, or the use of fasteners as an extra precaution.

Another important consideration is the number of continuous plies that should go around corners in the stiffener cross-section. The cross-section in Figure 12.5 gives only a minimum number of plies continuing from the web into the flanges. In a more detailed design sufficient load continuity around corners is ensured by performing additional analysis. For example, in the presence of out-of-plane shear loads, the shear flow distribution in the stiffener cross-section is calculated and the shear stresses at the flange/web intersections are computed. Then, the number of plies with continuous fibres around each corner is determined such that the local shear stresses cause no failure.

In the above design, most of the margins were quite high suggesting there is room for weight reductions. This can be done in subsequent design iterations. One interesting point is that if the stiffeners instead of being aligned with the long dimension of the panel were aligned with the short dimension, they would be more effective for the pressure load case. At the same time, the compression loads would have to be taken by the skin only. So there is potential for additional tradeoffs including the two possible stiffener directions, parallel or perpendicular to the loading.

The weight of the stiffened panel can now be determined:

$$W_{\text{stiffened}} = 1609$$

$$\left[(0.75 - 2(0.0276)) \, (1.5 - 2(0.0276)) \, 0.00217 + 6(1.5) \left(\begin{matrix} (0.027)(3)(0.0046) + \\ 0.057(0.00648) \end{matrix} \right) \\ + (2 \, (0.75 \times 0.0276) + 2(1.5 - 2(0.0276))0.0276) \, 0.00762 \right] = 15.73 \text{ kg}$$

13.3 Sandwich Design

For the sandwich design, a constant thickness sandwich is designed first. Then, depending on preferences, a full-depth close-out or rampdown configuration (see Section 10.6) can be defined for attachment to adjacent structure.

The design starts with the wrinkling requirement under combined compression and shear. From Table 10.3:

$$\frac{N_x}{N_{xwr}} + \left(\frac{N_{xy}}{N_{xywr}} \right)^2 = 1 \tag{13.28}$$

where N_{xwr} and N_{xywr} are the wrinkling failure loads when compression and shear act individually. If the left-hand side of Equation (13.28) is less than 1, there is no failure.

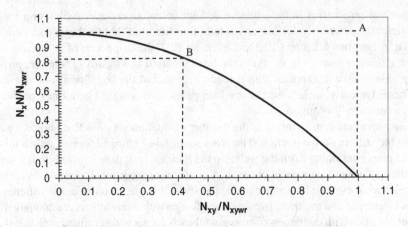

Figure 13.9 Interaction curve for wrinkling under compression and shear

By solving Equation (13.28) with respect to N_x, one can obtain the interaction curve:

$$\frac{N_x}{N_{xwr}} = 1 - \left(\frac{N_{xy}}{N_{xywr}}\right)^2 \tag{13.29}$$

which is plotted in Figure 13.9.

It is clear from Figure 13.9, that if a design is selected for which $N_x = N_{xwr}$ and $N_{xy} = N_{xywr}$, it will lie above the interaction curve (point A) and it will fail. This means that the design must be such that both the applied compression and shear loads are a fraction, less than one, of their corresponding wrinkling failure loads. In the present case, the applied shear load N_{xy} is half the applied compression load N_x. So somewhat arbitrarily, a design is sought for which

$$\frac{N_{xy}}{N_{xywr}} = \frac{N_x}{2N_{xwr}} \tag{13.30}$$

Using this assumption, substitution in Equation (13.29) leads to:

$$\frac{N_x}{N_{xwr}} = 1 - \left(\frac{N_x}{2N_{xwr}}\right)^2 \Rightarrow \frac{N_x}{N_{xwr}} = 0.828 \tag{13.31}$$

This means that a design is sought for which $N_x/N_{xwr} = 0.828$ and $N_{xy}/N_{xywr} = (0.828)/2 = 0.414$. This is point B in Figure 13.9. This design would just fail when the maximum applied compression and shear loads are reached.

In a sandwich panel, each facesheet carries half of the applied in-plane loads. So the facesheet must have

$$N_{xwr} = \frac{N_x}{2(0.39)} = \frac{350}{2(0.828)} = 211.3 \text{ N/mm}$$

$$N_{xywr} = \frac{N_{xy}}{2(0.78)} = \frac{175}{2(0.414)} = 211.3 \text{ N/mm}$$

It is anticipated that the buckling requirement under the in-plane load and the maximum deflection requirement under pressure will require the use of core that is sufficiently thick so that antisymmetric wrinkling is not possible and a fraction of the core will deform during wrinkling (see Section 10.3.1 and exercise 10.1). Therefore, Equation (10.28) is applicable. Note that, once the core thickness is determined, the validity of these assumptions can be checked and another design iteration can be performed if necessary.

In order to use Equation (10.28) a core material must be selected from the three proposed. Since the product $E_c G_{xz}$ appears in Equation (10.28), one can compare by how much this ratio increases as the density increases from one core material to the next. Substituting values from Table 13.1, the following Table can be created:

Table 13.3 Core comparisons on the basis of wrinkling load improvement

Core	Normalized $(E_c G_{xz})^{1/3}$	Normalized density
A	1	1
B	1.245	1.33
C	1.483	1.66

It can be seen from Table 13.3 that, as one switches from one core material to the next, the increase in density in column 3 (and thus the weight of the panel) is greater than the increase in wrinkling load in column 2. Therefore, as a rule, it is more efficient to use lower density cores than higher density cores. As a result, core A from Table 13.1 is selected. Then, substituting in Equation (10.28):

$$t_f E_f^{1/3} = \frac{N_{xwr}}{0.43 \, (E_c G_{xz})^{1/3}} \tag{13.32}$$

for wrinkling along the direction of compression loading and

$$t_f E_{45}^{1/3} = \frac{N_{45wr}}{0.43 \left(E_c \dfrac{G_{xz} + G_{yz}}{2} \right)^{1/3}} \tag{13.33}$$

where Section 10.3.2 was used and N_{45} is the compression load at 45° caused by the shear load N_{xy}. Using the second stress transformation Equation (3.35) with $\tau_{xy} = N_{xy}/t_f$, the compression load along a 45° line is determined as:

$$\frac{N_{45}}{t_f} = -2 \sin 45 \cos 45 \frac{N_{xy}}{t_f} \Rightarrow N_{45} = -N_{xy}$$

where the minus sign denotes compression.

Therefore, substituting numbers in Equations (13.32) and (13.33):

$$t_f E_f^{1/3} = \frac{211.3 \times 10^3}{0.43 \left(131 \times 10^6 \times 41.4 \times 10^6\right)^{1/3}} = 1298 \ m \left(\frac{N}{m^2}\right)^{1/3}$$

$$t_f E_{45}^{1/3} = \frac{211.3 \times 10^3}{0.43 \left(131 \times 10^6 \times \dfrac{41.4 + 20.7}{2} \times 10^6\right)^{1/3}} = 1429 \ m \left(\frac{N}{m^2}\right)^{1/3} \quad (13.34)$$

If the two Equations (13.34) are divided by each other and raised to the third power, the ratio of the stiffnesses of the facesheet in the 0 and 45° directions is obtained as

$$\frac{E_f}{E_{45}} = \left(\frac{1298}{1429}\right)^3 = 0.75$$

that is the stiffness in the 0 direction (aligned with the compression load) is 25% less than that in the 45° direction. This allows selection of the % of fibres in the different directions by picking a number for the 45° plies, taking 75% of that for the 0 plies and letting the remainder to be taken up by 90° plies. One such example that closely approximates this ratio is:

$$[(\pm 45)_2/(0/90)_2(\pm 45)_2]$$

with $E_f/E_{45} = 0.71$.

This facesheet consists of only fabric plies, has a thickness of 1.143 mm and

$$A_{11} = 58{,}787 \ \text{N/mm}$$
$$A_{12} = 25{,}812 \ \text{N/mm}$$
$$A_{22} = 58{,}787 \ \text{N/mm}$$
$$A_{66} = 27{,}388 \ \text{N/mm}$$
$$D_{11} = 5346 \ \text{N mm}$$
$$D_{12} = 3865 \ \text{N mm}$$
$$D_{22} = 5346 \ \text{N mm}$$
$$D_{66} = 4036 \ \text{N mm}$$
$$E_f = 41.5 \ \text{GPa}$$
$$E_{45} = 58.2 \ \text{GPa}$$

Note that this satisfies rules 1–5 and 17 in Chapter 12.
Double checking the wrinkling failure load, for wrinkling under compression only:

$$N_{xwr} = 0.43(1.143) \left(41.5 \times 10^3 \times 121 \times 41.4\right)^{1/3} = 291.1 \ \text{N/mm}$$

Also, for wrinkling under shear only:

$$N_{xwr} = 0.43(1.143) \left(58.2 \times 10^3 \times 121 \times \left(\frac{41.4 + 20.7}{2} \right) \right)^{1/3} = 296.1 \text{ N/mm}$$

The interaction curve, Equation (13.28) gives for each facesheet:

$$\frac{175}{291.1} + \left(\frac{87.5}{296.1} \right)^2 = 0.688$$

This is less than 1 and, therefore, there is no wrinkling failure. To calculate the margin of safety, the number R by which both loads must be multiplied so that the left-hand side of this equation equals exactly 1 is determined: $R = 1.39$. Then:

$$MS_{wr} = 1.39 - 1 = +0.39$$

It is important to note that the individual wrinkling loads (for compression and shear separately) determined for the selected facesheet do not match the target value of 211.3 N/mm. The reason for this is that even though the stiffness ratio E_f/E_{45} is close to the target value of 0.75, the facesheet thickness corresponding to the selected laminate is higher than it should be. Potentially, a lower facesheet thickness would still have a positive margin and, of course, lower weight.

The facesheet is also checked for material failure. The Tsai-Wu failure criterion (see Chapter 4) is used with applied loads (per facesheet), $N_x = 175$ N/mm and $N_{xy} = 87.5$ N/mm. The critical plies are plies 3 and 4 with an R value (number by which the applied loads must be increased to barely cause first-ply failure) of 1.54. Then, the safety margin is:

$$MS_{TW} = 1.54 - 1 = +0.54$$

The core thickness can now be selected such that there is no buckling of the panel under compression and shear and the maximum deflection requirement of 2 mm is met. Using Equation (10.1) to determine the bending stiffnesses D_{ij} for the sandwich and applying Equation (13.8), it is found that a core thickness $t_c = 24$ mm results in a maximum panel deflection of 1.97 mm.

The buckling load of this panel under pure compression when no transverse shear effects are accounted for is obtained from Equation (6.7):

$$N_{xEcrit} = 1547 \text{ N/mm}$$

corresponding to two half-waves along the length of the panel.

Using Equation (10.3) the buckling load under pure compression including transverse shear effects is determined as:

$$N_{xcrit} = \frac{1547}{1 + \dfrac{1547(1000)}{0.024(41.4 \times 10^6)}} = 605 \text{ N/mm}$$

Similarly, the buckling load under pure shear, neglecting transverse shear effects is given by Equation (6.27):

$$N_{xyEcrit} = (0.7)\frac{9\pi^4 b}{32a^3}\left(D_{11} + 2\left(D_{12} + 2D_{66}\right)\frac{a^2}{b^2} + D_{22}\frac{a^4}{b^4}\right) = 2216 \text{ N/mm}$$

Then, using Equation (10.6) to correct for transverse shear effects:

$$N_{xycrit} = \frac{(41.4 + 20.7) \times 10^6(0.024)}{\dfrac{(41.4 + 20.7) \times 10^6(0.024)}{1965 \times 10^3} + 2} = 540 \text{ kN/m} = 540 \text{ N/mm}$$

To check for buckling under the combined compression and shear, the interaction curve Equation (6.28) is used:

$$\frac{N_x}{N_{xcrit}} + \left(\frac{N_{xy}}{N_{xycrit}}\right)^2 = \frac{350}{605} + \left(\frac{175}{540}\right)^2 = 0.683$$

which is less than 1 and, therefore, there is no buckling failure.

The factor R by which the applied loads must be multiplied to just cause buckling failure turns out to be 1.38. Thus, the buckling margin of safety is:

$$MS_{buck} = 1.38 - 1 = +0.38$$

For crimping failure, there is no interaction curve available. Very conservatively one can use a straight-line curve as shown in Figure 13.10. The reason, a straight line is conservative, is that all interaction curves that represent or simulate failure under combined loads are concave

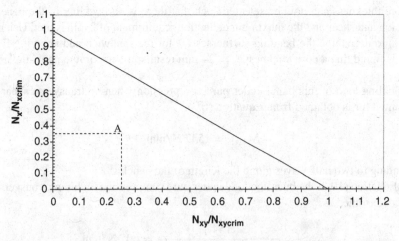

Figure 13.10 Conservative interaction curve for sandwich crimping under combined shear and compression

with respect to the origin and thus would lie above the straight line of Figure 13.10. For the present design,

$$N_{x\text{crimp}} = t_c G_{xz} = 993,600 \text{ N/m}$$

$$N_{xy\text{crimp}} = t_c \sqrt{G_{xz} G_{yz}} = 702,581 \text{ N/m}$$

Then, $N_x/N_{x\text{crimp}}$ and $N_{xy}/N_{xy\text{crimp}}$ define point A in Figure 13.10 which is below the conservative interaction curve and crimping is not a concern. The corresponding margin of failure is:

$$MS_{\text{crimp}} = 1.66 - 1 = +0.66$$

Strictly speaking a check for dimpling, see Section 10.5 would also have to be done. However, core cell sizes are not reported in Table 13.1. For typical core materials, the cell size is between 3 and 5 mm. For these core sizes, Equation (10.34) gives high positive margins.

At this point, it remains to check if there is any failure of the sandwich panel under pressure loading. Using Equations (13.10a) and (13.11), the maximum bending moments and shear forces are found to be:

$$M_{x\text{max}} = 378 \text{ Nm/m}$$

$$M_{y\text{max}} = 654 \text{ Nm/m}$$

$$M_{xy\text{max}} = -386 \text{ Nm/m}$$

$$Q_{x\text{max}} = 4654 \text{ N/m}$$

$$Q_{y\text{max}} = 6413 \text{ N/m}$$

For a sandwich, as already mentioned at the beginning of Chapter 10, the bending moments can be resolved as a force couple with moment arm $t_c + t_f$. Then, using the above values,

$$N_{x\text{max}} = \frac{M_{x\text{max}}}{t_c + t_f} = \frac{378}{24 + 1.143} = 15.0 \text{ N/mm}$$

$$N_{y\text{max}} = \frac{M_{y\text{max}}}{t_c + t_f} = \frac{654}{24 + 1.143} = 26.0 \text{ N/mm}$$

$$N_{xy\text{max}} = \frac{M_{xy\text{max}}}{t_c + t_f} = -\frac{386}{24 + 1.143} = -15.3 \text{ N/mm}$$

Clearly, these loads are much lower than the in-plane loads for the compression and shear case and there is no need to check for failure. The corresponding margins will be positive and high.

The shear forces $Q_{x\text{max}}$ and $Q_{y\text{max}}$ cause constant shear stresses in the core. The quadratic distribution of shear stresses is not applicable in sandwich with core which has no bending stiffness. This can be seen by applying the first of Equations (5.2) to the core with $\tau_{xy} = 0$ (no in-plane shear capability of the core):

$$\frac{\partial \sigma_x}{\partial x} + \frac{\partial \tau_{xz}}{\partial z} = 0 \tag{13.35}$$

with z in the out-of-plane direction. Since the core carries no in-plane stress σ_x, the first term in Equation (13.35) is zero. Then:

$$\frac{\partial \tau_{xz}}{\partial z} = 0 \tag{13.36}$$

and the shear stress in the core τ_{xz} is independent of z, that is, it is constant through the thickness. An analogous proof can be used for τ_{yz}.

Therefore, the shear stresses in the core caused by the applied Q_{xmax} and Q_{ymax} are:

$$\tau_{xz} = \frac{Q_{xmax}}{t_c} = \frac{4654}{0.024} = 0.194 \text{ MPa}$$

$$\tau_{yz} = \frac{Q_{ymax}}{t_c} = \frac{6413}{0.024} = 0.267 \text{ MPa}$$

These should be compared to the core allowable shear strengths, 2.413 and 1.723 MPa, respectively (from Table 13.1). Then the lowest margin is:

$$MS_{coreshear} = \frac{1.723}{0.267} - 1 = +5.453$$

Finally, for attaching the sandwich panel to the adjacent structure, the same edge design as for the monolithic and stiffened panels is used. This means that the sandwich panel must be ramped down near the edges to the monolithic laminate determined in Section 13.1 for the edge of the panel. This requires some careful considerations for determining the ramp angle and transitioning the facesheets to the monolithic laminate with judicious placement of plydrops along the flat and ramped portion of the rampdown (see Section 10.6.1). The details of the rampdown are not discussed here. It should be kept in mind that the rampdown and the associated transition region may end up adding as much as 30% to the weight of the constant thickness panel. This does not mean that rampdowns are not efficient. If a constant thickness sandwich is used even at the panel edges where the attachments are, the requirements for densification or other local reinforcement to be able to cope with fastener loads and the requirements for protecting the core from moisture (water-proof close-out of the edges) add up, in some cases, to about the same amount of weight (30%) as the rampdown does.

Assuming now that the facesheets are bonded to the core with the use of film adhesive that weighs 0.293 kg/m^2, the weight of the sandwich panel can be estimated:

$$\begin{aligned}
W_{sand} = \; & 1609\,[2(0.001143)(1.5 - 2(0.0276))(0.75 - 2(0.0276)] + \\
& 1609\,[(2\,(0.75 \times 0.0276) + 2(1.5 - 2(0.0276))0.0276)\,0.00762] + \\
& 2(0.293)(1.5 - 2(0.0276))(0.75 - 2(0.0276)) + \\
& 48.2\,[0.024(1.5 - 2(0.0276))(0.75 - 2(0.0276))] = 6.926 \text{ kg}
\end{aligned}$$

13.4 Cost Considerations

An estimate of the recurring cost for each of the three designs in the previous sections can be obtained by using the following equation:

$$C(hrs) = 0.4764 + 1.935 \times 10^{-4} A + 0.001097 P + 0.0111 P/s$$
$$+ 0.000107 A^{0.7706} + 0.001326 A^{0.5252} + \left(0.00233 + n_{plies} 0.000145\right) A^{0.6711}$$
$$+ 0.001454 n_{plies} L^{0.8245} + n_{pliesm} C_m L_m n_{bm}$$
$$+ n_{plies f} C_f L_f n_{bf} \tag{13.37}$$

where : A = Area of the ply or plies being laid-up (square inches)

P = Perimeter(inches)

n_{plies} = Number of plies

L = Longest dimension of part

C_m = 0.00007 for convex (male) mould corner

C_f = 0.00016 for concave (female) mould corner

L_{m}, n_{bm} = Length (inches) of n_{pliesm} plies bent around a (male) mould and number of male bends

L_{f}, n_{bf} = Length (inches) of $n_{plies f}$ plies bent into a (female) mould and number of (female) bends

s = Fastener spacing (inches)

This equation is based on a much simplified version of the labour cost equations in [2] augmented to include an estimate for the assembly time-assuming fasteners are used. The equation applies to hand layup and reflects a specific part number on a learning curve. So it would change if a factory had more or less experience than the part or unit number to which the above equation corresponds. Therefore, even though the predictions are good for comparing different designs, the cost estimates obtained should not be viewed as exact unless the factory capabilities to which they are applied are identical to the assumed factory implied in this equation.

Applying Equation (13.37) to the three different designs in Sections 13.1–13.3, the following cost numbers are obtained:

Monolithic panel cost = 15.3 hrs

Stiffened panel cost = 33.4 hrs

Sandwich panel cost = 15.1 hrs

Note that, for the stiffened panel which uses seven identical stiffeners, a learning curve of 85% was also assumed.

13.5   Comparison and Discussion

A summary of the most important margins of safety for the three designs is given in Table 13.4. Note that the deflection requirement is also translated to a margin with a 0 value implying matching exactly the maximum deflection of 2 mm.

The lowest margin of safety, excluding the deflection requirement, is highlighted in Table 13.4 for each of the three designs. It appears that there is room for further weight reductions since no margin is equal to zero. However, this will affect the maximum deflection and one must be very careful in removing the material to make sure the stiffness remains the same so that the deflection, which, for all three designs is almost equal to the maximum allowable, does not exceed that maximum allowable value. In particular, weight can be removed from the attachment region which is the same for all three designs. The bearing and inter-rivet buckling margins are high and a significant amount of the material can probably be removed. This will have a big impact on the total weight of the panel.

A breakdown of the weight distribution for the three designs is shown in Figure 13.11.

It can be seen from Figure 13.11 that the centre of the laminate is a very large percentage (almost 94%) of the total weight. This is a result of the deflection requirement and the lack of other means to stiffen the structure.

Table 13.4   Summary of safety margins for the three designs

Design	Failure mode	Margin
Monolithic panel	Deflection	0.004
	Tsai-Wu strength (pressure)	17.7
	Tsai-Wu strength (comp. + shear)	4.17
	Panel buckling	**1.137**
	Bearing	1.362
	Inter-rivet buckling	33.6
Stiffened panel	Deflection	0.550
	Crippling (web)	3.272
	Crippling (flanges)	5.245
	Column buckling (stiffener)	10.17
	Tsai-Wu flange strength	8.55
	Tsai-Wu web strength	3.269
	Tsai-Wu skin strength	0.510
	Panel buckling	**0.234**
	Bearing	1.362
	Inter-rivet buckling	33.6
Sandwich panel	Deflection	0.015
	Wrinkling	**0.39**
	Tsai-Wu facesheet strength	0.54
	Panel buckling	0.38
	Crimping	0.66
	Core shear strength	5.453
	Bearing	1.362
	Inter-rivet buckling	33.6

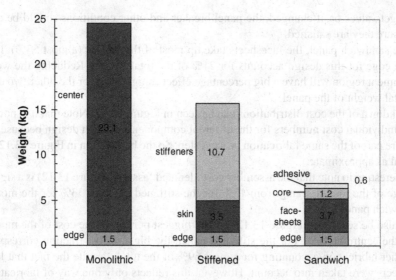

Figure 13.11 Weight distributions for the different designs (see Plate 30 for the colour figure)

For the stiffened panel, the biggest weight fraction corresponds to the stiffeners (68%). Subsequent design iterations would focus on reducing that weight. Reducing the number of stiffeners might be one way which would also benefit the cost (see Figure 13.12). However, this would increase the stiffener spacing and would require an increase in the skin thickness to delay buckling of the skin between stiffeners. Another option is keeping the number of stiffeners and reducing their thickness since the crippling margins are high (see Table 13.4) and there is room to do so. However, the EA and EI of the stiffeners should still be kept close

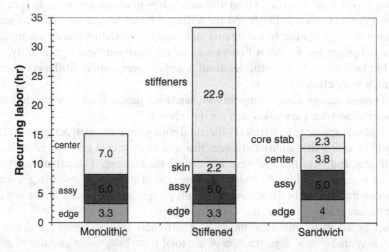

Figure 13.12 Cost distributions for the different designs (see Plate 31 for the colour figure)

to the target values or, if changed, the panel breaker and other conditions should be checked to make sure they are satisfied.

For the sandwich panel, the facesheets take up most of the weight (almost 53%). Interestingly, the edge for this design accounts for 21% of the total weight. Reducing the weight of the attachment region will have a big percentage effect (bigger than for the other two designs) on the total weight of the panel.

A breakdown of the cost distributions can be seen in Figure 13.12. Note that it is not easy to separate individual cost numbers for the different components of each design because several of them are part of the same fabrication process step. So the breakdown in Figure 13.12 should be viewed as approximate.

It is interesting to note that the assembly cost (denoted 'assy' in Figure 13.12) is a significant percentage of the total, ranging from 15% for the stiffened panel to 33% for the monolithic and sandwich panels.

It can also be seen from Figure 13.12 that the biggest portion of the cost of the monolithic panel is the centre at 46%. For the stiffened panel, the biggest cost fraction corresponds to the stiffener fabrication accounting for almost 69% of the total, despite the fact that learning curve effects were taken into account. However, this reflects only one way of fabricating this design with each stiffener fabricated as an individual part. If, instead, one long stiffener were fabricated and cut up to individual pieces, the cost would be significantly lower. Finally, for the sandwich, the biggest cost fraction corresponds to the assembly which accounts for 33% of the total. Note also that the edge cost in the sandwich is higher than the edge cost in the other two designs because of extra labour associated with the rampdown and transitioning the plies from the edge to the centre of the panel.

On the basis of Figures 13.11 and 13.12, several general conclusions can be drawn as well as recommendations on how to refine the individual designs to further reduce weight and cost.

- For flat rectangular panels, the cost of monolithic panels is 0.62 h/kg, that of stiffened panels is 2.12 h/kg and that of sandwich 2.18 h/kg. These numbers are expected to occur after fairly long production runs (more than 400 units). It is interesting to note that per kilogram of finished product, the sandwich and the stiffened panel have essentially the same cost. Automation through the use of automated tape layup or fibre/tow placement machines for the skins and pultrusion for the stiffeners will reduce these numbers significantly.
- Monolithic laminates are not efficient from a weight perspective. Stiffeners or core help create much more efficient designs.
- Stiffened panels cannot easily compete with sandwich panels for lower weight unless they are allowed to post-buckle (which was not done here).
- Sandwich panels appear to be the most efficient from a weight and cost perspective. However, one should be aware that additional issues that add to their cost and weight such as close-out treatments and densification were not taken into account. The extent to which these become an issue in a sandwich design is a function of the design itself (e.g. core density and rampdown angle used). Therefore, the trends presented here should be viewed as trends showing the biggest potential of the sandwich designs.
- Care must be taken to minimize the cost of stiffeners in a stiffened panel because the stiffeners may make up a large fraction of the total cost. Integration and use of automated processes such as pultrusion can significantly reduce the cost.

- Minimizing the number of stiffeners reduces the cost but may adversely affect the weight if the skin thickness must be increased to prevent or delay buckling.
- Minimizing the number of fasteners used for attachment reduces the cost. The spacing selected, if not governed by other requirements, must guarantee that the desired boundary conditions, for example simply supported, are enforced.

In addition, certain aspects that were not included but should be in a more detailed design to obtain a more accurate picture should be mentioned here: (a) The weight of fasteners was not included in the weight calculations. (b) The material cost (including scrap) is not included in the cost. (c) The thickness of the monolithic design may require special process steps such as provisions for good heat conducting material used for the mould, special heat-up and cool-down rates to avoid exothermic reactions and minimize creation of microcracks, extra dwell during cure for improved part quality. (d) The recurring cost trends are not necessarily representative of non-recurring cost trends. The tooling cost and the size of the production run should be factored in the design for a better tradeoff of the different concepts.

References

[1] Kassapoglou, C., Simultaneous Cost and Weight Minimization of Composite Stiffened Panels Under Compression and Shear, Composites A. *Applied Science and Manufacturing*, **28A**, 419–435(1997).
[2] LeBlanc, DJ., Kokawa, A., Bettner, T., Timson, F., and Lorenzara, J., Advanced Composites Cost Estimating Manual (ACCEM), Technical Report AFFDL-TR-76-87, volumes I and II, Northrop Corp., 1976.

Index

Design and Analysis of Composite Structures: With Applications to Aerospace Structures, Second Edition. Christos Kassapoglou.
© 2013 John Wiley & Sons, Ltd. Published 2013 by John Wiley & Sons, Ltd.